JSP+Servlet+Tomcat

应用开发从零开始学

刘华贞 编著

（第3版）

清華大学出版社
北 京

内 容 简 介

本书全面系统地介绍JSP+Servlet+Tomcat开发中涉及的相关技术要点和实战技巧。本书内容讲解循序渐进，结合丰富的示例使零基础的读者能够熟练掌握JSP+Servlet+Tomcat的应用开发和部署。本书配套示例代码、PPT课件、作者答疑服务。

本书共17章。第1~7章为Java Web基础开发，内容包括搭建Java Web开发环境、JSP基础语法、JSP内置对象、Servlet技术、请求与响应、会话管理、Servlet进阶API、过滤器、监听器等；第8~15章为Java Web高级开发，内容包括MySQL 8数据库开发、JSP与Java Bean、EL标签、JSTL标签库、自定义标签、JDBC详解、XML概述、资源国际化等；第16~17章为Java Web实战，分别讲解两个典型的系统，即家校通门户网站（JSP+HTML+CSS）和在线购物系统（JSP+Java Bean+MySQL）。

本书内容精练、结构清晰、注重实战，适合广大Java Web开发初学者学习，还可作为高等院校或高职高专计算机及相关专业的教材使用。

图书在版编目（CIP）数据

JSP+Servlet+Tomcat 应用开发从零开始学/刘华贞编著. —3 版. —北京：清华大学出版社，2023.5（2024.8重印）

ISBN 978-7-302-63617-5

Ⅰ. ①J… Ⅱ. ①刘… Ⅲ. ①JAVA 语言－程序设计 Ⅳ. ①TP312.8

中国国家版本馆 CIP 数据核字（2023）第 094055 号

责任编辑： 夏毓彦
封面设计： 王　翔
责任校对： 闫秀华
责任印制： 沈　露

出版发行： 清华大学出版社
网　　址： https://www.tup.com.cn，https://www.wqxuetang.com
地　　址： 北京清华大学学研大厦 A 座　　**邮　　编：** 100084
社 总 机： 010-83470000　　**邮　　购：** 010-62786544
投稿与读者服务： 010-62776969，c-service@tup.tsinghua.edu.cn
质量反馈： 010-62772015，zhiliang@tup.tsinghua.edu.cn
印 装 者： 三河市天利华印刷装订有限公司
经　　销： 全国新华书店
开　　本： 190mm×260mm　　**印　　张：** 23.75　　**字　　数：** 641 千字
版　　次： 2015 年 1 月第 1 版　　2023 年 7 月第 3 版　　**印　　次：** 2024 年 8 月第 2 次印刷
定　　价： 99.00 元

产品编号：102060-01

前　言

本书是面向Java Web开发初学者的一本高质量图书。Java是当今程序开发中最流行的编程语言之一，不但可以开发手机应用、桌面应用，而且越来越多地应用于Java Web开发中。Java优越的跨平台特性使它备受欢迎。近年来，Java Web框架技术层出不穷，跨浏览器、跨系统等要求更是体现了Java Web开发的强大生命力。

目前，市面上有关Java Web的书籍非常多，初学者常常不知道应该如何选择。本书从初学者的角度出发，用浅显的实例说明复杂的知识点，为那些想在Java Web开发中大展拳脚的开发人员精心编写，虽然所讲内容未涉及当前大型项目的主流框架，但都属于Java Web中的基础知识。只有夯实基础知识，才能更好地学习其他技术框架。本书从底层原理入手，并从实战的角度进行讲解，以便让想要学习Java Web开发的初学者快速掌握相关技术，并能够根据实际需求开发出有用的Web应用。

本书特点

（1）内容丰富，知识全面。本书内容几乎涵盖Java Web基础开发的各个方面。本书还涉及Servlet 5.0版本的知识与编写规范，并利用详细的实例进行说明。

（2）循序渐进，由浅入深。为方便读者学习，本书首先介绍如何搭建Java Web开发的基础环境，然后介绍JSP的基础语法与Servlet的基本概念。帮助读者掌握这些基础知识后，让读者逐渐学习请求与响应的过程、会话管理、Servlet 5.0以上的版本中进阶API以及过滤器、监听器、自定义标签的编写等，从而更深入地掌握Java Web开发技术。

（3）编码规范，讲解详细。书中每个知识点都给出了详尽的操作示例，以供读者参考，并对代码进行详细解释。实例中的代码是严格按照Java规范进行编写的，并配有详细的代码注释。

（4）易学易用，案例丰富。本书通过简单的实例讲解每个知识点，力求用简单的实例来诠释复杂的知识，使读者快速了解并掌握Web开发所需的知识。对于各种语法几乎都配有一个实例来说明其用法。

（5）案例精讲，图文并茂。对于难以理解的知识点，编者用图表的方式进行讲解，让读者更加直观地理解知识点。编者根据多年的项目经验，在每章中尽量用一个综合示例对知识点进行整合，使读者对每章的知识点有整体掌握。

进阶路线

第1~7章，Java Web基础开发：讲解Java Web开发环境的搭建、JSP基础语法、JSP内置对象、Servlet技术、请求与响应、会话管理、Servlet进阶API、过滤器、监听器等基础知识。

第8~15章，Java Web高级开发：讲解MySQL数据库开发、JSP与Java Bean、EL标签、JSTL标签库、自定义标签、JDBC详解、XML概述、资源国际化等Java Web高级开发所需的知识。

第16~17章，Java Web实战：讲解如何运用Java Bean、MySQL、JSP技术以及标签开发家校通门户网站和带数据库的在线购物系统，使读者能够快速掌握Java Web开发技术和编写规范。

第 3 版修订说明

随着Java Web技术的快速发展，所使用的技术也在不断更新，为了方便读者学习最新技术，本书在第2版的基础上进行相应的升级。JDK的版本更新为17.0.4，Servlet升级到5.0并修改了相应的章节内容，JSP版本升级到3.1，Tomcat服务器由Tomcat 9改为Tomcat 10。本书的更新都是为了让读者跟上当前技术发展的步伐，希望读者不要停下学习的脚步，努力向前。

示例代码、PPT 课件下载

本书示例源代码、PPT课件的下载地址可扫描右侧的二维码获得。如果阅读过程中发现问题，请用电子邮件联系booksaga@163.com，邮件主题务必写“JSP+Servlet+Tomcat应用开发从零开始学（第3版）”。

本书适合的读者

- JSP+Servlet 开发初学者
- Java Web 开发初学者
- Java Web 开发工程师
- Web 应用开发人员
- 高等院校、中职学校、培训机构的学生

本书第1版本由林龙主笔，第2版本由刘华贞修订，第3版由刘华贞修订，在此表示感谢。

编　者

2023年3月

目　录

第 1 章 搭建 Java Web 开发环境

正所谓“工欲善其事，必先利其器”，开发一个Web应用程序，首先必须搭建好开发环境，选择好开发工具，从而达到事半功倍的开发效果。现如今支持Web的应用服务器非常多，例如WebSphere、WebLogic、Tomcat等，它们的配置方法各不相同，本书选择Apache Tomcat 10.0.22作为服务器开发平台，JDK使用的是17.0.4版本。

本章主要涉及的知识点有：

- JSP支持的网络协议
- Web应用程序的运行环境和开发环境
- Tomcat软件的安装和配置
- JSP开发工具的选择

1.1 Web 开发背景知识

本节的重点是介绍Web开发的一些基础知识，首先简单介绍Web访问的基本原理，然后对超文本传输协议（Hyper Text Transfer Protocol，HTTP）进行简单介绍，最后介绍静态网页与动态网页的区别以及各种Web服务器的优缺点。通过本节的学习，读者可以掌握Java Web开发所需的背景知识。

1.1.1 Web访问的基本原理

Web访问可以简单划分为两个过程：客户端请求和服务器端响应。客户端的请求通过Servlet引擎传递给Servlet模块，Web服务器接收客户端的请求，并把处理的结果返回给客户端。客户端与服务器之间的通信协议就是超文本传输协议。客户端与服务器之间的请求模式如图1.1所示。

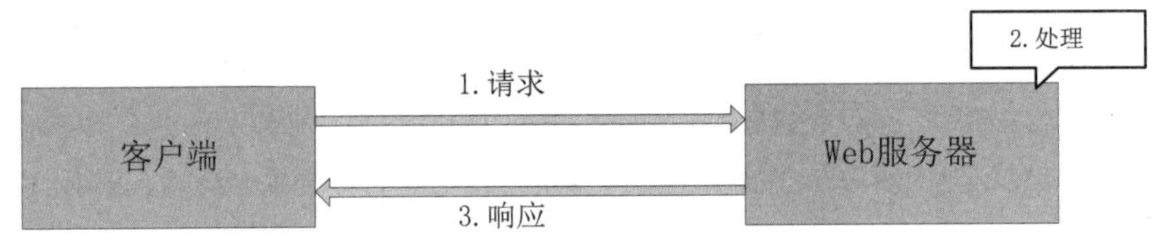

图1.1 Web访问原理

1.1.2 超文本传输协议

超文本传输协议是一种在互联网上应用最为广泛的网络协议，是一种无状态的协议。自1990年起，它就已经被应用于WWW全球信息服务系统。所有的WWW文件都必须遵守这个标准。

超文本传输协议的主要特点如下：

- 简单、快速：客户端向服务器请求服务时，只需发送请求方法和路径URL。通常请求的方法有POST、GET、PUT和DELETE。由于HTTP的协议简单，使得HTTP服务器的程序规模相对较小，因此传输速度较快。
- 灵活：HTTP允许传输任意类型的数据，例如普通文本、超文本、音频、视频等，主要由Content-Type控制。
- 无状态：无状态是指对于数据库事务处理没有记忆能力。后续的处理如果需要前面的信息，就需要重新发送。
- 无连接：无连接的含义是每次连接只处理一次请求，处理完当次请求后就断开连接。

1.1.3 静态网页和动态网页

在网站设计中，直接使用HTML标记语言编写的网页通常被称为“静态网页”。静态网页是标准的HTML文件，后缀名为.htm或.html。它所展示的内容一般是固定不变的，早期的网站一般都是静态网页。静态网页更新起来比较麻烦，需要将更新的HTML页面重新上传到网站服务器。显然，这样的网站缺乏灵活性，同时网站的维护成本也比较高。动态网页技术的出现改变了如此不灵活的状态，用户在不同时间或不同地点访问同一动态网页时显示的内容可以是不同的。

> **注意：**所谓静态网页与动态网页，是基于用户访问网页时页面的内容有无变化而言的，与页面的视觉效果没有关系。因为动态的视觉效果大多是通过 JavaScript 或其他基于 JavaScript 的框架技术实现的，与动态网页技术没有必然的联系。

动态网页中的变化内容大部分来自数据库中数据的变化，通过增加、删除、修改、查找数据库中存储的数据来显示内容的变化。例如，在微博中发布一条微博后，查看微博时，会将所发的微博即时显示出来，这在静态网页中是无法完成的。动态网页在被访问时，首先运行服务器端脚本，通过它生成网页内容。显然，动态网页的显示内容是在访问该网页的时候动态生成的，而静态网页是提前做好放在服务器中的，因此，当前网络上的网页大多是动态网页，很少有静态网页，除非一些固定不变的内容，例如发布公告等新闻内容。

目前比较流行的动态网页技术主要包含 ASP、PHP 以及 JSP（Java Server Pages）。

- ASP更精确地说应该是一个中间件，它将Web上的请求转入IIS解释器中，IIS将全部解析执行ASP中的Script脚本。其缺点就是不能跨平台，只能在Windows平台下使用，开发受到诸多限制。其优点是微软提供了强大的IDE，所以开发者容易上手且开发效率也较高。
- PHP是当前比较流行的动态网页技术，是一种HTML内嵌式的语言，其语法融合了Java、C以及Perl，能够比CGI（Common Gateway Interface）更加快速地执行动态网页。其优点是开源、跨平台，正因为它具有开源和跨平台特性，所以很多网站都采用PHP编写自身的网页；其缺点是安装复杂，需要添加很多的外部库来支持，如图形需要gd库等。

注意：CGI 也是一种动态网页技术，虽然它的功能比较强大，但是它的由于比较编程困难、效率低下、修复复杂等缺陷，逐渐被新技术淘汰。

- JSP采用Java语言作为服务器端脚本，页面由HTML和嵌入Java代码组成。随着Java的广泛应用，JSP的应用也越来越广泛。其优点是简单易用，完全面向对象，具有Java的平台无关性、安全性和可靠性。目前大多数的动态网站开发都采用JSP技术，它具有很高的市场占有率。

1.1.4 Web浏览器和Web服务器

1. Web 浏览器

浏览器是指Web服务的客户端浏览程序。它可以向服务器发送各种请求，并对从服务器中返回的各种信息（包括文本、超文本、音频、视频等各种数据）进行解释、显示和播放。现如今，Web浏览器遍地开花，国外的有Internet Explorer（IE）、Chrome、Firefox、Safari，以及近几年逐渐步入公众视野的Microsoft Edge，国内的有360浏览器、QQ浏览器、傲游浏览器、搜狗浏览器、猎豹浏览器等，它们都各自占据着一片江山。由于IE和Edge绑定在Windows操作系统中，因此IE和Edge在市场中占有明显较高的份额，但随着互联网的不断进步，Chrome浏览器大量普及，已逐渐展露出超越IE的势头。

国内的浏览器从利用IE支持的内核逐步发展为支持多种内核，例如新版本的360浏览器同时支持Trident+Blink内核；而国外的浏览器内核是自主研发的，例如IE的内核是Trident、Chrome的内核由Webkit转为Blink、Firefox的内核是Gecko、Safari的内核是Webkit等。

2. Web 服务器

浏览器与服务器的关系可谓是“唇齿相依”，浏览器发送请求，服务器处理请求并将结果返回给浏览器显示。Web 服务器的种类繁多，目前比较流行的有 WebSphere、WebLogic、Tomcat 等。它们的配置、启动方式各不相同，也有各自的优缺点。

- WebSphere服务器是IBM的产品，是一款功能很完善的Web服务器，基于Java程序研发，用于建立、部署和管理Web应用程序。WebSphere产品有开发版和商业版，但是WebSphere不开源。
- WebLogic服务器是一款多功能、基于标准的Web服务器。它遵从开放的标准、支持基于组件的开发。因为它性能比较稳定，所以国内有很多大公司在使用。

- Tomcat是一款开放源代码、基于Java的Web服务器，也是一款轻量型的Web服务器，是基于Apache许可证下开发的自由软件，根据JSP和Servlet技术标准实行。它安装简单、占有系统空间较小，却能支撑较大的Web应用系统，所以它是开发者、小型公司、学校网站建设者的首选软件。

1.2 JSP简介

上一节简单介绍了Web开发的一些背景知识，读者已经了解了Web访问的基本原理、超文本传输协议、静态网页与动态网页的区别，以及主流的浏览器和Web服务器。本节将介绍JSP的基本概念、执行过程等，让读者了解JSP是什么、能做些什么。

1.2.1 什么是JSP

JSP技术是由SUN公司（现被Oracle收购）提出、多家公司参与的，于1999年推出的一款建设动态网页的方法。它基于Java Servlet技术来开发动态的、高性能的Web应用程序。JSP的网页实际上是由HTML文件中加入的Java代码片段和JSP的特殊标记构成的。

因为JSP是Java的成员，所以JSP具有平台无关性，即实现了跨平台功能，实现了用户界面和程序代码的解耦合，使得业务逻辑与代码的耦合度更低，开发人员可以在不更改JSP程序的情况下修改用户的界面。

JSP即Java服务器界面，实质上是一个Servlet程序。JSP页面不仅包含HTML代码，还包含用于产生动态网页内容的Java代码，这些Java代码可以是Java Bean、SQL语句、RMI（远程方法调用）对象等。例如，一个JSP页面包含了用于产生静态网页的HTML代码，同时也包含了连接数据库的JDBC代码，正因为如此才能称之为动态网页。

1.2.2 JSP的优势

JSP可以看作Java Servlet的一种扩展，JSP在使用前必须被编译为Servlet，也就是Java类，然后被调用执行，Servlet所产生的Web页面是不能包含在HTML标签中的，因为它离不开Java类文件的支持。第一次访问JSP页面时，Web服务器会将此页面翻译成一个Java源文件，并编译成为.class扩展名的字节码文件，打开源代码可以发现该文件内的类是一个继承自HttpJspBase的Java类，而HttpJspBase这个类继承自HttpServlet类。随着学习的深入，用户将体验到JSP的很多优势。

1. 开发简单方便

在JSP中的编辑跟编写HTML文件基本一样，在处理表单方面极为方便。对于设置HTTP报头，JSP同样提供了丰富的方法，使得JSP开发者在编写通用功能时十分便捷，从而花费更多的时间在业务逻辑上。

2. 跨平台

Java本身就有跨平台的特性，因此JSP程序可以在支持Java的平台上开发运行，显然这对平台移植极为有利。当JSP在更换服务平台时，若不涉及数据库等相关操作，则几乎不做任何变动就可以完成服务平台的迁移。当需要更换Web服务器时，JSP同样可以做到不修改或者少量修改就在新的Web服务器中编译、运行。

3. 高效率和高性能

JSP可以是Servlet的扩展，因此Java虚拟机为每一个请求创建一个单独的线程而不是进程，如此系统就能很快地处理请求。同时JSP只会被编译一次，即在首次加载时需要编译，这样就加快了系统的响应速率。当一个请求处理结束之后，相关JSP映射的Java类并不会从内存中删除，而是被保留在内存中，当下次同样的请求发生时，系统会提供更快的响应速度。

4. 低成本

众所周知Java是开源的开发语言，JSP也是基于Java的开源环境开发的动态网页技术，所以就省去了商业的付费项目。再有，开发者可以从众多的Java IDE中选择一款适合自己的开发工具来进行项目研发，当然，也可以利用文本编辑器直接编写，只是这样比较耗时而且容易出错。还有许多的商业软件可以使用，但是通常来说JSP的开发总成本比采用其他技术要低一些。

综上所述，采用 JSP 动态网页技术是目前 Web 开发者的最佳选择。

1.2.3 JSP的执行顺序

在编写JSP程序时，我们要了解它的执行顺序，这对于后续的学习会有很大的帮助。JSP程序的执行过程大致如下：首先客户端向Web服务器提出请求，然后JSP引擎负责将页面转换为Servlet，此Servlet经过虚拟机编译生成类文件，再把类文件加载到内存中执行，最后由服务器将处理结果返回给客户端。整个流程如图1.2所示。

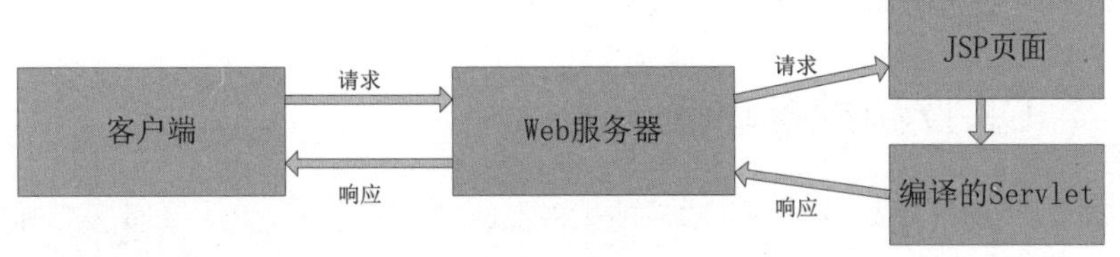

图1.2 JSP执行顺序

JSP页面代码会被编译成Servlet代码，所以从执行效率上来说肯定是没有Servlet快，但并不是每一次都需要编译JSP页面。当JSP被编译成类文件后，重复调用该JSP页面时，JSP引擎发现该JSP页面没有被改动过，那么就会直接使用编译后的类文件，而不会再次编译为新的Servlet。当然，如果页面被修改过，则需要重新加载编译。

1.2.4 一个JSP的简单实例

以下是 JSP 网页的一个简单实例，功能是输出 01~10 的相加结果：

```
-----------------------index.jsp--------------------------
01  <%@ page import="java.util.*" pageEncoding="UTF-8"%>
02  <!DOCTYPE HTML>
03  <html>
04    <head>
05      <title>JSP简单例子</title>
06    </head>
07    <body>
08      <%
09        int count=0;
10        for(int i=1;i<10;i++)
11        {
12          count+=i;
13        }
14        out.print("1到10的相加结果："+count);
15      %>
16    </body>
17  </html
```

注意：在第 01 行中，pageEncoding 标签可以设定字符类型，第 08~15 行为 Java 代码。

该程序的主要作用是利用JSP输出1~10的和，其中代码是由简单的HTML代码和JSP表达式构成的，JSP表达式中包括一段Java程序段。

1.3 安装开发环境

上一节简单介绍了JSP中的相关概念，让读者对JSP有一个初步的了解。本节将介绍Java开发环境的安装和搭建，使读者能了解JSP所需的开发环境。

1.3.1 安装JDK17和配置环境变量

JDK是由SUN公司（现被Oracle收购）提供的Java开发工具和API，首先从Oracle公司的官网下载Java SE。需要注意的是，JDK的版本较多，各版本之间还存在不兼容问题以及操作系统的兼容性问题。JDK版本分为长期支持版本（LTS）和短期支持版本（STS）。短期支持版本一般在半年内会被更新替换，所以尽量不要使用。目前JDK8、JDK11、JDK17都是官网支持的长期版本。本书使用的是Windows 10的64位操作系统，所以下载了jdk-17_windows-x64_bin.zip版本，读者可以根据自身的操作系统下载对应的版本。

注意：API 在 Java 世界中就是可以查找的参考内容，也就是 Java 方法的说明。

下载完成后解压到指定目录，笔者这里解压到F:\Program Files\Java\jdk-17.0.4路径下，解压完成后应该在bin目录下出现如图1.3所示的文件。

server	2022/7/26 14:59	文件夹	
api-ms-win-core-console-l1-1-0.dll	2022/6/28 15:30	应用程序扩展	12 KB
api-ms-win-core-console-l1-2-0.dll	2022/6/28 15:30	应用程序扩展	12 KB
api-ms-win-core-datetime-l1-1-0.dll	2022/6/28 15:30	应用程序扩展	12 KB
api-ms-win-core-debug-l1-1-0.dll	2022/6/28 15:30	应用程序扩展	12 KB
api-ms-win-core-errorhandling-l1-1-...	2022/6/28 15:30	应用程序扩展	12 KB
api-ms-win-core-file-l1-1-0.dll	2022/6/28 15:30	应用程序扩展	15 KB
api-ms-win-core-file-l1-2-0.dll	2022/6/28 15:30	应用程序扩展	12 KB
api-ms-win-core-file-l2-1-0.dll	2022/6/28 15:30	应用程序扩展	12 KB
api-ms-win-core-handle-l1-1-0.dll	2022/6/28 15:30	应用程序扩展	12 KB
api-ms-win-core-heap-l1-1-0.dll	2022/6/28 15:30	应用程序扩展	12 KB
api-ms-win-core-interlocked-l1-1-0.dll	2022/6/28 15:30	应用程序扩展	12 KB
api-ms-win-core-libraryloader-l1-1-0....	2022/6/28 15:30	应用程序扩展	13 KB
api-ms-win-core-localization-l1-2-0.dll	2022/6/28 15:30	应用程序扩展	14 KB
api-ms-win-core-memory-l1-1-0.dll	2022/6/28 15:30	应用程序扩展	12 KB
api-ms-win-core-namedpipe-l1-1-0.dll	2022/6/28 15:30	应用程序扩展	12 KB
api-ms-win-core-processenvironmen...	2022/6/28 15:30	应用程序扩展	13 KB
api-ms-win-core-processthreads-l1-1...	2022/6/28 15:30	应用程序扩展	14 KB
api-ms-win-core-processthreads-l1-1...	2022/6/28 15:30	应用程序扩展	12 KB
api-ms-win-core-profile-l1-1-0.dll	2022/6/28 15:30	应用程序扩展	11 KB
api-ms-win-core-rtlsupport-l1-1-0.dll	2022/6/28 15:30	应用程序扩展	12 KB
api-ms-win-core-string-l1-1-0.dll	2022/6/28 15:30	应用程序扩展	12 KB
api-ms-win-core-synch-l1-1-0.dll	2022/6/28 15:30	应用程序扩展	14 KB
api-ms-win-core-synch-l1-2-0.dll	2022/6/28 15:30	应用程序扩展	12 KB
api-ms-win-core-sysinfo-l1-1-0.dll	2022/6/28 15:30	应用程序扩展	13 KB
api-ms-win-core-timezone-l1-1-0.dll	2022/6/28 15:30	应用程序扩展	12 KB
api-ms-win-core-util-l1-1-0.dll	2022/6/28 15:30	应用程序扩展	12 KB
api-ms-win-crt-conio-l1-1-0.dll	2022/6/28 15:30	应用程序扩展	13 KB
api-ms-win-crt-convert-l1-1-0.dll	2022/6/28 15:30	应用程序扩展	16 KB

图1.3　JDK安装目录下的文件

该目录文件夹下包含了很多可执行文件，例如java.exe、javac.exe、javadoc.exe等。javac.exe是Java的编译器，用来编译Java文件，将它变为字节码。在DOC环境下，利用命令"javac test.java"的形式编译一个Java文件。java.exe是运行编译后的Java Class文件。java.exe也可以运行.jar文件，比如执行命令"java -jar test.jar"，就可以运行test.jar文件。

JDK11 之后默认不再安装 JRE，如果需要 JRE，可以进行手动配置。首先使用 Windows+R 组合键打开 cmd，切换到 JDK 安装目录。然后输入如下命令：

```
bin\jlink.exe --module-path jmods --add-modules java.desktop --output jre
```

执行完毕后没有任何输出，表示执行成功，如图 1.4 所示。生成成功后打开 JDK 目录，可以看到新增了 JRE 目录。

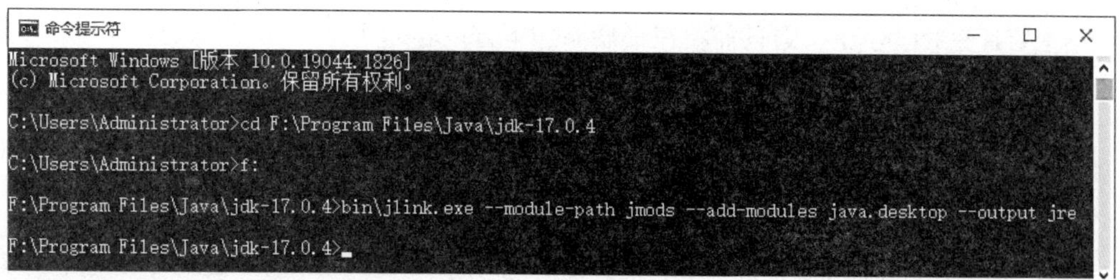

图1.4　生成JRE

接着就是配置Java的环境变量，配置的作用如下：

- 系统自动查找Java编译器路径。
- 服务器或其他需要依赖Java的程序在安装时需要知道Java路径。
- 编译和执行时指定Java路径。

配置环境变量的步骤如下：

01 以Windows为例，右击“此电脑”图标，在弹出的快捷菜单中单击“属性”命令，在打开的页面右侧单击“高级系统配置”选项，弹出“系统属性”对话框，在“系统属性”对话框中选择“高级”选项卡，单击“环境变量”按钮，如图1.5所示。

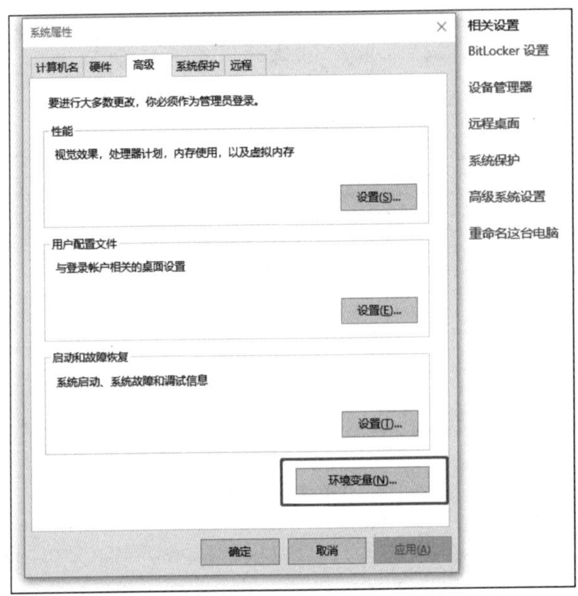

图1.5 配置环境变量

02 在弹出的“环境变量”对话框中单击“新建”按钮，分别设定JAVA_HOME、JRE_HOME、PATH 和 CLASSPATH 这 4 个新的环境变量。4 个变量值分别如下，其中%JAVA_HOME%、%JRE_HOME%表示引用值：

```
JAVA_HOME= F:\Program Files\Java\jdk-17.0.4
JRE_HOME= F:\Program Files\Java\jdk-17.0.4\jre
CLASSPATH= .;%JAVA_HOME%\lib;%JRE_HOME%\lib
PATH= ;%JAVA_HOME%\bin;%JRE_HOME%\bin
```

如果PATH变量已存在，则只需要在最后添加下方内容即可：

```
;%JAVA_HOME%\bin;%JRE_HOME%\bin
```

JDK9及以后的版本删除了tools.jar等JAR包，不再进行相关配置。配置完成后，打开cmd命令窗口，输入java -version命令，如果显示信息中JDK版本为17.0.4，则表示配置成功。

注意：CLASSPATH 的值前面有一个“.”符号。

1.3.2 安装IntelliJ IDEA开发工具

IntelliJ IDEA是当下流行的Java IDE（Integrated Developing Environment）开发工具，深受开发者的青睐，它能支持目前主流的技术和框架，擅长企业应用、移动应用和Web应用的开发，并且拥有丰富的插件。

从官网下载IntelliJ IDEA 2022.1.4，或者使用搜索引擎搜索IntelliJ IDEA2022.1.4，然后选择合适的链接下载。下载完成后双击安装包，安装界面如图1.6所示。

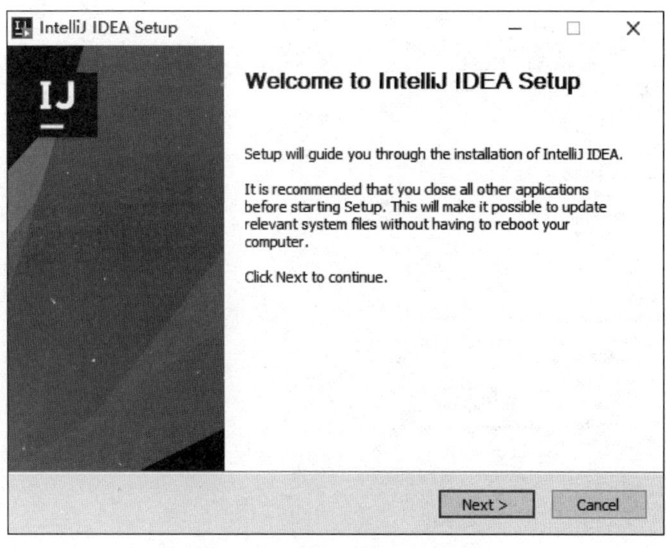

图1.6　IntelliJ IDEA的安装界面

每一步都选择默认选项，直接单击“Next”按钮，进入安装进程，安装完成后会生成桌面快捷方式，也可根据操作系统的位数双击安装目录内bin目录下的idea.exe或idea64.exe运行程序，例如D:\Program Files\JetBrains\IntelliJ IDEA 2022.1.4\bin\idea.exe目录。运行IntelliJ IDEA后会提示输入注册码，这时可以输入注册码或者选择试用30天。

IntelliJ IDEA安装成功后，在新建项目或导入项目时还需配置项目所需的JDK。例如，配置已安装的JDK时，可选择File | Project Structure| Platform Settings | SDKs选项来添加JDK，如图1.7所示。

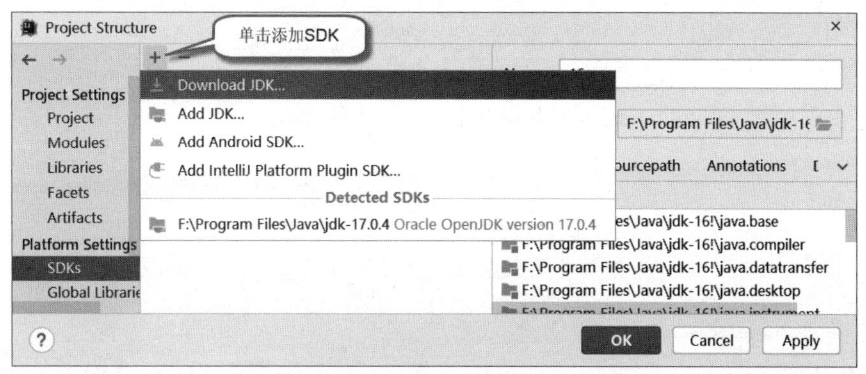

图1.7　IntelliJ IDEA查找JDK的选择界面

配置Tomcat时，可在Run | Edit Configurations中单击上方的+按钮，在左侧下拉框中选择Tomcat Server | Local选项，如图1.8所示，在弹出的窗口中配置Tomcat的安装路径，并在已安装的JDK版本中选择jdk17.0.4，如图1.9所示。

当然，还有很多Java IDE工具，例如MyEclipse、NetBeans、JCreator、JRun等。编辑JSP页面的工具还有Dreamveaver MX等。由于篇幅有限，这里就不一一介绍了。单从学习的角度来说，建议选择轻便型的IDE工具，如果是企业开发，那么人性化、智能化的IntelliJ IDEA将是最佳选择。

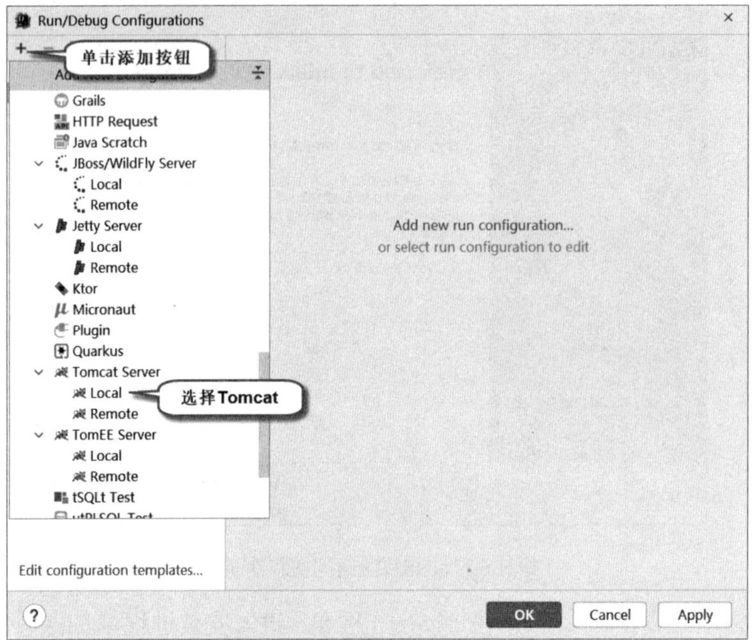

图1.8　打开配置Tomcat的界面

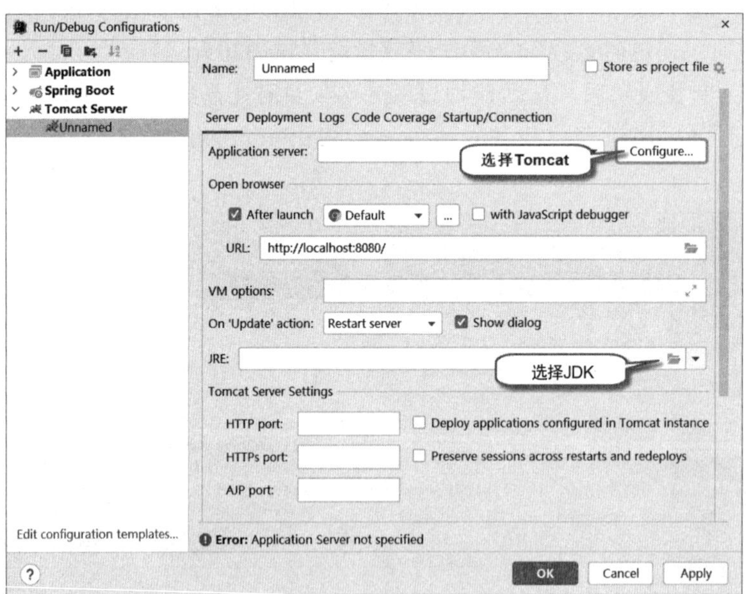

图1.9　选择Tomcat和JDK的界面

1.3.3　安装Tomcat 10服务器

1．目录结构

Tomcat 是轻量级的 Web 应用服务器。可以从官网下载最新的 Tomcat 服务器版本，本书使用的 Tomcat 10.0.22 版本。下载完成后，直接解压 Tomcat 文件到指定的目录下，例如 F:\Program Files\apache-tomcat-10.0.22 中。下面介绍一下 Tomcat 的目录结构：

- bin文件夹，包含Tomcat服务器启动和终止的批处理文件。例如startup.bat、startup.sh、shutdown.bat、shutdown.sh、catalina.bat、catalina.sh等。其中startup.bat、shutdown.bat、catalina.bat是Windows中的批处理文件，startup.sh、shutdown.sh、catalina.sh是Linux中的脚本文件。
- conf文件夹，包含Tomcat的配置信息，主要有server.xml和web.xml两个配置文件。在server.xml中可以更改服务器端口和Web默认的访问目录。
- lib文件夹，存放Tomcat运行中需要的JAR包文件，例如catalina.jar、servlet-api.jar、tomcat-dbcp.jar等JAR包，正因为有这些包的支持，Tomcat才可以运行Web应用程序。
- logs文件夹，存放执行Tomcat的日志文件。
- temp文件夹，存放Tomcat的临时文件信息。
- webapps文件夹，是Tomcat默认的Web文件夹。本身自带两个admin应用和manager应用。开发人员可以直接将Web应用存放在该文件夹中。
- work文件夹，存放Tomcat执行应用后的缓存。

2. 配置修改

在 Tomcat 服务器中，需要经常修改配置信息来满足系统的需求，例如在 server.xml 中可以更改服务器端口和 Web 默认的访问目录。

1）修改端口号

方法如下：

```
<Connector port="8080" protocol="HTTP/1.1" connectionTimeout="20000"
           redirectPort="8443" URIEncoding="UTF-8"/>
```

Tomcat默认的端口号为8080，也可以自定义其他端口，修改完毕后，保存server.xml，然后重启Tomcat服务器，这样服务器的端口就成功变更了。

2）修改Web默认的访问目录

方法如下：

```
<Host name="localhost"  appBase="webapps"
        unpackWARs="true" autoDeploy="true"
        xmlValidation="false" xmlNamespaceAware="false">
```

修改appBase中的文件夹地址。例如，将appBase的属性值webapps改为D:\test，修改后的文件如下：

```
<Host name="localhost"  appBase="D:\test"
        unpackWARs="true" autoDeploy="true"
        xmlValidation="false" xmlNamespaceAware="false">
```

这样就可以将Web默认的访问目录更改为D:\test，以后加载Web应用程序时就会在该目录下创建目录。

3）建立自身的Web目录

开发人员可以将应用部署在 Tomcat 服务器的默认 webapps 目录下，也可以部署在自己创建的目录下。方法如下：

（1）首先创建自身的目录 D:\test，其次配置 Web 目录，在 server.xml 文件的末尾</HOST>中加入如下语句：

```
<Context path="text" docBase="D:\test" debug="0" reloadable="true"></Context>
```

该语句的作用是将目录 D:\test 设置为 Tomcat 服务器的 Web 目录，将该文件的访问路径设置为 text。属性 docBase 的值为 D:\test，它是指应用的物理路径。修改后将 server.xml 文件进行保存。假设现在 test.jsp 页面位于 D:\test 目录下，那么页面的访问路径就为 http://localhost:8080/text/test.jsp。

（2）在 bin 文件夹下，可以修改 catalina.bat 或者 catalina.sh 来更改 Tomcat 的启动配置信息。例如增加 Java 的运行内存：

```
set JAVA_OPTS=-XX:PermSize=512M -XX:MaxPermSize=512m -Xms512m -Xmx1024m
```

更多的修改内容可参见 Tomcat 官网说明。

1.4 小　结

本章为读者介绍了搭建Web开发环境以及Web开发的基础知识，这些都是学习JSP开发技术之前必须掌握的知识。读者不仅需要理解和掌握本章内容，还需要亲自动手安装JDK、Tomcat、IntelliJ IDEA以及制作一个简单的JSP测试程序。

1.5 习　题

（1）JSP动态网页技术有哪些优势？

（2）JSP引擎的作用有哪些？

（3）JSP页面执行的顺序是什么？

（4）如何修改Tomcat的服务器端口？

（5）如何创建一个自己的Web目录？

第 2 章

JSP 基础语法：与编写 HTML 一样容易

上一章介绍了Web开发所需的基础知识以及如何安装JDK、IntelliJ IDEA以及Tomcat应用服务器，本章将要介绍JSP的基本语法、如何在JSP页面中嵌套Java以及JSP的指令等。从本章开始，读者将正式开始学习JSP技术。

本章主要涉及的知识点有：

- JSP中的注释表达式
- JSP中的声明表达式
- JSP中指令标签的作用和使用方法
- 运用HTML页面的元素、Java代码段、JSP标签创建JSP实例

2.1 JSP 注释

在编写规范的代码时，合理和必要的注释是至关重要的，这也是对开发人员的一项基本要求。编程语言的注释起到了说明、解释的作用，在实际执行过程中并不会被执行。在JSP页面中，注释可以归纳为两种：一种是原有的HTML的注释，另外一种是JSP的注释。

1. HTML 的注释

HTML的注释语法如下：

```
<!--  注释的内容 -->
```

例如：

```
<!--  注释的内容会被照搬到浏览器中 -->
```

2. JSP 的注释

JSP 的注释语法如下：

```
<%-- 注释的内容 --%>
```

例如：

```
<%-- 注释的内容不会被照搬到浏览器中 --%>
```

在实际应用中，JSP 通常会引入 Java 的注释，二者混合使用。因为 Java 的注释是在脚本内，而 JSP 的注释是在脚本外。

【例2.1】 JSP中不同的注释用法。

```
<%@ page  pageEncoding="UTF-8"%>
<!DOCTYPE HTML>
<html>
  <head>
    <meta content="text/html;charset=utf-8" http-equiv="Content-Type">
    <title>JSP注释例子</title>
  </head>
      <body>
        <!-- 该JSP注释可以在浏览器文件中看到 -->
        <br/>
        <br/>
         <table align="center">
              <tr>
                   <th>
                        <b>JSP注释</b>
                   </th>
              </tr>
         </table>
         <br/>
         <table>
              <tr>
                   <th>
                        <b>JSP注释</b>
                   </th>
              </tr>
         </table>
         <table align="right">
              <tr>
                   <th>
                        <b>JSP注释</b>
                   </th>
              </tr>
         </table>
        <%--该JSP注释无法在浏览器中看到 --%>
        <%
        //  这是脚本中的Java注释
        /*
              这也是脚本中的Java注释
        */
         %>
      </body>
    </html>
```

上述代码介绍了 JSP 的各种注释格式，其中第 09 行是“<!-- -->”格式，第 34 行是“<%-- --%>”格式。因为都是注释，所以在界面运行的结果中看不到注释中的内容，通过浏览器中的“查看源文件内容”功能可以看到 HTML 注释的内容。本例运行的结果如图 2.1 所示。

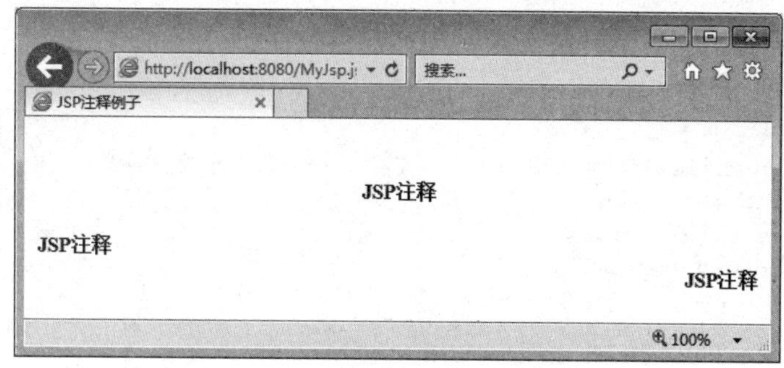

图2.1 JSP注释运行结果界面

注意：第 36 行的代码说明在 JSP 页面中可以嵌套 Java 注释。

2.2 JSP 声明

JSP声明用于定义JSP中的变量、方法以及静态方法，实际上它与在Java中定义一个全局变量、共用方法一样，JSP声明部分的变量和方法是可以直接在JSP页面中被调用的。

JSP 声明的基本语法有如下两种：

```
<%! 变量定义/方法定义/类 %>
```

或

```
<jsp:declaration>变量定义/方法定义/类</jsp:declaration>
```

注意：第 2 种语法已经过时，一般采用第 1 种声明语法。

JSP声明的结尾跟Java的结尾一样都以“;”结束，可以一次定义多个变量，利用“,”分隔。声明部分的变量和方法只在当前的页面中有效。

【例2.2】 演示变量、方法和类的声明。

```
------------------------declar.jsp---------------------------
<%@ page  import="java.util.*" pageEncoding="UTF-8"%>
<!DOCTYPE HTML>
<html>
  <head>
    <meta content="text/html;charset=utf-8" http-equiv="Content-Type">
    <title>JSP声明例子</title>
  </head>
  <%!
     int x,y=60,z;                          //多个声明以","分隔
     String name="John";
     Date date = new Date();
```

```
    %>
    <%!
      int add(int m,int n){             //计算两个数的和
          int result=0;
          result = m+n;
          return result;
      }
     %>
     <%!
      int chengji(int m,int n){              //计算两个数的乘积
          int result=0;
          result = m*n;
          return result;
      }
     %>
     <%!
      class Circle{                          //计算圆的面积
          double r;
          Circle(double r){
              super();                       //继承空的构造方法
              this.r = r;
          }
          double area(){                     //取整
              return  Math.floor(Math.PI*r*r);
          }
      }
      %>
   <body>
     <%
         out.println("我的名字: "+name);
         out.println("<br/><br/>");
         out.println("x的值为: "+x);
         out.println("<br/><br/>");
         out.println("y的值为: "+y);
         out.println("<br/><br/>");
         out.println("z的值为: "+z);
         out.println("<br/><br/>");
         out.println("现在的时间为: "+date);
         out.println("<br/><br/>");
         out.println("10与20的和: "+add(10,20));
         out.println("<br/><br/>");
         out.println("10与20的积: "+chengji(10,20));
      %>
     <br/>
     <br/>
     <%
         Circle c = new Circle(6);
         out.println("半径为6的圆面积为: "+c.area());
      %>
   </body>
 </html>
```

页面效果如图 2.2 所示。

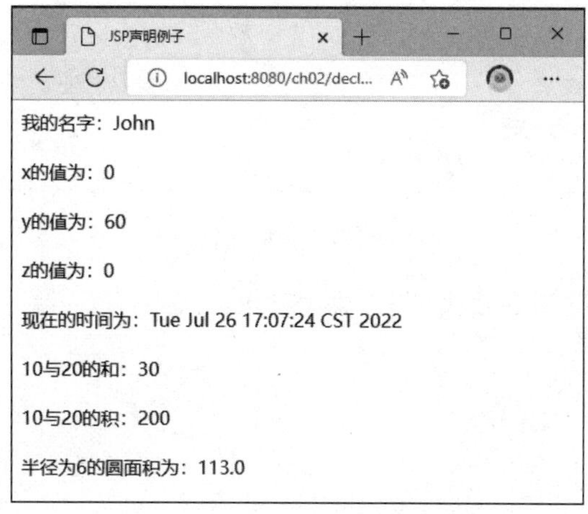

图2.2　declar.jsp页面的运行结果

2.3　JSP 表达式

JSP 表达式的作用是将动态信息显示在页面中，它的语法形式也有以下两种：

```
<%=变量或者表达式%>
```

或者

```
<jsp:expression>变量或者表达式</jsp:expression>
```

注意： 第 2 种形式已经很少使用，在一般的 IDE 工具中也不提供这种形式的表达式；第 1 种形式是目前主流的写法，本书实例也是使用该种形式书写的。

表达式的值由服务器负责计算，计算结果以字符串的形式发送到客户端。

下面看一个实例，JSP 页面利用 Date 类输出当前的时间。

```
01  <%@ page import="java.util.*" pageEncoding="UTF-8"%>
02  <!DOCTYPE HTML>
03  <html>
04    <head>
05      <meta content="text/html;charset=utf-8" http-equiv="Content-Type">
06      <title>JSP表达式例子</title>
07    </head>
08
09    <body>
10        当前时间为：</br>
11        <%=new Date()%>
12    </body>
13  </html>
```

在上述代码中，第 01 行导入 Date 类库，第 11 行直接引用 Date 对象，页面效果如图 2.3 所示。从页面的显示结果来看，文本“当前时间为：”被正常显示，其后的“</br>”（HTML 标记）使得后面显示的内容换行，只有“<%=new Date()%>”一行被替换成当前的时间。

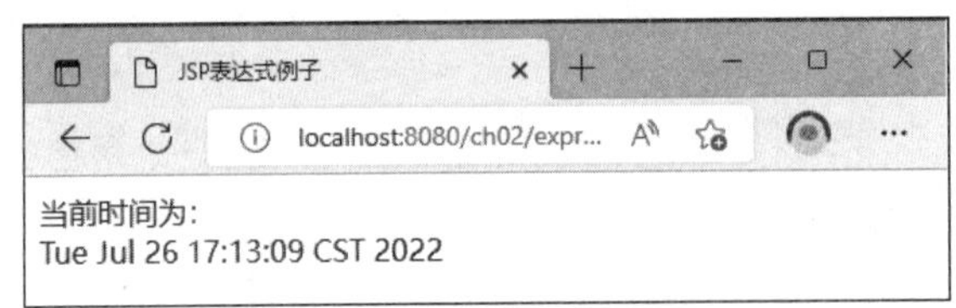

图2.3　JSP表达式运行结果

以下是上面 JSP 页面生成的源代码，对比 JSP 页面代码可以发现，只有第 01 行和第 11 行不同，其他的代码都一样。

```
01  <!DOCTYPE HTML>
02  <html>
03    <head>
04      <meta content="text/html;charset=utf-8" http-equiv="Content-Type">
05      <title>JSP表达式例子</title>
06    </head>
07
08    <body>
09        当前时间为：</br>
10        Tue Jul 26 17:13:09 CST 2022
11    </body>
12  </html>
```

通过对比可以发现，JSP 页面中的 HTML 元素在源代码中被原样保留，只有 JSP 代码会发生改变。因此，JSP 页面中的静态代码都是利用 HTML 模板来编写的。

注意：JSP 页面生成的源代码一般是通过浏览器来查看的。对于 IE 浏览器或者是 IE 内核的浏览器，可以在待查看的页面上右击，然后在弹出的快捷菜单中选择“查看源文件”命令。

2.4　JSP 指令

上一节介绍了JSP表达式的用法以及示例，通过学习，读者已经可以应用前面章节的知识来编写简单的JSP动态网页了，但要编写出功能强大、复杂的JSP页面还需要学习更多的内容。本节将介绍JSP的另一元素：指令。JSP指令包括page指令、include指令、taglib指令。

2.4.1　与页面属性相关的page指令

page 指令用来设置 JSP 页面的属性和相关功能，基本语法形式有如下两种：

```
<%@ page attribute1="value1" […attribute n="value n"]%>
```

或者

```
<jsp:directive.page attribute1="value1" […attribute n="value n"] />
```

注意：现在很少使用第 2 种形式，大多使用第 1 种表达形式。

page指令有多种属性，使用最多的是language、import、pageEncoding这3个。其中language

是必须设置的，由于目前JSP页面使用Java语言，所以其默认值是java；import用于声明需要导入的包；pageEncoding用于设置页面的编码。在所有的属性中除import可以声明多个外，其他的属性都只能出现一次。page指令中的其他属性如表2.1所示。

表 2.1　page 指令的常用属性

属性和属性值	说　明
session="true|false"	限定 session 对象是否可用，默认为 true
autoFlush="true|false"	指明缓冲区域是否自动清除，默认为 true
info="text"	描述该 JSP 页面的相关信息
errorPage="URL"	当页面产生异常时跳转的路径
isErrorPage="true|false"	指定该 JSP 页面是否为处理异常错误的页面，当设定为 true 时才能使用 exception 对象，默认为 false
isThreadSafe="true|false"	是否允许多线程使用，默认为 true
buffer="8kb"	输出流是否有缓冲区，默认为 8KB
contentType="text/html; charset=UTF-8"	设定 MIME 类型和编码属性。编码属性一般设置为 UTF-8，MIME 类型还有很多，比如 application/vnd.ms-excel 表示 Excel 电子表格，image/gif 表示 GIF 图像等
extends="class"	指明由该 JSP 页面产生的 Servlet 所继承的父类

2.4.2　引入文件的include指令

include 指令用于在 JSP 页面生成 Servlet 时引入需要包含的页文件。这个页文件既可以是 HTML 文件，也可以是 JSP 文件，还可以是其他文件（例如.js 文件）。include 指令的作用是在标签插入的位置插入静态的文件内容，使它与 JSP 文件组合成新的 JSP 页面，然后由 JSP 引擎翻译成 Servlet 文件。这样做有如下两个好处：

- 页面的代码可以复用，因为被引入的文件是静态文件，所以在其他的JSP页面中也可以导入。
- JSP页面的代码结构清晰易懂，维护也比较简单。

include指令的基本语法如下：

```
<%@include file="URL"%>
```

其中 file 属性指向要包含的文件，一定要注意引入的路径是否正确，一旦路径出错，在编译的时候就不能通过。

include指令经常用来包含网站中经常出现的相同页面。例如，一般情况下网站为每个页面都设置了导航栏，把它放在页面的顶端或者左边，这部分代码在每个页面都重复，可以用include来解决，为开发者省去重复动作。下面的实例将对include指令进行演示。

【例 2.3】　演示 include 指令。

index_include.jsp 页面演示 include 指令的用法，其源代码如下：

```
----------------- index_include.jsp -----------------
<%@ page pageEncoding="UTF-8"%>
<!DOCTYPE HTML>
<html>
<head>
```

```
<meta content="text/html;charset=UTF-8" http-equiv="Content-Type">
<title>JSP include指令演示</title>
</head>
<body>
<%@include file="John.html" %>
<br/>
<div align="center">JSP include指令演示</div>
<%@include file="copyRight.jsp" %>
</body>
</html>
```

John.html是内嵌页面，用来显示内容，其源代码如下：

```
----------------- John.html ----------------
<table align="center" border="1" width="600">
    <tr>
        <td align="center" height="18" width="100%">
            <span>This is top page.</span>
        </td>
    </tr>
</table>
```

copyRight.jsp是显示版权信息的页面，其源代码如下：

```
----------------- copyRight.jsp ----------------
<%@ page import="java.util.*,java.text.SimpleDateFormat"
pageEncoding="UTF-8"%>
<%
    Date d = new Date();
    SimpleDateFormat sdf = new SimpleDateFormat("yyyy");
    String t = sdf.format(d);
    String copyRightsMess = "John 版权所有 2010-"+t;
%>
<br/>
<div align="center" ><%=copyRightsMess%></div>
```

本例的页面效果如图 2.4 所示。

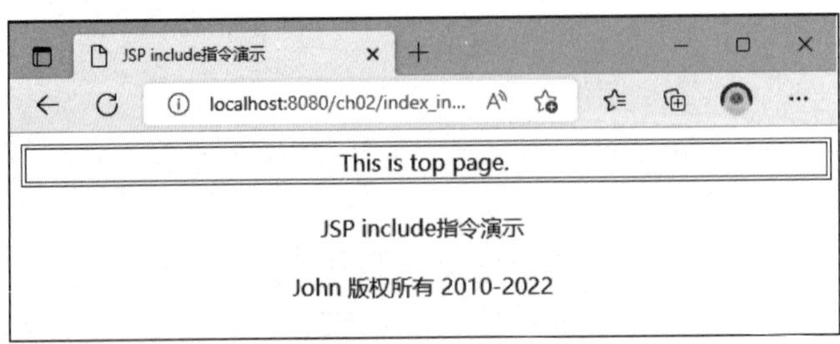

图2.4　index_include.jsp 页面的运行结果

注意：include 指令在 JSP 页面被转换成 Servlet 时才将文件导入，这与<jsp:include>动作不同。

2.4.3　与标签相关的taglib指令

taglib 指令（又名标签指令）是 JSP 新增的一个指令，通过该指令用户可以自定义新的标签在页面中执行。taglib 指令的语法如下：

```
<%@taglib  uri="tagliburl" prefix="tagPre" %>
```

其中uri属性用来表示自定义标签库的存放位置；prefix属性是用来区分不同标签库的标签名，在页面中引用标签也是以prefix开头的。

现在比较流行的JSTL、EL标签是如何在JSP中使用的呢？下面通过实例来说明如何利用JSTL标签输出数字01~10。

注意： JSTL标签库会在后续的章节中进行详细说明，这里先演示它的基本用法。

【例2.4】 演示taglib指令。

```
----------------- index.jsp ----------------
<%@ page import="java.util.*" pageEncoding="UTF-8"%>
<%@ taglib uri="http://java.sun.com/jstl/core_rt" prefix="c"%>
<!DOCTYPE HTML>
<html>
  <head>
    <meta content="text/html;charset=UTF-8" http-equiv="Content-Type">
    <title>JSP taglib指令演示</title>
  </head>
  <body>
       <table>
           <tr>
               <td>输出值</td>
           </tr>
           <c:forEach begin="1" end="10" var="num">
              <tr>
                 <td><c:out value="${num}"></c:out></td>
              </tr>
           </c:forEach>
       </table>
  </body>
</html>
```

从上述代码可以看出，JSTL标签使得JSP页面十分简洁，它不需要定义或初始化对象、方法。但凡事都不能只看表面，应该注重本质，标签的定制在页面显示时是如此简单，但其在编制时却是一个复杂的过程，它通过一定的编程步骤将JSP代码和Java代码联系起来。标签定制的最大好处就是使得开发者的职责分工更加明细：标签定制者无须关注业务逻辑的实现，页面编程人员直接使用标签即可。这样两者就不会冲突，分工明确。下面简单介绍定制标签的过程。

1. 标签库

标签库是用于定义标签的XML文件。该文件中包含了标签属性、标签名称、JSP在处理标签时所需的类文件等信息。在使用自定义标签时，必须指明标签库所在的目录；在执行标签功能时，通过定义标签属性来决定显示的内容。下面看一段标签的源代码，其中包括了标签名定义、属性定义以及标签对应的Java类。

```
<?xml version="1.0" encoding="UTF-8" ?>
<!DOCTYPE taglib
  PUBLIC "-//Sun Microsystems, Inc.//DTD JSP Tag Library 1.2//EN"
  "http://java.sun.com/dtd/web-jsptaglibrary_1_2.dtd">
<taglib>
```

```
06      <tlib-version>1.0</tlib-version>
07      <jsp-version>1.2</jsp-version>
08      <short-name>lms</short-name>
09      <!--  页面引用的url地址 -->
10      <uri>/lms-tags</uri>
11
12      <tag>
13          <!--  标签名称 -->
14          <name>page</name>
15          <!--  标签处理类 -->
16          <tag-class>com.common.model.PageTag</tag-class>
17          <body-content>JSP</body-content>
18          <!--  标签描述 -->
19          <description>分页</description>
20          <!--  标签属性 -->
21          <attribute>
22              <!--  标签属性名称 -->
23              <name>object</name>
24              <!--  标签属性是否为必需，true是必需，false不是必需 -->
25              <required>true</required>
26              <!--  标签属性对应的处理类 -->
27              <type>com.common.model.PageObject</type>
28              <!--  标签属性描述 -->
29              <description>PageObject对象</description>
30          </attribute>
31      </tag>
32
33      <tag>
34          <!--  标签名称 -->
35          <name>substring</name>
36          <!--  标签处理类 -->
37          <tag-class>com.gsta.common.tag.SubStringTag</tag-class>
38          <description>substring</description>
39          <attribute>
40              <name>content</name>
41              <required>true</required>
42              <type>java.lang.String</type>
43          </attribute>
44      </tag>
45  </taglib>
```

在上述代码中，第 06~08 行代码分别定义了标签的版本、JSP 的版本、标签的简称，第 12~31 行代码定义了 page 标签（第 21~30 行代码定义 page 标签的属性）。在定义属性名称的时候，可以定义该标签的数据类型以及该属性是否为必需的。例如第 41 行，JSP 页面就会根据其定义在执行时检查相关的错误信息。

2．标签处理类

其实标签处理类就是一个 Java 类，该 Java 类实现标签接口（jakarta.servlet.jsp.Tag），当自定义标签被 JSP 处理时即可被执行。每个标签中都需要执行的方法是：

```
Public int doStartTag() throws JspException
```

在标签中定义属性后，在Java类中就必须有这些属性并且设定其get/set方法。例如，在上面标签的源代码中与substring标签对应的是一个用于返回结果的Java类，在定义substring标签时，必须定义其content属性并给出get/set方法，代码如下：

```
private String content;
public void setContent(String content) {
    this.content = content;
}

public void setLength(int length) {
    this.length = length;
}
```

上述代码给出了属性 content 的 get/set 方法。

2.5　JSP 动作

上一节介绍了JSP的page、include、taglib这3个指令，这些指令不仅可以帮助JSP编程人员编写出复杂的JSP页面，而且能简化JSP页面代码，为后期维护带来便利。本节将介绍JSP中的另外一组元素：JSP动作。在JSP中一共有13个动作：<jsp:include>、<jsp:forward>、<jsp:plugin>、<jsp:param>、<jsp:useBean>、<jsp:getProperty>、<jsp:setProperty>、<jsp:output>、<jsp:attribute>、<jsp:element>、<jsp:body>、<jsp:params>、<jsp:fallback>。其中<jsp:include>、<jsp:forward>、<jsp:param>使用较多，下面将详细介绍这3个动作。

2.5.1　<jsp:include>动作

<jsp:include>动作与 include 指令十分相似，它们的作用都是引入文件到目标页面。<jsp:include>的语法格式如下：

```
<jsp:include page="relative URL" flush="true"/>
```

例如：

```
<jsp:include page="top.html" flush="true"/>
```

其中，page 是<jsp:include>动作的一个必选属性，它指明了需要包含的文件的路径，该路径可以是相对的，也可以是绝对的。flush 属性用于指定输出缓存是否转移到被导入文件中，如果指定为 true，则包含在被导入文件中；如果指定为 false，则包含在源文件中。

注意： page 属性所指的绝对路径不是文件系统的绝对路径，而是以应用系统目录为根目录的绝对路径。通常情况下，Web 应用下的 WEB-INF 目录就是根目录，一般在设定 URL 时指定相对路径，这样简单一些。

使用<jsp:include>动作可以达到与include指令相同的界面效果。例如，例2.3中介绍的使用include指令演示界面页脚的例子，直接将include指令替换成<jsp:include>动作，它们在页面显示结果中没有什么不同。

使用<jsp:include>包含页脚信息的代码如下：

```
----------------- index_include.jsp ----------------
01  <%@ page import="java.util.*" pageEncoding="UTF-8"%>
02  <!DOCTYPE HTML>
03  <html>
04      <head>
05        <meta content="text/html;charset=UTF-8" http-equiv="Content-Type">
06        <title>JSP include动作演示</title>
07      </head>
08      <body>
09          <jsp:include page=" John.html " />
10          <br/>
11          <div align="center">JSP include指令演示</div>
12          <jsp:include  page="copyRight.jsp"  flush="true"/>
13      </body>
14  </html>
```

在上述示例代码中，将第09行代码换成<jsp:include>动作。可以看出，两种不同的包含方式达到了相同的效果。

但是，<jsp:include>动作与include指令还是有些不同的：首先，<jsp:include>动作是在页面被访问时导入的，而include指令是由JSP引擎在编译时导入的；其次，在include指令中，被包含的文件会同主页面一起被编译为一个Servlet类文件，而<jsp:include>动作包含的文件跟主页面会是相对独立的两个文件，在编译时会被编译成两个Servlet类文件，因此<jsp:include>在效率上稍微慢些。

【例2.5】 演示<jsp:include>动作。

```
<%@ page  import="java.util.*" pageEncoding="UTF-8"%>
<!DOCTYPE HTML>
<html>
  <head>
     <title>新闻摘要</title>
     <meta http-equiv="pragma" content="no-cache">
     <meta http-equiv="cache-control" content="no-cache">
     <meta http-equiv="expires" content="0">
     <meta http-equiv="keywords" content="keyword1,keyword2,keyword3">
     <meta http-equiv="description" content="This is my page">
  </head>
  <body>
      <br/>
      <table align="center">
          <tr>
              <th>中华新闻网</th>
          </tr>
      </table>
      <br/>
      <p>最新新闻摘要：
         <ul>
             <li><jsp:include page="news1.html" /></li>
             <li><jsp:include page="news2.html"/></li>
             <li><jsp:include page="news3.html"/></li>
         </ul>
      </p>
  </body>
</html>
```

在上述代码中，第 22~24 行代码应用了 JSP 动作，页面运行效果如图 2.5 所示。

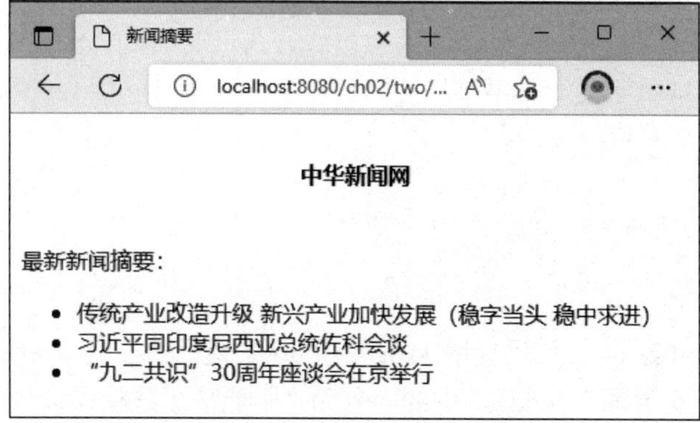

图2.5　<jsp:include>运行结果

2.5.2　<jsp:forward>动作

<jsp:forward>动作的作用是转发请求到另外一个页面中，在请求过程中会连同请求的参数数据一起被转发到目标页面中，目标页面通过 request.getParameter 方法获得参数值进行进一步处理。<jsp:forward>的基本语法如下：

```
<jsp:forward page="relative URL">
```

<jsp:forward>只有一个page属性，其值为相对的URL。例如，执行如下代码将跳转到错误页面：

```
<jsp:forward page="/error.jsp"/>
```

下面来看一个<jsp:forward>的实例：如果随机数能被 2 整除，则跳转到 even.jsp，否则跳转到 odd.jsp。

【例2.6】　演示<jsp:forward>动作。

```
----------------------- forwardE.jsp--------------------------
<%@ page  import="java.util.*" pageEncoding="utf-8"%>
<%
    String url = "";
    int random = (int)(Math.random()*10);//产生10以内的随机数
    int m = random%2;
    switch(m){
        case 0:
            url = "even.jsp";
            break;
        case 1:
            url = "odd.jsp";
            break;
    }
 %>
<!DOCTYPE HTML>
<html>
  <head>
    <title>My JSP 'forwardE.jsp' starting page</title>
```

```
19        <meta http-equiv="pragma" content="no-cache">
20        <meta http-equiv="cache-control" content="no-cache">
21        <meta http-equiv="expires" content="0">
22        <meta http-equiv="keywords" content="keyword1,keyword2,keyword3">
23        <meta http-equiv="description" content="This is my page">
24     </head>
25
26     <body>
27        <jsp:forward page="<%=url %>"/>
28     </body>
29  </html>
```

在上述代码中，第 02~14 行代码使用 Java 判断 url 的返回值，其中 even.jsp 中的内容为“能被 2 整除，跳转到偶数页界面”，odd.jsp 中的内容为“不能被 2 整除，跳转到奇数页界面”，效果如图 2.6 所示。

图2.6　<jsp:forward>的运行结果

从上述代码的运行结果可以看出，界面已经被重定向到odd.jsp页面，但是浏览器中的地址仍显示的是跳转前的地址，这也是<jsp:forward>动作的一个特点，即相对于请求者而言，所看到的响应仍然是原先请求的页面给出的，请求者并不会获得转发后的页面地址，因此相对来说，请求具有隐蔽性。

注意：<jsp:forward>动作中的 URL 页面只能是该 Web 应用中的文件，而不能是该 Web 应用之外的文件。

2.5.3　<jsp:param>动作

<jsp:param>动作用来传递参数信息，它经常与其他动作一起使用，例如与<jsp:forward>、<jsp:include>等结合使用，传递主页面的参数到目标页面。<jsp:param>动作的基本语法如下：

```
<jsp:param name="参数名称" value="参数值">
```

例如：

```
<jsp:param name="username" value="李四">
```

例 2.7 和例 2.8 分别是<jsp:param>和<jsp:include>、<jsp:forward>结合使用的实例，这两个实例都由主页面和子页面构成，子页面从主页面中获得参数值。

【例2.7】　<jsp:param>和<jsp:include>结合使用。

```
----------------------- main.jsp--------------------------
<%@ page  import="java.util.*" pageEncoding="utf-8"%>
<%request.setCharacterEncoding("utf-8"); //设定页面传递参数的编码格式%>
```

```
03  <!DOCTYPE HTML>
04  <html>
05    <head>
06      <title>主页面</title>
07    </head>
08
09    <body>
10
11            <table align="center">
12                <tr>
13                    <th>主页面</th>
14                </tr>
15            </table>
16
17        <jsp:include page="subPage.jsp" >
18             <jsp:param value="John" name="userName"/>
19             <jsp:param value="10086" name="passwd"/>
20             <jsp:param value="北京丰台" name="address"/>
21        </jsp:include>
22    </body>
23  </html>
```

在上述代码中，第17行代码使用<jsp:include>动作，第18~20行代码使用<jsp:param>参数。

```
----------------------- subPage.jsp--------------------------
<%@ page import="java.util.*" pageEncoding="utf-8"%>
<%
    String userName= request.getParameter("userName");
    String passwd= request.getParameter("passwd");
    String address= request.getParameter("address");
    System.out.println(address);
 %>
<!DOCTYPE HTML>
<html>
  <head>
    <title>子页面</title>
  </head>
  <body>

        <table align="center">
            <tr>
                <th>子页面：人员信息</th>
            </tr>
        </table>


        <table align="center">
              <tr>
                 <td>用户名：<%=userName%></td>
              </tr>
              <tr>
                 <td>密    码：<%=passwd%></td>
              </tr>
              <tr>
                 <td>用户地址：<%=address%></td>
              </tr>
```

```
32              </table>
33
34      </body>
35    </html>
```

在上述代码中，第 22~32 行代码用于获取主页面传递的参数值。

注意：所有的页面编码格式要统一，否则在传递中文时会出现乱码问题。request.setCharacterEncoding 方法用于设定传递参数的编码格式，在传递中文时要特别注意。在 URL 中包含参数传递时，可能会有中文乱码或者特殊字符问题，需要对这些参数进行编码处理。在 JavaScript（简称 JS）中，通过调用函数 encodeURI 或者 encodeURIComponent 实现编码；在 JSP 脚本中，通过方法 URLEncoder.encode 实现转码。

页面效果如图2.7所示。

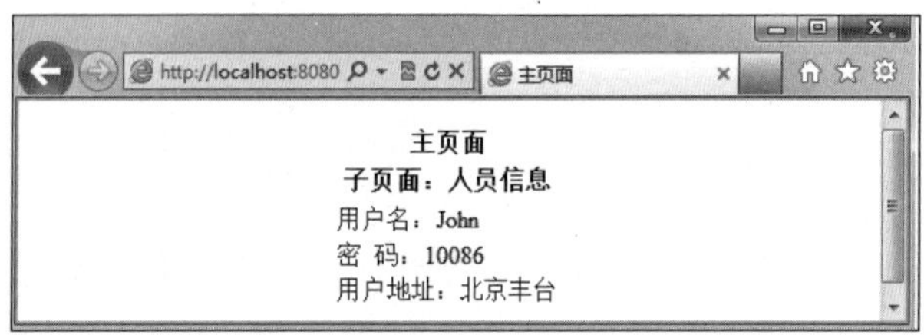

图2.7 <jsp:param>和<jsp:include>结合使用的运行结果

【例2.8】 <jsp:param>和<jsp:forward>结合使用。

```
------------------------ main.jsp---------------------------
01  <%@ page  import="java.util.*" pageEncoding="utf-8"%>
02  <%request.setCharacterEncoding("utf-8"); //设定页面传递参数的编码格式%>
03  <!DOCTYPE HTML>
04  <html>
05    <head>
06      <title>主页面</title>
07    </head>
08
09    <body>
10
11            <table align="center">
12                <tr>
13                    <th>主页面</th>
14                </tr>
15            </table>
16
17        <jsp:forward page="subPage.jsp" >
18             <jsp:param value="Smith" name="userName"/>
19             <jsp:param value="10086" name="passwd"/>
20             <jsp:param value="北京丰台" name="address"/>
21        </jsp:forward>
22    </body>
23  </html>
```

在上述代码中，第02行设置请求编码，第17行代码使用<jsp:forward>动作，第18~20行代码使用<jsp:param>参数。

```
----------------------- subPage.jsp--------------------------
01  <%@ page import="java.util.*" pageEncoding="utf-8"%>
02  <%
03     String userName= request.getParameter("userName");
04     String passwd= request.getParameter("passwd");
05     String address= request.getParameter("address");
06     System.out.println(address);
07   %>
08  <!DOCTYPE HTML>
09  <html>
10    <head>
11      <title>子页面</title>
12    </head>
13    <body>
14
15            <table align="center">
16                <tr>
17                    <th>子页面：人员信息</th>
18                </tr>
19            </table>
20
21
22            <table align="center">
23                 <tr>
24                    <td>用户名：<%=userName%></td>
25                 </tr>
26                 <tr>
27                    <td>密　码：<%=passwd%></td>
28                 </tr>
29                 <tr>
30                    <td>用户地址：<%=address%></td>
31                 </tr>
32            </table>
33
34     </body>
35   </html>
```

在上述代码中，第 02~07 行代码用于获取主页面传递的参数值，第 24~30 行代码用于显示获取的参数值，页面效果如图 2.8 所示。

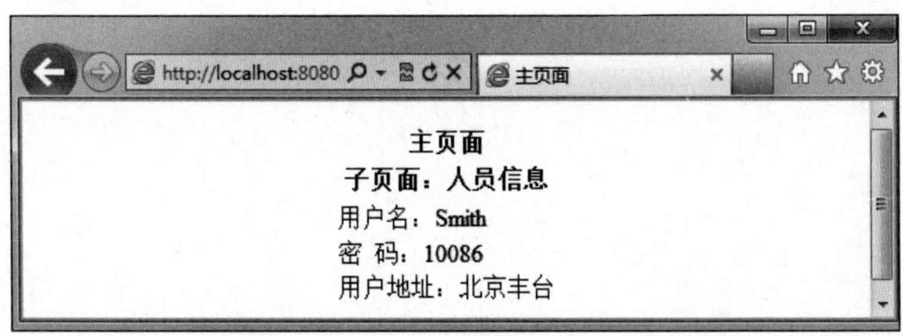

图2.8　<jsp:param>与<jsp:forward>结合使用的运行结果

从上述实例可以看出，<jsp:param>在<jsp:include>、<jsp:forward>中的用法基本一样，只要将<jsp:include>替换成<jsp:forward>动作就可以了。

2.6 小　结

本章介绍了JSP注释、JSP声明、JSP表达式、JSP指令以及JSP动作，这些都是编写JSP页面所必须了解和掌握的知识，其中掌握JSP指令是编写JSP页面的必要条件，JSP动作能简化JSP页面编程。在JSP动作中<jsp:include>和<jsp:forward>会被经常用到，读者必须掌握它们的用法。

2.7 习　题

（1）在JSP中应该如何声明一个变量？

（2）JSP的注释有哪几种？

（3）JSP的表达式如何表达？

（4）JSP的指令有哪几种？

（5）JSP的动作有哪些？include指令和<jsp:include>动作有什么不同？

（6）JSP页面由哪三类元素组成？各有什么作用？

（7）编写一个JSP页面，网页有上部、中间、下部三部分，通过由一个主JSP页面包含三个子页面的方式实现页面效果。要求利用两种方式实现，效果如图2.9所示。

图2.9　例图

第 3 章 JSP 内置对象

JSP内置对象是指可以直接在JSP页面中使用的对象，使用前不需要声明它们。若能熟悉并了解JSP内置对象，则可以方便读者更好地操作页面、开发页面，完成更复杂的业务流程。

本章主要涉及的知识点有：

- 讲解8个内置对象request、response、session、application、out、page、config、pageContext的作用和使用方法
- 了解JSP的4个作用域

3.1 request 对象

本节将要介绍request对象的作用范围及其常用的方法。用户每访问一个页面，就会产生一个HTTP请求。这些请求中一般都包含了请求所需的参数值或者信息，如果将request对象看作客户端请求的一个实例，那么这个实例就包含了客户请求的所有数据。因此，可以通过request来获取客户端和服务器端的信息，如IP地址、传递的参数名和参数值、应用系统名称、服务器主机名称等。

3.1.1 request对象的常用方法

request对象的常用方法如表3.1所示。

表 3.1 request 对象的常用方法

方 法	方法说明
getParameter()	取得请求中指定的参数值，返回 String 类型。如果有必要，需要将取得的参数值转换为合适的类型
getParameterValues()	将同名称的参数一次性地读入 String 类型的数组中

（续表）

方　法	方法说明
getParameterNames()	获取参数名称，返回枚举类型
getMethod()	获取客户提交信息的方式，即 post 或 get
getServletPath()	获取 JSP 页面文件的目录
getHeader()	获取 HTTP 头文件中的指定值，例如 accept、user-agent、content-type、content-length 等
getRemoteAddr()	获取客户的 IP 地址
getServerName()	获取服务器的名称
getServerPort()	获取服务器的端口号
getContextPath()	获取项目名称，如果项目为根目录，则得到空的字符串
getHeaders()	获取表头信息，返回枚举类型

3.1.2 使用request对象接收请求参数

request对象有两种方法获得请求参数值：一个是getParameter()，另一个是getParameterValues()。下面将演示如何使用getParameter()方法获取页面请求参数。

【例 3.1】 获取网页请求参数。

getParameter.jsp 用于获取页面参数值，代码如下：

```
------------------ getParameter.jsp ----------------
01  <%@ page  pageEncoding="UTF-8"%>
02  <!DOCTYPE HTML>
03  <html>
04    <head>
05      <title>' getParameter.jsp'</title>
06      <meta http-equiv="Content-Type" content="text/html;charset=utf-8">
07    </head>
08    <body style="text-align: center;">
09
10        <%
11          String name= request.getParameter("name");
12          String city = request.getParameter("city");
13          if(name!=null&&city!=null)
14          {
15        %>
16          <p>Welcome  <%=name %>,您所在的城市是<%=city %></p>
17        <%
18          }else{
19        %>
20          <p>欢迎访问本页面！</p>
21        <%
22          }
23        %>
24
25    </body>
26  </html>
```

在上述代码中，第 11、12 行代码分别用于获得 name、city 的参数值。

在浏览器地址栏上输入"http://localhost:8080/ch03/getParameter.jsp?name=John&city= Beijing"，页面效果如图3.1所示。

注意：在传递 URL 参数时，页面地址后使用"？"连接请求参数；参数由"="相连，表示参数名和参数值；一个请求携带多个参数时使用"&"连接。

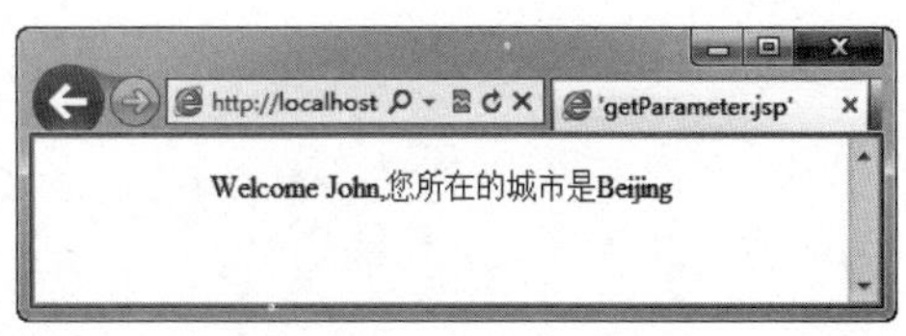

图3.1　运行结果

实际上，上述传递参数的方法在实际的开发中相比表单提交参数方法要较少使用，因为这种提交参数的方法完全将参数暴露出来，不利于隐私保护，所以我们一般采取表单提交的方式传递参数。

在表单form中，有两个非常重要的属性：action和method。属性action指明表单提交后的数据跳转到指定页面并处理参数。method有两个值，分别是get和post，通常指定为post，因为指定为get时，在表单中设定的参数和参数值将附加到页面地址的末尾以参数的形式提交，这样会暴露参数值，不安全；而指定为post并提交表单时，表单中的参数将被作为请求头中的信息发送，不会在目标地址中添加任何参数，这样也就不会暴露参数值，所以出于安全考虑，一般会选择post提交。

【例 3.2】　指定为 post 并提交，以获取网页请求参数。

ex3_1.jsp页面将用户填写的内容提交给getParameter.jsp，页面关系如图3.2所示。

图3.2　页面关系图

ex3_1.jsp 是用户提交信息的页面，源代码如下：

```
----------------- ex3_1.jsp ----------------
<%@ page  pageEncoding="UTF-8"%>
<!DOCTYPE HTML>
<html>
  <head>
    <title>'ex3_1.jsp'</title>
     <meta http-equiv="Content-Type" content="text/html;charset=utf-8">
  </head>

  <body >

      <form action="getParameter.jsp" method="post">
          <table align="center">
                 <tr>
                     <td>姓名</td>
                     <td><input type="text" name="name" value=""/></td>
                 </tr>
                 <tr>
                     <td>城市</td>
                     <td><input type="text" name="city" value=""/></td>
                 </tr>
```

```
21                  <tr>
22                      <td><input type="submit" value="提交"/></td>
23                      <td><input type="reset" value="重置"/></td>
24                  </tr>
25              </table>
26          </form>
27
28      </body>
29  </html>
```

在上述代码中，第 11~26 行代码用于提交表单 form，其中 action 为 getParameter.jsp、method 为 post，第 12~25 行代码用于输入姓名和城市。此处的 getParameter.jsp 与例 3.1 中的 getParameter.jsp 代码相同。

在浏览器中访问ex3_1.jsp并在“姓名”和“城市”输入框中输入值“John”和“Beijing”，单击“提交”按钮后，可以看到显示的结果与例3.1的结果是一样的。

注意：在测试时，读者可以设定 method 的两个值分别进行测试，并查看地址栏的变化情况。

3.1.3 请求中的中文乱码的处理

在上述实例中，请求信息中的参数值都是英文，但在实际开发中，请求信息通常包含中文。在上述的例子中，若是使用中文，例如在表单的“姓名”输入框中输入中文字符，则在提交之后显示的值不是中文而是“？”或者其他字符，如图3.3所示。

图3.3 显示奇怪的字符

产生这种错误的原因是请求信息所使用的字符集与页面使用的字符集不同，有如下 3 种方法可以解决这个问题。

- 第1种方法是在接收请求的页面中规定请求字符编码的代码，例如在例3.2的getParameter.jsp页面中添加语句“request.setCharacterEncoding("utf-8")”即可。
- 第2种方法是在取得参数值后，通过转码的方式将值转为合适的字符集。例如，将例3.1中获得参数值的代码修改为如下形式：

```
    String name= new String(request.getParameter("name").getBytes
("ISO-8859-1"), "utf-8");
    String city= new String(request.getParameter("city").getBytes
("ISO-8859-1"), "utf-8");
```

注意：通过上述两种方法都可以解决中文乱码问题，但是在代码维护方面却有很大麻烦，因为它不但增加了代码量，而且移植性也差。

- 第3种方法，通过编写一个Servlet过滤器来解决中文乱码问题，并可以通过配置让过滤器解决所有请求处理字符集的问题，这样请求处理页面就不用关心字符集处理了。第3种方法可以有效防止中文乱码问题且移植性强。

3.1.4 获取请求的头部信息

请求的头部信息在实际应用中也是有其重要作用的，这些信息有时候对服务器的响应特别有用，而且也能查看到服务器中的相关信息。

请求头部信息的相关方法除了第 3.1.1 节提到的 getHeaders()外，还有以下方法：

- String getHeader(String name)：获取字符串型的表头信息。
- int getIntHeader(String name)：获取整型的表头信息。

【例 3.3】 获取请求中所有的头名称和对应的值。

getHeaders.jsp 页面用于获取请求中的所有头名称和对应的值，源代码如下：

```
------------------ getHeaders.jsp ----------------
01  <%@ page  pageEncoding="UTF-8"%>
02  <!DOCTYPE HTML>
03  <html>
04    <head>
05      <title>'getHeaders.jsp'</title>
06      <meta http-equiv="Content-Type" content="text/html;charset=utf-8">
07    </head>
08
09    <body >
10        <%
11              Enumeration<String> enumeration = request.getHeaderNames();
12              while(enumeration.hasMoreElements()){
13                  String name = enumeration.nextElement();
14                  String value = request.getHeader(name);
15                  if(value==null||"".equals(value)){
16                        value="空字符串";
17                  }
18        %>
19              <p>表头名称：<%=name %>  对应的值：<%=value %></p>
20        <%   } %>
21    </body>
22  </html>
```

在上述代码中，第11~17行代码获取请求的Header信息，并循环遍历输出。在浏览器中直接输入页面地址，效果如图3.4所示。对这些头信息的含义说明如下：

- accept：客户端能接收的MIME类型。
- accept-language：浏览器的首选语言。
- user-agent：客户端程序的相关信息，例如浏览器版本、操作系统类型等。
- host：表示服务器的主机名和端口号。
- connection：判断客户端是否可以持续性地连接HTTP。
- accept-encoding：指明客户端能够处理的编码类型有哪些。
- cookie：会话的信息，每个会话都有一个会话ID或者其他信息。

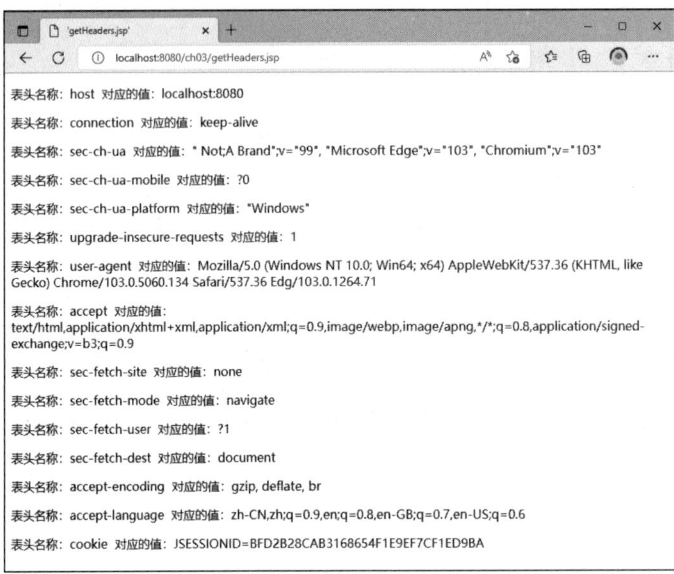

图3.4 网页的请求头信息

以上是通过直接在地址栏中输入页面地址获得的头文件信息。如果通过 post 方式跳转到页面，那么头文件信息会稍微多些。例如，在一个表单中将 method 的属性指定为 post，然后将 action 指定为例 3.3 的跳转页面，则显示结果如图 3.5 所示。

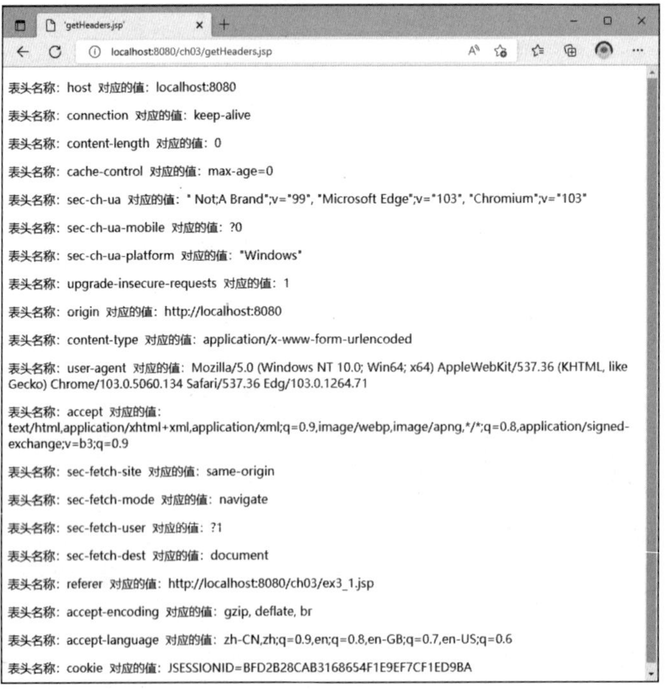

图3.5 利用post方式获取的请求头信息

从结果对比来看，使用 post 方式增加了以下几个信息。

- referer：向页面发起请求的地址。

- content-type：指明表单的MIME编码类型，一般值为text/plain、multipart/form-data、application/x-www-form-urlencoded。默认为application/x-www-form-urlencoded。如果需要表单上传一个文件，则表单中的enctype设置为multipart/form-data。
- content-length：用于设置提交请求中的数据长度，表单中提交的数据越多，长度就越长。
- cache-control：网页缓存控制。默认值为no-cache，表明每次访问都从服务器中获取页面。
- origin：指明当前请求来自于哪个站点。

在一般的开发中，我们不需要关注请求的头信息，针对不同的需求，在网页中设置相关的属性值就可以了。但是，当我们了解它们之后，在解决某些问题上将起到关键的作用。

3.1.5 获取主机和客户机的信息

获取主机和客户机的信息，一般通过以下几个方法实现。

- getRemoteAddr()：获取客户机的IP。
- getRemoteHost()：获取客户机的名称。
- getLocalAddr()：获取本地主机的IP。
- getLocalHost()：获取本地主机的名称。
- getServerName()：获取服务器主机的名称。
- getServerPort()：获取服务器端口。

【例 3.4】 获取客户机和主机信息。

getHostInfo.jsp 是获取客户机和本地主机信息的页面，源代码如下：

```
----------------- getHostInfo.jsp ----------------
<%@ page  pageEncoding="UTF-8"%>
<!DOCTYPE HTML>
<html>
  <head>
    <title>'getHostInfo.jsp'</title>
     <meta http-equiv="Content-Type" content="text/html;charset=utf-8">
  </head>

  <body >
          <p>
               本地机器IP：<%=request.getLocalAddr() %><br>
               本地机器名称：<%=request.getLocalName() %><br>
               本地机器端口：<%=request.getLocalPort() %><br>
          </p>
          <p>
               客户主机IP：<%=request.getRemoteAddr() %><br>
               客户主机名称：<%=request.getRemoteHost() %><br>
               客户主机端口：<%=request.getRemotePort()%><br>
          </p>
          <p>
               服务器IP：<%=request.getServerName() %><br>
               服务器端口：<%=request.getServerPort() %><br>
          </p>
     </body>
</html>
```

在上述代码中，第 11~13 行代码用于获取本地机器信息，第 16~18 行代码用于获取客户机信息，第 21~22 行代码用于获取服务器信息。页面效果如图 3.6 所示。

图3.6　获取客户机和主机信息

注意：一般而言，服务器主机与本地机器的 IP 应该是相同的，端口号也应该一样。

3.2　response 对象

本节将介绍response对象的常用方法。当用户访问一个页面时，就会产生一个HTTP请求，服务器做出响应时调用的是response响应包。response响应包实现的接口是jakarta.servlet.http.HttpServletResponse。

3.2.1　response对象的常用方法

response对象的常用方法如表3.2所示。

表 3.2　response 对象的常用方法

方　法	方法说明
addHeader(String name,String value)	向页面中添加头和对应的值
addCookie(Cookie cookie)	添加 Cookie 信息
sendRedirect(String uri)	实现页面重定向
setStatus(int code)	设定页面的响应状态代码
setContentType(String type)	设定页面的 MIME 类型和字符集
setCharacterEncoding(String charset)	设定页面响应的编码类型
setHeader(String name,String value)	设置响应的头信息

3.2.2　设置头信息

设置头信息包括设置页面返回的 MIME 类型、返回的字符集、页面中的 meta 等信息。其中设置 MIME 类型和返回的字符集尤为重要，因为它们关系到页面的显示是否会出现乱码。有两种方法用于设置 MIME 类型和返回的字符集，分别如下：

- response.setContentType(String type)：其中type的值为“text/html;charset=utf-8”，当然也可以是其他的MIME类型和字符集。
- page指令：基本固定格式为“<%@ page contentType="text/html;charset=utf-8" %>”。

【例 3.5】　setContentType 的用法示例。

setContentType.jsp 用来设置 MIME 类型和字符集，MIME 设置为“text/html”，字符集设置为“UTF-8”，源代码如下：

```
-----------------setContentType.jsp-----------------
<%
   response.setContentType("text/html;charset=UTF-8"); //设置字符集和MIME类型
   String str = new String("这是测试例子".getBytes("ISO-8859-1"),"utf-8");
%>
<!DOCTYPE HTML>
<html>
  <head>
    <title>setContentType.jsp</title>
  </head>

  <body>
      <p>这里是一段中文字符</p>
      </br>
      <%=str %>
  </body>
</html>
```

页面效果如图 3.7 所示。

从运行结果可以看出，脚本中经过转码的文字显示正常，但在HTML中的文字显示的却是乱码。

使用page指令设定字符集时，在写法和处理上都相对简单些，HTML中的中文字不需要转码，而且脚本中的中文字也不需要转码。

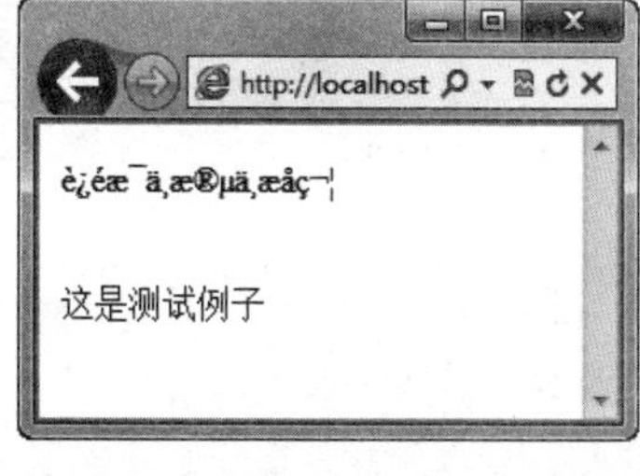

图3.7　setContentType运行结果

【例3.6】　利用page指令设置页面字符集。

```
<%@ page  pageEncoding="UTF-8"%>
<%
    String str = "这是测试例子";
 %>
<!DOCTYPE HTML>
<html>
  <head>
    <title>page 指令setContentType.jsp</title>
  </head>

  <body>
      <p>这里是一段中文字符</p>
      </br>
      <%=str %>
```

```
15      </body>
16    </html>
```

在上述代码中，第 01 行代码利用 page 指令设置页面字符集，页面效果如图 3.8 所示。

注意：在 JSP 中设置页面字符集会破坏 JSP 容器自身的页面编码处理，所以不建议设置，但可以在 Servlet 中设定。利用 page 指令设置字符集相对简单，所以在开发中一般都采用这种模式。

图3.8　利用page指令设置字符集

meta信息是指在HTML页面中存在于<head></head>之间的meta标签信息。例如：

- <meta http-equiv="pragma" content="no-cache">：设定禁止浏览器从本地缓存中调取页面内容，设定后离开网页就不能从Cache中再调用。
- <meta http-equiv="cache-control" content="no-cache">：请求和响应遵循的缓存机制策略。
- <meta http-equiv="expires" content="0">：用于设定网页的到期时间，一旦过期就必须到服务器上重新调用。
- <meta http-equiv="keywords" content="keyword1,keyword2,keyword3">：向搜索引擎说明网页的关键字。
- <meta http-equiv="description" content="This is my page">：向搜索引擎说明网页的主要内容。

【例 3.7】　设定 meta 头信息。

页面 setMeta.jsp 用于设置 meta 中的信息，源代码如下：

```
-----------------setMeta.jsp----------------
01  <%@ page import="java.util.*" pageEncoding="UTF-8"%>
02  <!DOCTYPE HTML>
03  <html>
04    <head>
05      <title>My JSP 'setMeta.jsp' starting page</title>
06    </head>
07    <body>
08       <div class="aa" style="text-align: center;">
09          <p class="bb">
10             现在的时间为：<br/>
11             <%
12                 out.print(""+ new Date());
13                 response.setHeader("refresh","1");
14                 response.setHeader("description","实时显示当前时间");
15                 response.setHeader("keywords","实时,显示,显示当前时间");
16                 response.setHeader("cache-control","no-cache");
17              %>
18            <br/><br/>
19            copyright:2018
20       </div>
21    </body>
22  </html>
```

在上述代码中，第 12~16 行代码用于设置头信息，页面效果如图 3.9 所示。

图3.9　设置meta头信息

从运行结果可以看出在response中设置meta信息的效果，但在实际开发中是直接在HTML标记中写定这些值，而不是在响应中设定。

3.2.3　设置页面重定向

重定向是指一个页面在收到一个访问请求后，根据请求的 URL 重新跳转到其他的页面。设置重定向的方法如下：

```
response.sendRedirect(String url);
```

其中 url 代表了跳转路径，路径可以是相对路径，也可以是绝对路径。

【例 3.8】　设置页面重定向。

在 sendRedirect.jsp 页面中设置跳转到一个不存在的页面，源代码如下：

```
-----------------sendRedirect.jsp----------------
<%@ page  pageEncoding="UTF-8"%>
<%
    response.sendRedirect("sendPageError.jsp");
 %>
<!DOCTYPE HTML>
<html>
  <head>
    <title>My JSP 'sendRedirect.jsp' starting page</title>
  </head>

  <body>
    This is my JSP page. <br>
  </body>
</html>
```

运行结果如图 3.10 所示。

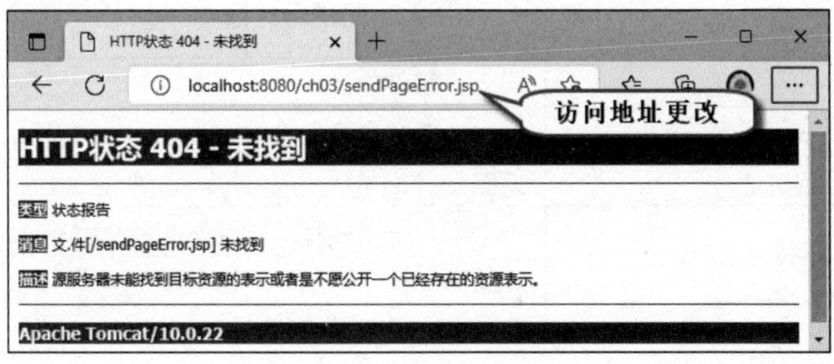

图3.10　设置页面重定向

从运行结果可以看出跳转页面时，URL地址改变了且不显示sendRedirect.jsp中的页面信息。重定向的运行过程如图3.11所示。

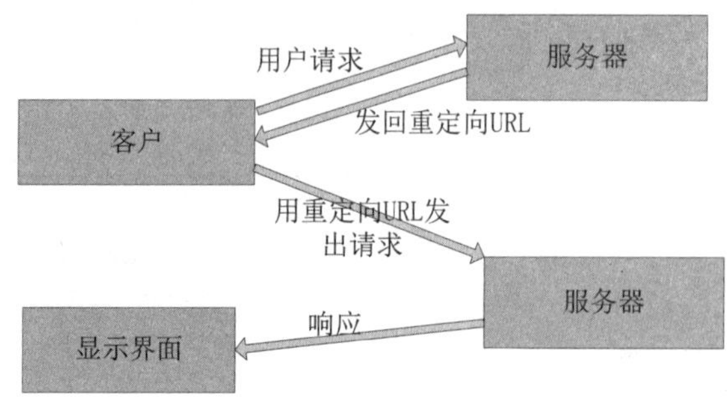

图3.11 重定向的运行过程

3.3 session对象

在Web开发中，session对象同样占据着极其重要的位置，它是一个非常重要的对象，可以用来判断是否为同一用户，还可以用来记录客户的连接信息等。HTTP是一种无状态的协议（不保存连接状态的协议），每次用户请求在接收到服务器的响应后，连接就关闭了，服务器端与客户端的连接被断开，因此，如果用户的浏览器还没关闭时又发起请求，那么网站就应该识别出该用户的情况。在这种情况下，session对象就起到了关键作用。session的相关概念如表3.3所示，常用方法如表3.4所示。

表 3.3 session 的相关概念

名　称	说　明
会话	从用户打开浏览器连接到一个 Web 应用或者是某个界面，直至关闭浏览器这个过程称为一个会话。其实打开一个浏览器就意味着打开了一个会话对象
session 对象生命周期	从用户访问某个页面到关闭浏览器这段时间称为 session 对象的生命周期，也可以说从会话开始到结束这段时间为 session 对象的生命周期
session 对象与 Cookie 对象	session 对象与 Cookie 对象是一一对应关系。JSP 引擎会将创建好的 session 对象存放在对应的 Cookie 中

表 3.4 session 的常用方法

方　法	说　明
void setAttribute(String name, Object value)	将参数名和参数值存放在 session 对象中
Object getAttribute(String name)	返回 session 中与指定参数绑定的对象，如果不存在就返回 null
Enumeration getAttributeName()	一个用户一个线程，从而保证多个用户单击同一页面时 session 对象的唯一性
String getId()	获取 session 对象的 ID 值
void removeAttribute(String name)	移除 session 中指定名称的参数

（续表）

方　法	说　明
long getCreationTime()	获取对象创建的时间，返回结果是 long 型的毫秒数
int getMaxInactiveInterval()	获取 session 对象的有效时间
void setMaxInactiveInterval()	设置 session 对象的有效时间
boolean isNew()	用于判断是否为一个新的客户端
void invalidate()	使 session 对象不合法，即失效

3.3.1　获取session ID

获取session对象ID可以判断会话是否为同一会话，用户可以通过会话中的信息来进行相关的操作。

【例 3.9】　获取 session ID 值。

创建两个Web目录并部署应用，页面之间通过超链接联系起来。

假设我们有3个页面：ex3_2.jsp、ex3_3.jsp、ex3_4.jsp，其中ex3_2.jsp和ex3_3.jsp部署在同一应用下，ex3_4.jsp部署在另外一个应用中。ex3_2.jsp通过表单提交到ex3_3.jsp，ex3_3.jsp通过超链接指向ex3_4.jsp，ex3_4.jsp通过表单提交指向ex3_2.jsp，ex3_2.jsp可以通过超链接指向ex3_4.jsp，ex3_3.jsp也可以通过超链接指向ex3_2.jsp。它们之间的关系如图3.12所示。

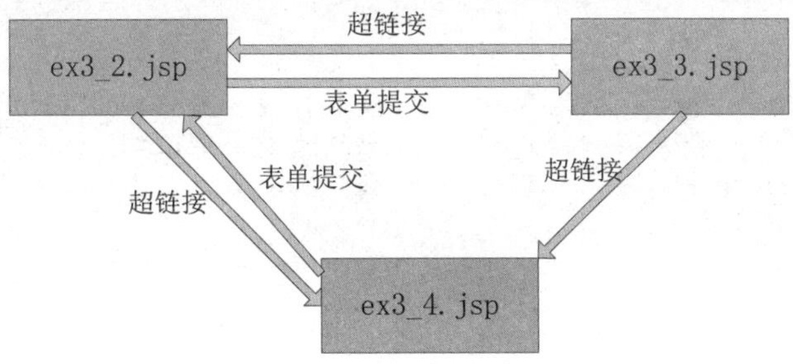

图3.12　页面之间的关系图

ex3_2.jsp 发送请求和超链接到 ex3_3.jsp、ex3_4.jsp，源代码如下：

```
------------------ex3_2.jsp-----------------
<%@ page import="java.util.*" pageEncoding="UTF-8"%>
<!DOCTYPE HTML>
<html>
  <head>
    <title>My JSP 'ex3_2.jsp' starting page</title>
     <meta http-equiv="pragma" content="no-cache">
     <meta http-equiv="cache-control" content="no-cache">
  </head>
  <body>
  <%
      String sessionID = session.getId();
      session.setAttribute("name","John");                //存储参数name
      String author = (String)session.getAttribute("author");
      long time = session.getCreationTime();
      Date date = new Date(time);
```

```
16    %>
17    <div style="text-align: center;">
18            <p>您访问的是ex3_2.jsp页面</br>
19               <%=author %>, 您的session对象ID为: </br>
20               <%=sessionID%></br>
21               session对象创建时间: <%=date %>
22               </br>
23            </p>
24            <form action="ex3_3.jsp" method="post">
25                <input type="submit" value="转向ex3_3.jsp"/>
26            </form>
27            <a href="../ch03/ex3_4.jsp">欢迎到ex3_4.jsp页面</a>
28     </div>
29    </body>
30  </html>
```

在上述代码中，第 10~16 行代码用于获取 session 中的 author 参数值，并设置当前日期；第 19~21 行代码用于显示获取的参数值。页面效果如图 3.13 所示。

从图3.13可以看出，页面输出了session对象的ID值和创建的时间，但是没有获得从页面ex3_4.jsp传来的参数值author。

ex3_3.jsp 超链接到 ex3_2.jsp 和 ex3_4.jsp，源代码如下：

```
-----------------ex3_3.jsp----------------
<%@ page import="java.util.*" pageEncoding="UTF-8"%>
<!DOCTYPE HTML>
<html>
  <head>
    <title>My JSP 'ex3_2.jsp' starting page</title>
     <meta http-equiv="pragma" content="no-cache">
     <meta http-equiv="cache-control" content="no-cache">
  </head>
  <body>
  <%
      String sessionID = session.getId();
      String name = (String)session.getAttribute("name");
      long time = session.getCreationTime();
      Date date = new Date(time);
  %>
  <div style="text-align: center;">
          <p>您访问的是ex3_3.jsp页面</br>
             <%=name %>,您的session对象ID为: </br>
             <%=sessionID%></br>
             session对象创建时间: <%=date %>
             </br>
          </p>
          <a href="ex3_2.jsp"><%=name %>,欢迎到ex3_2.jsp页面</a></br>
          <a href="../ch03/ex3_4.jsp"><%=name %>,欢迎到ex3_4.jsp页面</a>
   </div>
  </body>
</html>
```

运行结果如图 3.14 所示。从中可以看出，页面 ex3_3.jsp 获得的 session 对象 ID 和创建时间与 ex3_2.jsp 中的 session 对象 ID 和创建时间是一样的，且获得了参数 name 的值。

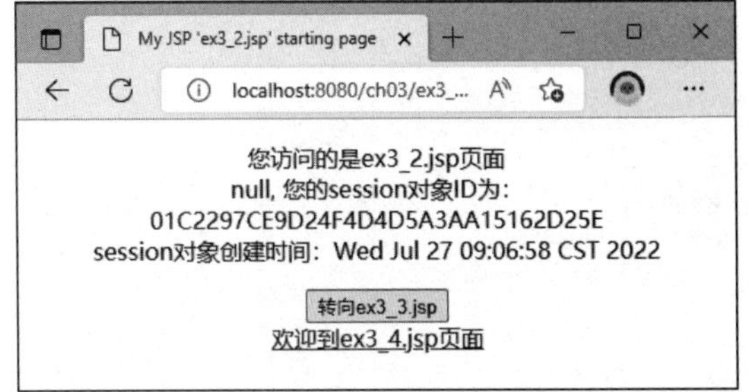

图 3.13　ex3_2.jsp 运行结果

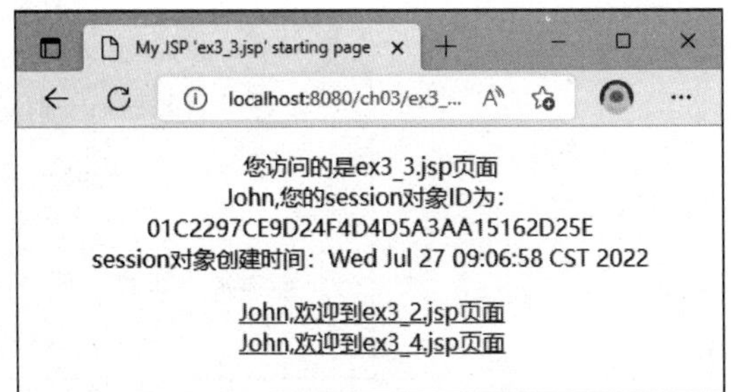

图 3.14　ex3_3.jsp 运行结果

ex3_4.jsp 发送请求到 ex3_2.jsp，其源代码如下：

```
------------------ex3_4.jsp-----------------
01  <%@ page import="java.util.*" pageEncoding="UTF-8"%>
02  <!DOCTYPE HTML>
03  <html>
04    <head>
05      <title>My JSP 'ex3_4.jsp' starting page</title>
06       <meta http-equiv="pragma" content="no-cache">
07       <meta http-equiv="cache-control" content="no-cache">
08    </head>
09    <body>
10    <%
11           String sessionID = session.getId();
12           String name = (String)session.getAttribute("name");
13           session.setAttribute("author","Smith");        //保存参数author
14           long time = session.getCreationTime();
15           Date date = new Date(time);
16     %>
17      <div style="text-align: center;">
18            <p>您访问的是ex3_4.jsp页面</br>
19               <%=name %>,您的session对象ID为: </br>
20               <%=sessionID%></br>
21               session对象创建时间: <%=date %>
22               </br>
23            </p>
24            <form action="../ch03/ex3_2.jsp" method="post">
25                 <input type="submit" value="转向ex3_2.jsp"/>
26            </form>
27      </div>
28    </body>
29  </html>
```

在上述代码中，第 11~15 行代码用于获取 session 中的用户名参数，第 18~23 行代码用于显示参数值，页面效果如图 3.15 所示。

从图3.15中可以看出，页面输出了session对象的ID值和创建时间，但是它和页面ex3_2.jsp、ex3_3.jsp的session对象ID值不同，并且页面没有获得从ex3_2.jsp传来的参数值name。

从以上的结果可以看出：一个Web应用的session对象的ID值是唯一的，并且两个应用之间的参数利用session对象是获取不到值的。

注意：读者务必要理解 session 对象的生命周期。

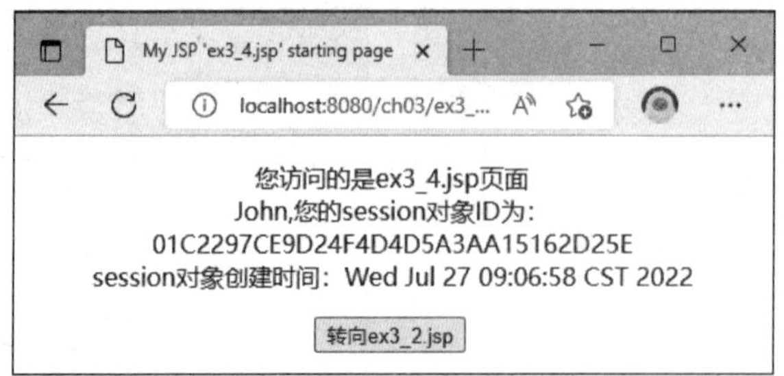

图3.15　ex3_4.jsp运行结果

3.3.2　用户登录信息的保存

【例3.10】　用户登录信息的保存。

login.jsp是用户登录页面，validate.jsp是验证用户合法性页面，class.jsp是登录成功后显示班级管理的页面，logout.jsp是退出登录页面。它们之间的关系如图3.16所示。

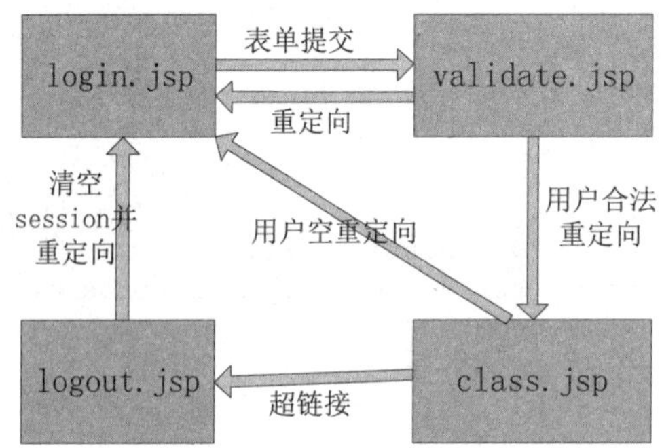

图3.16　页面之间的关系图

login.jsp 是用户登录页面，其源代码如下：

```
------------------login.jsp-----------------
<%@ page pageEncoding="UTF-8"%>
<%
  String path = request.getContextPath();
  String basePath = request.getScheme()+"://"+request.getServerName()
+":"+ request. getServerPort()+path+"/";
%>
<!DOCTYPE HTML>
<html>
  <head>
    <base href="<%=basePath%>">
    <title>用户登录</title>
     <meta http-equiv="pragma" content="no-cache">
     <meta http-equiv="cache-control" content="no-cache">
     <meta http-equiv="expires" content="0">
     <meta http-equiv="keywords" content="keyword1,keyword2,keyword3">
     <meta http-equiv="description" content="This is my page">
  </head>

  <body>
```

```
20        <div style="text-align: center;">
21              <span style="font-size: 26px;">用户登录</span>
22              <hr/>
23              <form action="validate.jsp" method="post">
24                   用户名称：<input type="text" name="username"/>
25              <br/>
26                   用户密码：<input type="password" name="password"/>
27              <br/>
28              <input type="submit" value="登录"/>
29              </form>
30        </div>
31     </body>
32  </html>
```

在上述代码中，第 23~29 行代码利用 form 表单提交登录信息。

validate.jsp 是验证用户合法性页面，其源代码如下：

```
-----------------validate.jsp----------------
<%@ page pageEncoding="UTF-8"%>
<!DOCTYPE HTML>
<html>
  <head>
    <title>success.jsp</title>
     <meta http-equiv="pragma" content="no-cache">
     <meta http-equiv="cache-control" content="no-cache">
     <meta http-equiv="expires" content="0">
     <meta http-equiv="keywords" content="keyword1,keyword2,keyword3">
     <meta http-equiv="description" content="This is my page">
  </head>
    <%!
      //声明一个用户集合，模拟从数据库中取出用户集
      Map<String, String> map =new HashMap<String, String>();
      //声明验证的标识
      boolean flag = false;
    %>
    <%
      //向集合添加数据
      map.put("John","123456");
      map.put("Smith","222222");
      map.put("Bob","333333");
      map.put("Bruth","666666");
    %>
    <%!
      //声明验证方法
      boolean validate(String username,String password){
           String passwd = map.get(username);
           if(passwd!=null&&passwd.equals(password)){
               return true;
           }else{
               return false;
           }
      }
    %>
    <%
      //获得页面提交的用户名与密码
```

```
38        String username = request.getParameter("username");
39        String password = request.getParameter("password");
40        if(username==null||username==""||password==null||password==""){
41            response.sendRedirect("login.jsp");
42        }
43        flag = validate(username,password);
44        if(flag){
45            //保存在session对象中
46            session.setAttribute("username",username);
47            session.setAttribute("password",password);
48            response.sendRedirect("class.jsp");
49        }
50      %>
51    <body>
52          <div style="text-align: center;">
53              <span style="font-size: 26px;">用户登录</span>
54          </div>
55          <br/>
56          <div style="text-align: center;">
57              <%if(!flag){ %>
58                  <a href="login.jsp">重新登录系统</a>
59              <%} %>
60          </div>
61    </body>
62  </html>
```

在上述代码中，第 18~24 行代码用于设置用户集合；第 25~35 行代码用于声明验证用户合法性的方法；第 38~49 行代码用于获取参数值，并判断其合法性，合法的保存在 session 中，否则跳转到 login.jsp 中。

class.jsp 是登录成功后显示班级管理的页面，其源代码如下：

```
------------------class.jsp-----------------
<%@ page pageEncoding="UTF-8"%>
<%
    String name = (String)session.getAttribute("username");
    if(name==null){
        response.sendRedirect("login.jsp");
    }
%>

<!DOCTYPE HTML>
<html>
  <head>
    <title>My JSP 'score.jsp' starting page</title>
     <meta http-equiv="pragma" content="no-cache">
     <meta http-equiv="cache-control" content="no-cache">
     <meta http-equiv="expires" content="0">
     <meta http-equiv="keywords" content="keyword1,keyword2,keyword3">
     <meta http-equiv="description" content="This is my page">
  </head>

  <body>
       <div  style="text-align: center;">
            <span style="font-size: 24px;">班级管理</span>
```

```
23          <hr/>
24          <h3>学生：<%=name %></h3>
25          <table>
26              <tr>
27                  <td>
28                      <a href="addClass.jsp">班级录入</a>
29                  </td>
30                  <td>
31                      <a href="modifyClass.jsp">班级修改</a>
32                  </td>
33                  <td>
34                      <a href="queryClass.jsp">班级查询</a>
35                  </td>
36                  <td>
37                      <a href="delClass.jsp">班级删除</a>
38                  </td>
39              </tr>
40          </table>
41          <a href="logout.jsp">退出登录</a>
42          </div>
43      </body>
44  </html>
```

在上述代码中，第 27~38 行代码用于列出班级的各项方法。

logout.jsp 是退出登录页面，其源代码如下：

```
-----------------logout.jsp-----------------
01  <%@ page pageEncoding="UTF-8"%>
02  <%
03     String username = (String)session.getAttribute("username");
04     session.removeAttribute("John");
05     session.invalidate();
06     response.sendRedirect("login.jsp");
07  %>
08
09  <!DOCTYPE HTML>
10  <html>
11    <head>
12      <title>My JSP 'logout.jsp' starting page</title>
13    </head>
14    <body>
15    </body>
16  </html>
```

在 login.jsp 页面中，第 23 行代码利用 form 表单的 post 方式提交，action 指向 validate.jsp 页面。在 validate.jsp 页面中，第 12~17 行代码用于声明 map 集合和成功标识 flag；第 18~24 行代码向 map 中添加数据；第 25~35 行代码是验证的方法；第 36~50 行代码用于获得页面传递的参数以及验证学生是否存在，如果存在，则存放在 session 中并重定向到 class.jsp；第 57~59 行代码用于验证是否登录成功，若失败则显示重新登录。在 logout.jsp 页面中，第 02~07 行代码用于获得 session 中的 name 参数值并移除该学生，使 session 失效。

运行结果如图3.17所示。

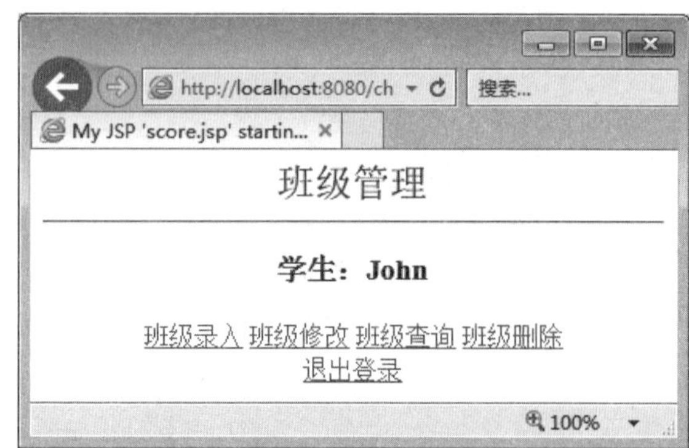

图3.17 运行结果

注意：该实例中的用户退出操作，只是利用简单的 session.invalidate 方法来实现，在开发中经常是结合 Struts 框架一起操作。

3.4 application 对象

application对象实现的接口为jakarta.servlet.ServletContext，它的生命周期是从application对象创建开始到应用服务器关闭为止，也就是说当服务器关闭时application对象才消失。可以将它视为Web应用的全局变量，当服务器运行时有效，如果关闭服务器，其中保存的信息也就消失了。

3.4.1 application对象的常用方法

application对象的常用方法如表3.5所示。

表 3.5 application 对象的常用方法

方　法	方法说明
getAttribute(String name)	获得存放在 application 中的含有关键字 name 的对象
setAttribute(String name,Object obj)	将关键字 name 的指定对象 obj 放进 application 对象中
Enumeration getAttributeNames()	获取 application 中所有参数的名字，返回值是枚举类型
removeAttribute(String name)	移除 application 对象中 name 指定的参数值
getServletInfo()	获取 Servlet 的当前版本信息
getContext(String uripath)	获取 uripath 指定路径的 context 内容
gctRcalPath(String path)	获取指定文件的实际路径
getMimeType(String file)	获取指定的文件格式

3.4.2 获取指定页面的路径

【例 3.11】 获取指定页面的实际路径、相对路径和当前应用程序路径。

application.jsp 用于指定页面输出它所在的三种路径，源代码如下：

```
-----------------application.jsp----------------
01  <%@ page pageEncoding="UTF-8"%>
02  <!DOCTYPE HTML>
03  <html>
04    <head>
05      <title>My JSP 'application.jsp' starting page</title>
06    </head>
07    <body>
08        <h3>指定页的实际路径、相对路径和当前应用程序路径</h3>
09        <hr/>
10        <table border="1" bordercolor="black">
11            <tr>
12                <td>当前服务器的名称和版本</td>
13                <td><%=application.getServerInfo() %></td>
14            </tr>
15            <tr>
16                <td>页面application.jsp的实际路径</td>
17                <td><%=application.getRealPath("application.jsp") %></td>
18            </tr>
19            <tr>
20                <td>页面application.jsp的URL</td>
21                <td><%=application.getResource("application.jsp") %></td>
22            </tr>
23            <tr>
24                <td>当前Web程序的路径</td>
25                <td><%=application.getContextPath() %></td>
26            </tr>
27        </table>
28    </body>
29  </html>
```

在上述代码中，代码第 11~26 行分别输出 application 中指定页面的实际路径、相对路径和当前应用程序路径等信息，页面效果如图 3.18 所示。

图3.18　application.jsp的运行结果

3.4.3　设计一个网站计数器

application对象还可以用于保存访问网站的人数，也就是我们常说的网站计数器，下面通过一个例子来演示。

【例 3.12】 网站计数器。

applicationCount.jsp 是网站计数器页面，源代码如下：

```
----------------- applicationCount.jsp----------------
01  <%@ page pageEncoding="UTF-8"%>
02  <%
03      Integer count =(Integer) application.getAttribute("count");
04      if(count==null){
05       count=1;
06      }else{
07          count++;
08      }
09      application.setAttribute("count",count);
10  %>
11
12  <!DOCTYPE HTML>
13  <html>
14    <head>
15      <title>网站计数器</title>
16    </head>
17    <body>
18        欢迎访问本网站，您是第<%=count %>位访问客户！
19    </body>
20  </html>
```

第 02~10 行代码从页面中获得计数值，如果为空就设定初始值为 1，如果不为空就加 1。程序运行结果如图 3.19 所示。

图3.19 网站计数器

注意：application 对象在 Web 应用运行时一直存在于服务器中，保存这种全局变量相对来说比较占用资源，因此不推荐使用。在实际开发中，一般都是让对象存在于必要的时间段中，否则当访问量加剧时，会造成内存不足等情况。

3.5 out 对象

out对象是继承jakarta.servlet.jsp.JspWriter类的一个输出流对象。它包含很多IO流中的方法和特性，最常用的方法就是输出内容到HTML中。

3.5.1 out对象的常用方法

out对象的常用方法如表3.6所示。

表 3.6　out 对象的常用方法

方　法	方法说明
append(char c)	将字符添加到输出流中
clear()	清空页面缓存中的内容
close()	关闭网页流的输出
flush()	网页流的刷新
println()	将内容直接打印在 HTML 标记中
write()	与 println()方法相似，区别在于 println()方法可以输出各种类型的数据，而 write 方法只能输出与字符相关的数据，例如字符、字符数组、字符串等

3.5.2　out对象的使用示例

out对象中的方法相对比较简单，而且也较少使用，下面就通过几个简单的例子来演示。

【例 3.13】　演示 out 对象的 println 方法。

outprintln.jsp 直接在 JSP 脚本中引用 HTML 标记，源代码如下：

```
------------------ outprintln.jsp-----------------
01  <%@ page pageEncoding="UTF-8"%>
02  <!DOCTYPE HTML>
03  <html>
04    <head>
05      <title>演示out对象</title>
06    </head>
07
08    <body>
09       <div style="text-align: center;">
10           <hr>
11           <h4>以下就是一个表格</h4>
12           <%
13            out.println("<table border='2' align='center'>");
14            out.println("<tr>");
15            out.println("<td width='60'>"+"姓名"+"</td>");
16            out.println("<td width='40'>"+"性别"+"</td>");
17            out.println("<td width='80'>"+"出生年月"+"</td>");
18            out.println("<td width='60'>"+"城市"+"</td>");
19            out.println("</tr>");
20            out.println("<tr>");
21            out.println("<td width='60'>"+"Smith"+"</td>");
22            out.println("<td width='60'>"+"Male"+"</td>");
23            out.println("<td width='60'>"+"1984.8"+"</td>");
24            out.println("<td width='60'>"+"NewYork"+"</td>");
25            out.println("</tr>");
26            out.println("</table>");
27           %>
28       </div>
29    </body>
30  </html>
```

在上述代码中，第 13~26 行代码利用 out 对象输出 HTML 格式，页面效果如图 3.20 所示。

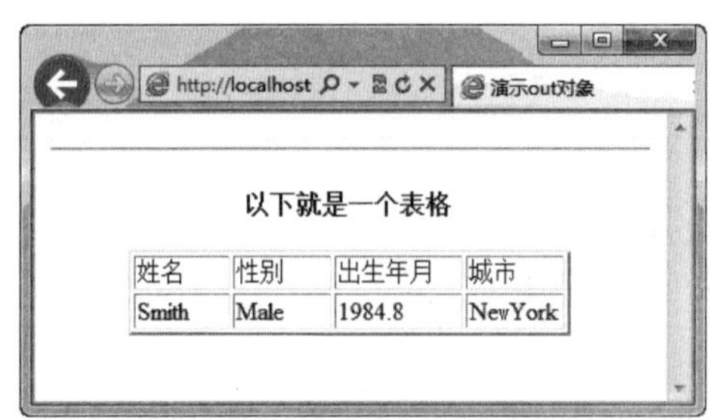

图3.20　outprintln.jsp页面的运行结果

注意：现在在开发中很少使用这种方法输出 HTML 标记，因为比较烦琐而且容易出错。

【例 3.14】　演示 out 对象的 clear 方法。

outclear.jsp 用于清空缓冲区中的内容，源代码如下：

```
----------------- outclear.jsp----------------
01  <%@ page pageEncoding="UTF-8"%>
02  <!DOCTYPE HTML>
03  <html>
04    <head>
05      <title>演示out对象的clear方法</title>
06    </head>
07
08    <body>
09       <h4>这是HTML中的内容</h4>
10       <%
11          out.print("<h4>这是out对象输出的信息</h4>");
12          out.clear();
13       %>
14       <h4 style="text-align: center;">这是HTML中的信息</h4>
15    </body>
16  </html>
```

在上述代码中，第 12 行代码用于清除缓冲区中的内容，页面效果如图 3.21 所示。

从运行结果可以看出，若在JSP页面中调用clear方法，那么以前向客户输出流中写入的数据都将被清除。

图3.21　outclear.jsp页面运行结果

注意：在调用 clear 之前，不要调用 flush 方法，否则会抛出 IO 异常。

3.6　page 对象

page对象的实质是java.lang.Object对象，它代表转译后的Servlet。page对象是指当前的JSP页面本身，在实际开发中并不常用。

3.6.1 page对象的常用方法

page对象的常用方法如表3.7所示。

表 3.7 page 对象的常用方法

方　法	方法说明
getClass()	返回当时被转译的 Servlet 类
hashCode()	返回此时被转译的 Servlet 类的哈希代码
toString()	将此时被转译的 Servlet 类转换成字符串
equals(Object obj)	比较此时的对象是否与指定的对象相等
clone()	将此时的对象复制到指定的对象中
copy(Object obj)	对指定对象进行克隆

3.6.2 page对象的使用示例

下面就通过简单的例子来演示page中的方法。

【例 3.15】 演示输出 page 对象的 toString()方法和 hashCode()方法。

通过 page.jsp 页面演示 toString()和 hashCode()的用法，源代码如下：

```
----------------- page.jsp-----------------
01  <%@ page pageEncoding="UTF-8"%>
02  <!DOCTYPE HTML>
03  <html>
04    <head>
05      <title>演示page对象</title>
06    </head>
07
08    <body>
09      <%
10          int code = page.hashCode();//hashcode
11          String str = page.toString();
12          out.println("page对象的hash码:"+code);
13          out.println("page对象的值:"+str);
14      %>
15    </body>
16  </html>
```

在上述代码中，第 10~13 行代码用于获取 page 对象中的值，并输出在页面中，页面效果如图 3.22 所示。

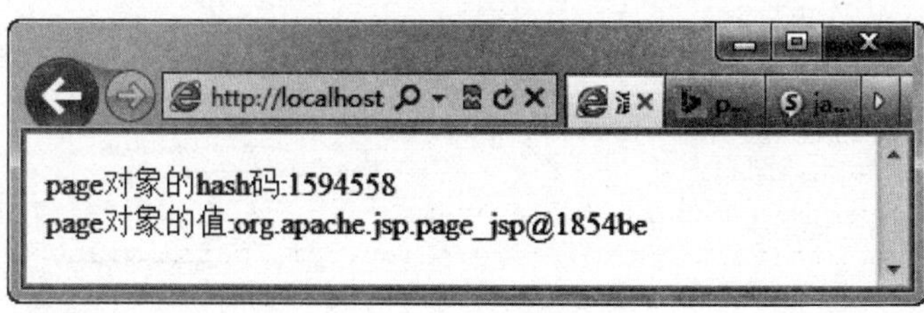

图3.22 page.jsp的运行结果

3.7 config 对象

config对象实现了jakarta.servlet.ServletConfig接口，它一般用于在页面初始化时传递参数。

3.7.1 config对象的常用方法

config对象的常用方法如表3.8所示。

表 3.8 config 对象的常用方法

方　法	方法说明
getInitParameter(String arg0)	获得指定的初始化值
getServletName()	获得 Servlet 名字
getServletContext()	获得 ServletContext 值
equals(Object obj)	比较此时的对象是否与指定的对象相等
getInitParameterNames()	获得初始化值的枚举值
toString()	获得此对象的值

3.7.2 config对象的使用示例

下面就通过简单的实例来演示config中的方法。

【例 3.16】 演示输出 config 对象的 getInitParameter()方法。

通过 config.jsp 页面演示 getInitParameter()方法的用法，在此假设 WEB-INF 文件夹下面存在 web.xml 文件，内容如下：

```
<?xml version="1.0" encoding="UTF-8"?>
<web-app version="5.0"
    xmlns="https://jakarta.ee/xml/ns/jakartaee"
    xmlns:xsi="http://www.w3.org/2001/XMLSchema-instance"
    xsi:schemaLocation="https://jakarta.ee/xml/ns/jakartaee
    https://jakarta.ee/xml/ns/jakartaee/web-app_5_0.xsd">
  <welcome-file-list>
   <welcome-file>index.jsp</welcome-file>
  </welcome-file-list>

  <servlet>
      <servlet-name>
          jspconfigdemo
      </servlet-name>
      <jsp-file>/config.jsp</jsp-file>
      <init-param>
          <param-name>url</param-name>
          <param-value>http://www.baidu.com</param-value>
      </init-param>
  </servlet>
  <servlet-mapping>
```

```
22       <servlet-name>
23              jspconfigdemo
24       </servlet-name>
25       <url-pattern>/config.jsp</url-pattern>
26    </servlet-mapping>
27  </web-app>
```

在上述代码中，第11~26行代码用于在web.xml中配置Servlet，包括其初始化参数等信息。config.jsp页面用于显示配置内容，源代码如下：

```
01  <%@ page import="java.util.*" pageEncoding="UTF-8"%>
02  <!DOCTYPE HTML>
03  <html>
04    <head>
05      <title>演示config对象</title>
06    </head>
07
08    <body>
09       <%
10          String url = config.getInitParameter("url");
11          String str = config.toString();
12          out.config("config对象的initParameter方法: "+url+"</br>");
13          out.config("config对象的toString方法: "+str);
14       %>
15    </body>
16  </html>
```

在上述代码中，第 10 行代码利用 config 对象获取初始化参数值的信息，页面效果如图 3.23 所示。

图3.23 config.jsp页面的运行结果

注意：一般而言很少在页面中使用 config 对象，因为 JSP 页面的实质是 Servlet。

3.8 pageContext 对象

pageContext对象是jakarta.servlet.jsp.PageContext 类的实例，用来代表整个JSP页面。这个对象主要用来访问页面信息，同时过滤掉大部分实现细节，存储了request对象和response对象的引用。application对象、config对象、session对象、out对象可以通过访问这个对象的属性来导出。pageContext对象也包含了传递给JSP页面的指令信息，包括缓存信息、ErrorPage URL、页面scope等。

3.8.1 pageContext对象的常用方法

pageContext对象的常用方法如表3.9所示。

表 3.9 pageContext 对象的常用方法

方　法	方法说明
getOut()	返回当前客户端响应被使用的 JspWriter 流（out）
getSession()	返回当前页面中的 HttpSession 对象（session）
getPage()	返回当前页面的 Object 对象（Object）
getRequest()	返回当前页面的 ServletRequest 对象（request）
getResponse()	返回当前页面的 ServletResponse 对象（response）
setAttribute(String name,Object attribute)	设置属性及属性值
getAttribute(String name,int scope)	在指定范围内获取属性的值
getAttributeScope(String name)	返回某属性的作用范围
forward(String relativeUrlPath)	使当前页面重定向到另一页面
include(String relativeUrlPath)	在当前位置包含另一文件
getException()	获取当前页面的异常对象

3.8.2 pageContext对象的使用示例

下面通过简单的示例来演示pageContext中的方法。

【例 3.17】 演示输出 pageContext 对象的常用方法。

演示 pageContext 对象的常用方法，源代码如下：

```
<%@ page pageEncoding="UTF-8"%>
<!DOCTYPE HTML>
<html>
  <head>
    <title>演示pageContext对象</title>
  </head>
  <body>
     <%
         int code = pageContext.hashCode();//hashcode
         String str = pageContext.toString();
         out.println("pageContext对象的hash码:"+code+"  </br>");
         out.println("pageContext对象的值:"+str+"</br>");
         ServletRequest sReq = pageContext.getRequest();
         out.println("使用getRequest获得的对象:"+ sReq.toString()+"</br>");
         ServletResponse sRes = pageContext.getResponse();
         out.println("使用getResponse获得的对象:"+ sRes.toString()+"</br>");
         pageContext.setAttribute("name","Hazel");
         out.println("使用getAttribute获取设置的name对象值:"+ pageContext.getAttribute("name"));
     %>
  </body>
</html>
```

在上述代码中，第 09~16 行代码用于获取 pageContext 对象中的值，第 17、18 行代码是属性设置与获取。页面效果如图 3.24 所示。

图3.24 pageContext.jsp页面的运行结果

【例 3.18】 演示 pageContext 对象的 getException 方法。

Exception 对象可以用于配置 JSP 页面的全局异常处理。这里就简单地演示打印错误信息，需要如下两个步骤：

01 在需要捕获异常的JSP页面中，通过page指令的errorPage属性指定错误页面的相对URL。

02 在错误页面中使用pageContext的getException方法获取异常信息。

首先查看抛出异常的testException.jsp页面的源代码：

```
<%@ page pageEncoding="UTF-8" errorPage="exception.jsp"%>
<!DOCTYPE HTML>
<html>
  <head>
    <title>演示exception对象--抛出异常界面</title>
  </head>
  <body>
    <%
        int code = 1/0;
    %>
  </body>
</html>
```

在上述代码中可以看到errorPage页面指向了exception.jsp，然后在页面中声明了一个肯定抛出异常的表达式。当我们访问该页面时，就会跳转到exception.jsp页面，源代码如下：

```
<%@ page pageEncoding="UTF-8"%>
<!DOCTYPE HTML>
<html>
  <head>
    <title>演示exception对象</title>
  </head>
  <body>
    <%
        Exception exception = pageContext.getException();
        out.println("异常消息提示：" +exception.getMessage()+"  </br>");
    %>
  </body>
</html>
```

上述代码使用 pageContext 的 getException()获取到了前一页面的异常信息，异常信息如图 3.25 所示。

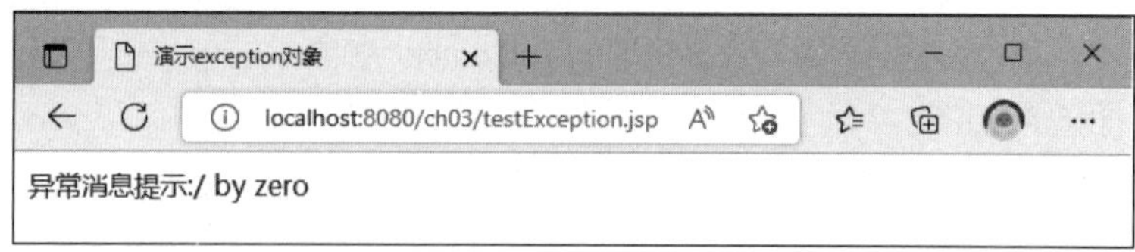

图3.25 testException.jsp页面的运行结果

3.9 小　结

本章主要介绍了JSP页面中的常用内置对象，在Web开发中要充分利用这些隐藏对象提供的方法来实现Web应用的强大功能。在开发过程中，要充分理解各个对象的生命周期、作用范围，这样就可以方便地操作页面中的属性和行为，实现页面与页面之间、页面与应用环境之间的通信等操作。本书后面的章节也会经常用到这些内置对象。

3.10 习　题

（1）JSP的内置对象有几种作用域？分别是什么？

（2）表单向JSP提交数据的方式有哪些？

（3）请用session对象统计访问页面的客户数。

（4）请编写一个会员注册程序，要求用户在注册页面中填写信息后提交请求，接收注册请求的页面可以显示出信息。

第 4 章 Servlet 技术

在Web应用中，Servlet是一项重要的技术。Servlet是利用Java类编写的服务器端程序，与平台架构、协议无关。JSP的实质是Servlet，因为JSP在执行第一次后，会被编译成Servlet的类文件（即.class），当再重复调用执行JSP时，就直接执行第一次所产生的Servlet，而不再重新把JSP编译成Servelt，所以Servlet至关重要。

本章主要涉及的知识点有：

- Servlet的基本概念和技术特点
- 一个Servlet的生命周期
- 如何编写和部署一个Servlet程序
- Servlet与JSP之间的关联与区别

4.1 Servlet 是什么

本节首先介绍Servlet的基本概念，Servlet是利用Java类编写的服务器端应用程序，顾名思义，它通常是在服务器端运行的程序，打开浏览器即可调用一个。它可以被看作位于客户端和服务器端的一个中间层，负责接收和请求客户端用户的响应。

Servlet使用了很多Web服务器都支持的API，可以调用和扩展Java中提供的大量程序设计接口、类、方法等功能。

Servlet可以提供以下功能。

1. 对客户端发送的数据进行读取和拦截

客户端在发送一个请求时，一般而言都会携带一些数据（例如URL中的参数、页面中的表单、Ajax提交的参数等），当一个Servlet接收到这些请求时，Java Servlet中的类通过所提供的方法就能得到这些参数（例如，方法request.getParameterName(name)用于获得名为name的参数值），也正因为这个原因，Servlet可以对发送请求起到拦截作用，它在某些请求发出前先做一个预处理分析，从而判断客户端是否可以做某些请求（例如检查访问权限、设定程序的字符集、检查用户角色等），当Servlet具有如上功能时，一般可以被称为拦截器。

2. 读取客户端请求的隐含数据

客户端请求的数据可以分为隐含数据和显式数据：隐含数据一般不直接跟随于URL中，它存在于请求的来源、缓存数据（Cookie）、客户端类型中；显式数据显然是用户可以直观看到的，例如表单数据和URL参数。Servlet不但可以处理显式数据，而且可以处理隐含数据，是一个“多面手”。

3. 运行结果或者生成结果

当一个Web应用程序对客户端发出的请求做出响应时，一般需要很多中间过程才能得到结果。Servlet起到这个中间角色的作用，协调各组件、各部分完成相应的功能，根据不同的请求做出相应的响应并显示结果。

4. 发送响应的数据

Servlet在对客户端做出响应并经过处理得出结果后，会对客户端发送响应的数据，以便让客户端获取请求的结果数据。在Web应用程序中，Servlet的这个功能相当突出，无论现有的技术多么先进，都是基于这个功能出发的。

综上所述，Servlet的程序运行顺序大致如图4.1所示。

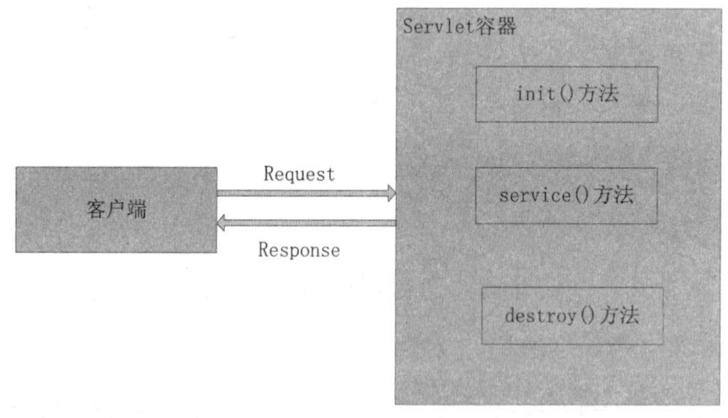

图4.1 Servlet的运行顺序

4.2 Servlet 的技术特点

上一节介绍了Servlet的概念和功能，从而让读者对Servlet有一个整体的印象和认识，并了解其运行的顺序等，本节将介绍Servlet的一些技术特点，让读者了解Servlet的优点。

Servlet 在开发中带来的优点就是能及时响应和处理 Web 端的请求，使得一个不懂网页的 Java 开发人员也能编写出 Web 应用程序，只是在开发/修改一个 Web 程序时比较麻烦，因为代码的可读性比较差，也比较难以维护，但是它却具有以下特点。

1. 高效率

Servlet本身就是一个Java类，在运行的时候位于同一个Java虚拟机中，可以快速地响应客户端的请求并生成结果。在Web服务器中处理一个请求使用的都是线程而非进程，也就是说在性能开销方面就小很多，无须大量的启动进程时间，在高并发量访问时，一个进程可以有多个线程，并发时线程在CPU中的开销代价要远小于进程的开销。

2. 简单方便

开发一个 Web 程序，从开发顺序上来说比较简单：首先定义一个 Servlet 类，然后在系统（web.xml）中配置程序，继而发布程序，这样一个 Web 程序就完成了。在开发的过程中，系统提供了大量的实用工具和方法，可以处理复杂的 HTML 表单数据、处理 cookie、跟踪网页会话等。

4.3 Servlet 的生命周期

每个生命都有特定的生命周期，例如人的生命周期为"婴儿→少年→青年→壮年→老人"，开发项目的生命周期为"立项→开发→运维→消亡"。同样，Servlet 也不例外，它有 3 个阶段，分别是：初始化（包括装载和初始化）、运行、消亡。

- 初始化阶段：初始化阶段可以分为装载和初始化两个子阶段。装载就是由Servlet容器装载一个Servlet类，把它装载到Java内存中，Servlet容器可创建一个Servlet对象并与web.xml中的配置对应起来；初始化子阶段是调用Servlet中的init()方法，在整个Servlet生命周期中init()方法只被调用一次。
- 运行阶段：在这个阶段中会实际响应客户端的请求，当有请求时Servlet会创建HttpServletRequest和HttpServletResponse对象，然后调用service(HttpServletRequest request, HttpServletResponse response)方法。serivce()方法通过request对象获得请求对象的信息并加以处理，再由response对象对客户端做出响应。
- 消亡阶段：当Servlet应用被终止后，Servlet容器会调用destroy()方法对Servlet对象进行销毁。在消亡的过程中，Servlet容器将释放被它所占的资源，例如关闭流、关闭数据库连接等。同样，在整个Servlet生命周期中destroy()方法也只被调用一次。

下面通过编写一个Servlet类来说明它的生命周期，完整的代码如下：

```
package com.etch.edu;

import java.io.IOException;
import java.io.PrintWriter;

import jakarta.servlet.ServletException;
import jakarta.servlet.ServletRequest;
import jakarta.servlet.ServletResponse;
import jakarta.servlet.http.HttpServlet;
import jakarta.servlet.http.HttpServletRequest;
import jakarta.servlet.http.HttpServletResponse;

```

```
public class HelloServlet extends HttpServlet {

    //序列化，可以自动生成或者自行定义
    private static final long serialVersionUID = 1L;

    public void init() throws ServletException{
        System.out.println("初始化 init 方法");
    }

    public void service(ServletRequest request,
ServletResponse response)  throws  ServletException, IOException{
        System.out.println("调用 public service 方法");
        response.setContentType("text/html;charset=gbk");
        PrintWriter out = response.getWriter();
        out.println("收到service请求");
    }
    protected void service(HttpServletRequest request,
        HttpServletResponse response) throws ServletException{
        System.out.println("调用protected service 方法");
    }

    public void doGet(HttpServletRequest request,
HttpServletResponse response)throws ServletException, IOException{
        System.out.println("调用doGet()方法");
        //设置响应的页面类别与页面编码
        response.setContentType("text/html;charset=gbk");
        PrintWriter out = response.getWriter();
        out.println("收到HelloServlet doGet()请求");
    }

    public void doPost(HttpServletRequest request,
HttpServletResponse response)throws ServletException, IOException{
        System.out.println("调用doPost()方法");
        //设置响应的页面类别与页面编码
        response.setContentType("text/html;charset=gbk");
        PrintWriter out = response.getWriter();
        out.println("收到HelloServlet doPost()请求");
    }

    public void destroy(){
        System.out.println("调用destroy()方法");
    }
}
```

可以看出，上述实例中的各个方法都只是执行打印功能，它们只是为了说明一个Servlet的生命周期的执行过程。

完成上述Servlet编译后，还需要配置一下web.xml，具体配置如下：

```
<servlet>
    <servlet-name>helloServlet</servlet-name>
    <servlet-class>com.etch.edu.HelloServlet</servlet-class>
</servlet>
<servlet-mapping>
    <servlet-name>helloServlet</servlet-name>
    <url-pattern>/HelloServlet</url-pattern>
</servlet-mapping>
```

注意：配置 servlet-name 时应区分字母的大小写。

除了在 web.xml 中配置 Servlet 外，在 Servlet 3.0 及以后的版本中还可以通过直接注入的方式进行配置，代码如下：

```
01  package com.etch.edu;
02
03  import java.io.IOException;
04  import java.io.PrintWriter;
05
06  import jakarta.servlet.ServletException;
07  import jakarta.servlet.ServletRequest;
08  import jakarta.servlet.ServletResponse;
09  import jakarta.servlet.http.HttpServlet;
10  import jakarta.servlet.http.HttpServletRequest;
11  import jakarta.servlet.http.HttpServletResponse;
12  @WebServlet(
13      urlPatterns = { "/HelloServlet " },
14      name = " helloServlet "
15  )
16  public class HelloServlet extends HttpServlet {
17
18      //序列化，可以自动生成或者自行定义
19      private static final long serialVersionUID = 1L;
20
21      public void init() throws ServletException{
22          System.out.println("初始化 init 方法");
23      }
24      public void doGet(HttpServletRequest request, HttpServletResponse
25  response){
26          ...
27      }
28      ...
29  }
```

在上述代码中，第 12~15 行就是利用注入声明的方式表示这是一个 Servlet 类，@WebServlet 中的参数如表 4.1 所示。从上述配置可以看出，这种方法比较简单，也是现在主流的开发形式，在随后的章节中还会继续介绍和使用这种方法。如果利用注入的方式进行了配置，那么 web.xml 就不用配置 Servlet。

表 4.1 @WebServlet 中的主要属性列表

属 性 名	描 述
name(String name)	指定 Servlet 的 name 属性，等价于<servlet-name>。如果没有指定，则该 Servlet 的取值为类的全名
urlPatterns(String[] urls)	指定 Servlet 的 URL 匹配模式，等价于<url-pattern>标签

配置完web.xml之后，可以通过以下步骤查看Servlet的生命周期执行过程。

01 启动Tomcat，将项目工程放在webapps文件夹下。

02 在浏览器地址栏中输入与该Servlet配置相对应的URL，在浏览器中可以看到“收到service请求”内容，在控制台中会输出：

```
初始化 init 方法
调用 public service 方法
```

03 再在浏览器中输入URL，在浏览器中看到“收到service请求”内容，在控制台中会输出：

```
调用 public service 方法
```

通过上述过程，验证了 Servlet 的生命周期过程。图 4.2 进一步说明了生命周期的不同阶段。

图4.2 Servlet的生命周期

从程序的运行结果还可以知道，当重写了service()方法之后，doPost()方法和doGet()方法是不会被处理的，由service()来管理转向对应的方法。

4.4 编写和部署 Servlet

上一节讲解了Servlet的生命周期过程，本节将介绍如何编写一个Servlet类和部署Servlet工程，编写和部署Servlet是开发一个Web工程的基础，也是读者必须掌握的内容。

4.4.1 编写Servlet类

本小节将讲述如何编写一个简单的Servlet类。

本书编写Servlet的开发工具为IntelliJ IDEA，在IntelliJ IDEA的主界面中依次单击File | New | Project命令，然后在弹出的窗口左侧选择Java，在右侧上方选择JDK17.0.4，进入下一步，输入项目名称，选择项目路径，这样就完成了初始工程的创建。

由于 Java EE 已正式更名为 Jakarta EE，而目前的 IDEA 只能支持 Java EE，因此创建完成后我们需要手动接入 Jakarta EE 的支持。首先在已创建的工程上右击鼠标，在弹出的快捷菜单中选择 Add FrameWork Support 命令，再在弹出的窗口中选择 Java EE | Web Application 选项，再单击 OK 按钮，如图 4.3 所示。然后打开项目工程目录，修改 web-app 的版本为 5.0，代码如下：

```
<?xml version="1.0" encoding="UTF-8"?>
<web-app xmlns="https://jakarta.ee/xml/ns/jakartaee"
        xmlns:xsi="http://www.w3.org/2001/XMLSchema-instance"
        xsi:schemaLocation="https://jakarta.ee/xml/ns/jakartaee
        https://jakarta.ee/xml/ns/jakartaee/web-app_5_0.xsd"
        version="5.0">
</web-app>
```

整个工程的目录结构如图 4.4 所示。

图 4.3 接入 Jakarta EE 的支持

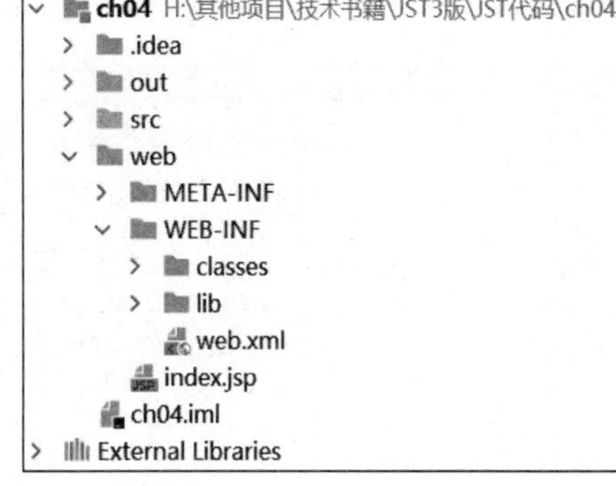

图 4.4 Servlet 工程目录

为了使目录结构更加具有层次感，新建包com.yxtech，然后右击该文件，在弹出的快捷菜单中选择New | Java Class命令，然后在Name文本框中输入自己定制的Servlet程序文件的名字。创建Servlet的对话框如图4.5所示。

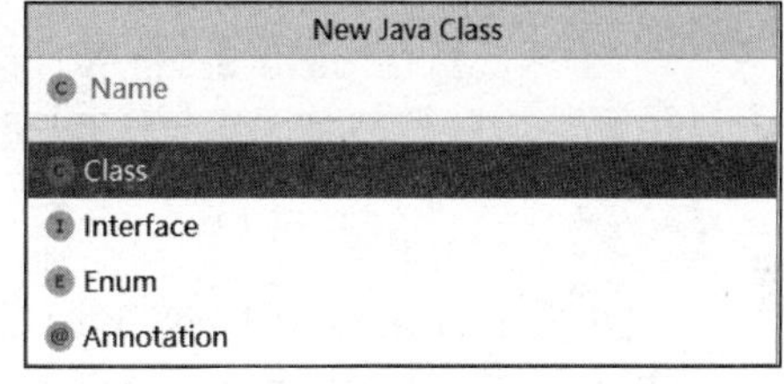

图4.5 创建一个Servlet类

单击 Finish 按钮后出现主编辑界面，在编写 Servlet 类时需要继承 HttpServlet 类，这样才可以开发具体的功能，假设想在页面中循环输出数字 0~10，那么程序代码如下：

```
package com.yxtech;

import java.io.IOException;
import java.io.PrintWriter;

import jakarta.servlet.ServletException;
import jakarta.servlet.http.HttpServlet;
import jakarta.servlet.http.HttpServletRequest;
import jakarta.servlet.http.HttpServletResponse;
@WebServlet(urlPatterns = "/FirstServlet",name = "firstServlet")
public class FirstServlet extends HttpServlet {

    //序列化，可以自动生成或者自行定义
    private static final long serialVersionUID = 1L;

    public void init() throws ServletException{
        System.out.println("初始化 init()");
    }

    public void doGet(HttpServletRequest request,
         HttpServletResponse response)
         throws ServletException,IOException
```

```
    {
        System.out.println("调用 doGet()方法");
        //设置响应的页面类别与页面编码
        response.setContentType("text/html;charset=gbk");
        PrintWriter out = response.getWriter();
        out.println("<html>");
        out.println("<head><title>测试0~10的循环结果</title>");
        out.println("<body>");
        out.println("开始执行……");
        int count=0;
        for(int i=0;i<=10;i++){
            count+=i;
        }
        out.println("程序执行结果："+count);
        out.println("</body>");
        out.println("</html>");
        out.flush();
        out.close();
    }
    public void doPost(HttpServletRequest request,
                        HttpServletResponse response)
                throws ServletException,IOException{
        doGet(request,response);
    }
    public void destroy(){
        System.out.println("调用 destroy()方法");
    }
}
```

这里采用的是注解@WebServlet 的方式注入 Servlet。

现在分析一下创建 Servlet 的步骤：

01 引入相应的包，例如jakarta.servlet包或者jakarta.servlet.http包（这两个包的区别在于前者是与协议无关的，后者是与HTTP协议相关的）。在平时开发的过程中，一般都是继承自HttpServlet类，因为它封装了很多基于HTTP的Servlet功能。当然要想自己开发一个协议，还可以继承GenericServlet类。FirstServlet中有两个处理请求的方法：一个是doGet()方法，响应HTTP Get请求；另一个是doPost()方法，响应HTTP Post请求。这里之所以使用jakarta包。而不是原来的javax包，原因在于Java EE在2017年由甲骨文转让给Eclipse基金会，2018年3月Eclipse基金会正式宣布Java EE更名为Jakarta EE。

02 创建一个扩展类，例如本例中的FirstServlet类。

03 重构doGet()或者doPost()方法。例如，本例中重构了doGet()方法，在该方法中完成处理请求，并输出到HTML页面中。

04 配置web.xml或添加@WebServlet注解。

4.4.2 部署Servlet类

在IntelliJ IDEA中部署一个Web工程有多种方法，本书只介绍两种方法：一种是原始的编译部署，另一种是直接利用在IntelliJ IDEA中配置的Tomcat服务器部署。

1. 原始的编译部署

利用Java编译器将具体的Java类进行编译，例如本例中运行编译命令“javac FirstServlet.java”，在当前目录中生成FirstServlet.class文件，将FirstServlet.class文件复制到WEB-INF文件夹下的classes文件夹中，如图4.6所示。

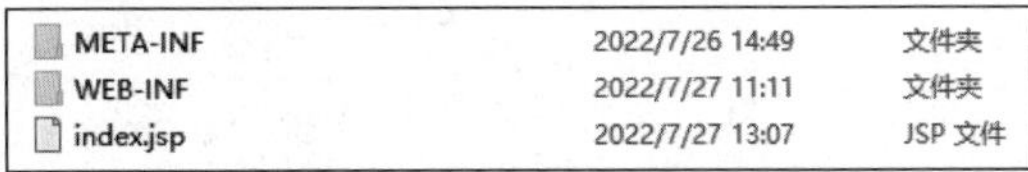

图4.6　Servlet文件夹路径

再将web.xml存放在WEB-INF目录下，接着将整个Servlet工程文件夹存放在webapps目录下，如图4.7所示，再启动Tomcat服务器。

图4.7　Servlet放置目录

2. 利用在 IntelliJ IDEA 中配置的 Tomcat 服务器部署

操作步骤如下：

01 首先单击右上角的按钮打开项目结构窗口，在窗口中选择Artifacts，单击加号，在弹出菜单中选择Web Application：Exploded，单击OK按钮完成配置，如图4.8所示。

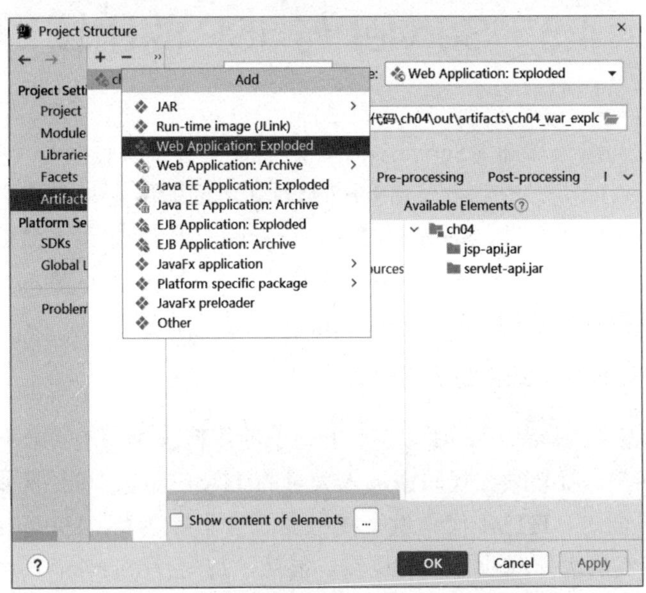

图4.8　在IntelliJ IDEA中部署Servlet工程

02 在项目界面选择Run | Edit Configurations打开配置界面，或者通过右上角的下拉框选择Edit Configurations打开，找到已配置的Tomcat，如果不存在已配置的Tomcat，请参照第1章有关IntelliJ IDEA工具下载部分配置Tomcat。单击对应的Tomcat，选择Server选项卡，配置URL和HTTP port端口号，单击OK按钮完成项目部署，如图4.9所示。配置完成后单击右上角的▶（运行）按钮启动项目即可。

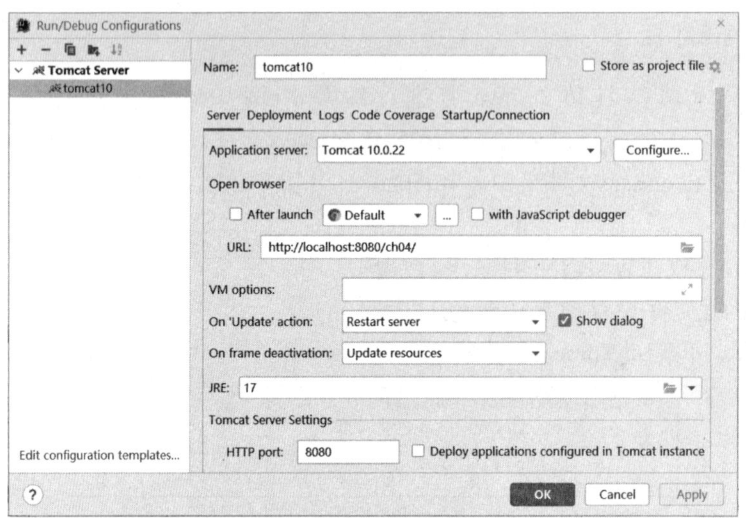

图4.9　部署Tomcat服务器

03 打开IE浏览器，在地址栏中输入地址“http://localhost:8080/ch04/FirstServlet”然后按回车键，如果在页面中显示信息“开始执行……程序执行结果：55”，那么恭喜你，一个Servlet成功部署并运行。

4.5 Servlet 与 JSP 的比较

上一节介绍了如何编写和部署Servlet程序，从而使读者了解编写一个Servlet的基本步骤，并能编写出简单的Servlet类。本节将介绍Servlet与JSP之间的区别与联系，包括它们之间的内在联系是什么。其实从本质上讲，Servlet与JSP是一样的，因为JSP页面最后在运行时会被转换成一个Servlet，但是从开发者的视角看，运用这两种技术还是有些区别的，这些区别也决定了在开发中应该如何选择使用它们。JSP与Servlet的主要区别说明如下。

1. Servlet 是 Java 代码，JSP 是页面代码

编写Servlet就是编写Java代码，所以应用Java中的规范去编写Servlet类就可以了，但是若想在客户端中响应结果，就必须在代码中加入大量的HTML代码。可想而知，当想要得到一个比较美观、复杂的界面时，HTML代码量会相当多而且非常烦琐。JSP以HTML代码为主，在页面中适当嵌入Java代码来处理业务上的逻辑。显然，JSP会比Servlet较易编写并更直观。

基于此的差异也是选择Servlet或者JSP技术的考量标准之一。如果业务中主要是以页面为主，就选择JSP技术；反之，则选择Servlet技术（适合服务器端开发）。

2. Servlet 的运行速度快过 JSP

Servlet本身就是一个Java类，编译的时候会直接被转换为类文件；JSP需要先被编译为Java类，而后再运行，所以Servlet的运行速度较快。

3. Servlet 需要手动编译，JSP 由服务器自动编译

Servlet类被编译成为类文件后，需要手动复制到Web应用程序目录下。JSP页面部署到Web应用则简单很多，只需要将JSP页面复制到指定的目录下即可，当它第一次被访问时，Web服务器自动将JSP代码转换为Servlet（Java代码）并自动编译。

4. 编辑 HTML 工具不支持编辑 Servlet

当前有很多制作网页的工具，例如Dreamweaver、Golive等，利用这些工具可以快速地开发网页、编写出复杂的界面，而且具备可视化效果，极大地提高了网页开发的效率。但对于大多数的网页工具而言，在HTML页面中加入Java代码是可以的，但是要在Java代码中加入HTML代码却不行，而且不能及时纠错。因此，初学者利用不同的编辑器工具编写JSP和Servlet也是可以理解的。

了解了 Servlet 与 JSP 的差别之后，在编写 Web 应用程序时，就要根据当前的需要权衡是使用 JSP 还是 Servlet。开发者应尽量使 JSP 和 Servlet 都发挥出最大的作用，同时又能够便于日后的代码维护工作。一般而言，Servlet 大多用于负责对客户端的请求进行处理和调用 Java Bean，由 Java Bean 负责提供可复用的数据以及访问数据等；而 JSP 页面主要负责页面的展示，将动态数据展现给客户，这就是开发者提出的简易 MVC 模式，这样分工大大减少了 JSP 页面中 Java 程序和 HTML 代码的耦合度，对维护工作具有重大的意义。

4.6 小　　结

本章主要介绍了Servlet的基础知识，包括它的基本概念、技术特点等。通过本章的学习，读者可掌握Servlet的基本编写方法和步骤、生命周期的意义，以及在生命周期各阶段调用的不同方法，最后需要了解并掌握Servlet与JSP的相同点和不同点。

4.7 习　　题

（1）从本质上说，Servlet是什么？

（2）当Servlet第一次被加载时，服务器端是如何创建Servlet对象的？

（3）简述Servlet与JSP的异同点。

（4）简述Servlet的生命周期。

（5）编写一个Servlet程序，用于输出一个九九乘法表。

第 5 章 请求与响应

在JSP开发中，请求与响应是最基本的两个内置对象。一个Web应用系统，必须得有请求和响应才能构建一个完整的程序。了解请求与响应的原理和方法可以有效提高前端人员的开发效率。

本章主要涉及的知识点有：

- 掌握请求和响应的基本概念
- 如何读取Body内容，取得上传文件、调派请求
- 如何输出字符、二进制流对象
- 如何正确跳转

5.1 从容器到HttpServlet

在第3、4章中，我们介绍了JSP的内置对象和Servlet的相关知识，以及如何部署和开发一个Servlet，但是并没有详细介绍如何将Servlet与JSP结合起来使用。Web容器是JSP唯一可以识别的HTTP服务器，所以必须了解Web容器如何生成请求和响应对象。本节将介绍Web容器与HttpServlet之间的“来龙去脉”。

5.1.1 Web容器用来做什么

Web容器的作用就是创建一个Servlet实例，并完成Servlet注册以及根据web.xml中的URL进行响应。当请求来到容器时，Web容器会转发给对应的Servlet来处理请求。

当客户端请求HTTP服务器时，会使用HTTP来传递请求、标头、参数等信息。HTTP协议是无意识的协议，通过文本信息传递消息，而Servlet是Java对象，运行在Web容器中。当HTTP服务器将请求转给Web容器时，Web容器会创建一个HttpServletRequest和HttpServletResponse

对象，将请求中的信息传递给HttpServletRequest对象，而HttpServletResponse对象则作为对客户端响应的Java对象，这一过程如图5.1所示。

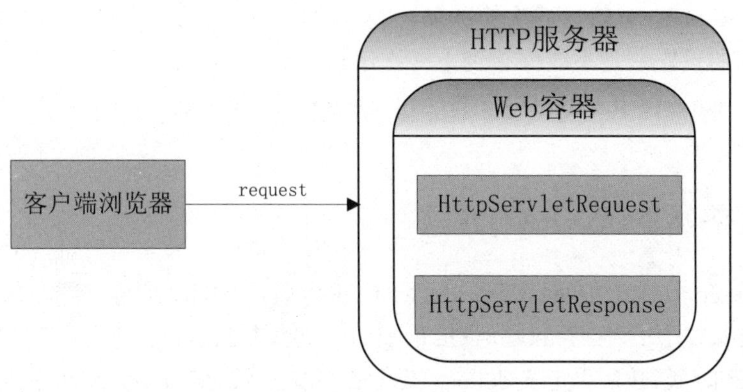

图5.1　Web容器收集请求信息并创建请求和响应对象

Web容器会根据配置信息（例如web.xml或者@WebServlet）查找相对应的Servlet并调用它的service()方法。service()方法会根据HTTP请求的方式决定是调用doPost()方法还是doGet()方法。例如，HTTP请求的方式为post，则调用doPost()方法。

在doPost()方法中，可以使用HttpServletRequest对象、HttpServletResponse对象。例如，使用getParameter()取得请求参数值，使用getWriter()取得输出流对象PrintWriter并进行响应处理。最后由Web容器转换为HTTP响应，由HTTP服务器对浏览器做出响应。随后，Web容器将HttpServletRequest对象、HttpServletResponse对象销毁回收，请求响应结束，过程如图5.2所示。

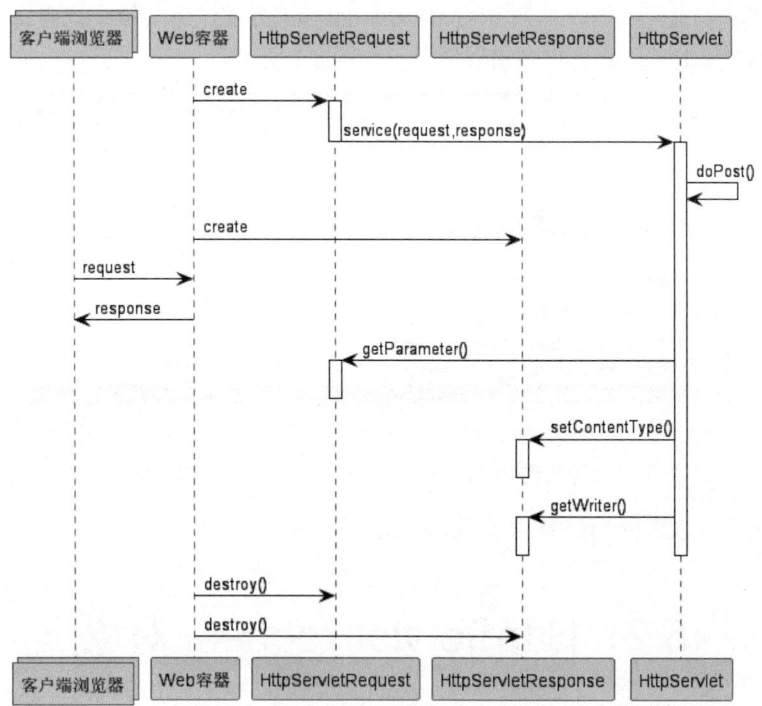

图5.2　从请求到响应流程的示意图

前面提到过HTTP协议是一种无状态的协议，每一次的请求、响应后，浏览器不会记得客户端的信息。Web容器每次都会根据请求创建新的request、response对象，做出响应后就销毁该次的request、response对象。下次的请求、响应就与上一次的请求、响应对象无关了，所以对于请求、响应的设置，是不能保存到下一次请求的。

类似于这种request、response对象的创建、服务、销毁，就是Web容器提供的请求、响应的生命周期管理。

5.1.2 令人茫然的doXXX()方法

Servlet 中的 service()方法包括多种：doGet()、doPost()、doHead()等。当 Method 是 POST 时，请求会调用 doPost()方法；当 Method 是 GET 时，请求会调用 doGet()方法。在定义 Servlet 时，一般是继承 HttpServlet 类，然后再定义 doPost()或者 doGet()方法。在 HttpServlet 中 doGet()和 doPost()方法的实现过程如下：

```
    protected void doGet(HttpServletRequest req, HttpServletResponse resp) throws
ServletException, IOException {
        String msg = lStrings.getString("http.method_get_not_supported");
        this.sendMethodNotAllowed(req, resp, msg);
    }
    protected void doPost(HttpServletRequest req, HttpServletResponse resp) throws
ServletException, IOException {
        String msg = lStrings.getString("http.method_post_not_supported");
        this.sendMethodNotAllowed(req, resp, msg);
    }
```

假如，在自定义 Servlet 时不实现 doGet()或者 doPost()方法，那么程序就会调用以上方法。当客户端发出 POST 请求时，就会收到错误消息，如图 5.3 所示。

图5.3 不实现doGet()或者doPost()方法时显示的错误信息

因此，在定义完Servlet后，需要实现doGet()或者doPost()方法。

5.2 HttpServletRequest 对象

HttpServletRequest对象是请求封装对象，由Web容器生成，使用该对象可以取得HTTP请求中的信息。在Servlet中，也是使用该对象进行请求处理的。如果要共享request中的属性值，可以将请求对象设置到该对象中，那么Servlet在同一请求中就可以共享其对象。在第3章中，曾介绍过JSP的内置对象request实质上就是HttpServletRequest。有关参数的请求和设置标头等都可以参考该章节。本节将要讨论的是如何读取Body内容、取得上传文件、调派请求等。

5.2.1　使用getReader()、getInputStream()读取Body内容

在HttpServletRequest对象中，可以运用getReader()方法来获取Body内容，而使用getInputStream()方法来获得上传文件的内容。下面利用一个示例来说明如何获取Body内容。

【例 5.1】　利用 getReader()方法获取 Body 内容。

首先，编写 GetReaderBody 的 Servlet 类，用于获取 BufferedReader 对象，并且逐行遍历其内容，随后输出到浏览器中。源代码如下：

```
-----------------------GetReaderBody.java--------------------------
01  package com.eshore;
02
03  import java.io.BufferedReader;
04  import java.io.IOException;
05  import java.io.PrintWriter;
06  import jakarta.servlet.ServletException;
07  import jakarta.servlet.http.HttpServlet;
08  import jakarta.servlet.http.HttpServletRequest;
09  import jakarta.servlet.http.HttpServletResponse;
10  @WebServlet(name = "getReaderBody", urlPatterns = { "/getReaderBody.do" })
11  public class GetReaderBody extends HttpServlet {
12
13      public void doPost(HttpServletRequest request, HttpServletResponse
14  response) throws ServletException, IOException {
15          BufferedReader br = request.getReader();    //获取BufferedReader对象
16          String input ="";
17          String body = "";
18          while((input = br.readLine())!=null){       //遍历body内容
19              body+=input+"<br/>";
20          }
21          response.setContentType("text/html;charset=UTF-8"); //设置响应的类型和
编码
22          PrintWriter out = response.getWriter();  //取得PrintWriter()对象
23          out.println("<!DOCTYPE HTML>");
24          out.println("<HTML>");
25          out.println("  <HEAD><TITLE>A Servlet</TITLE></HEAD>");
26          out.println("  <BODY>");
27          out.print(body);
28          out.println("  </BODY>");
29          out.println("</HTML>");
30          out.flush();
31          out.close();
32      }
33  }
```

在上述代码中，第 15 行代码用于获得 BufferedReader 对象，第 18~20 行循环遍历 body 内容，第 21~31 行代码用于向浏览器输出内容。

随后，编写 getReaderBody.jsp 页面，用于提交 form 表单信息到该 Servlet 中，源代码如下：

```
------------------------getReaderBody.jsp---------------------------
01  <%@ page pageEncoding="UTF-8"%>
02  <!DOCTYPE HTML>
03  <html>
04    <head>
05      <title>获取body内容</title>
06    </head>
07
08    <body>
09        <form action="<%=request.getContextPath()%>/getReaderBody.do"
10                method="POST">
11        用户名：<input name="username"/><br/>
12        密  码：<input name="password" type="password"/><br/>
13                <input type="submit" name="user_submit" value="提交" />
14        </form>
15    </body>
16  </html>
```

在上述代码中，第 09~14 行代码定义 form 表单及提交的 Servlet，并在 input 中分别定义 name 值。部署项目后，在浏览器中输入该页面地址并提交，提交后程序的运行效果如图 5.4 所示。

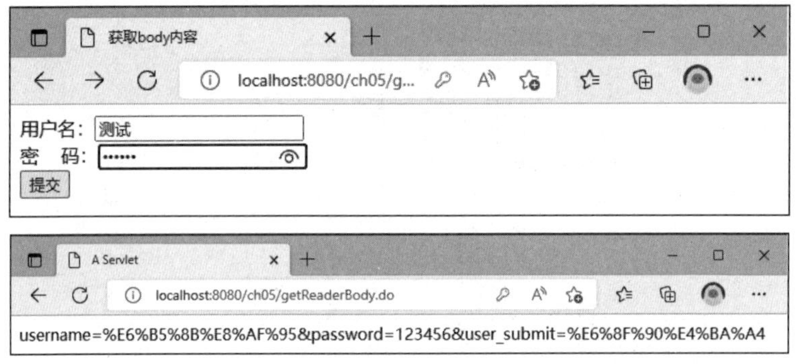

图5.4 浏览器发出的请求Body内容

从图5.4中可以看出，页面用的是UTF-8编码，所以在Servlet中都统一被转换为UTF-8编码，并且参数值都是一一对应的。以上是用getReader()方法获取Body内容。如果在form中设置enctype的值为multipart/form-data，则表示从客户端上传文件，获取Body内容时就会获得一堆很奇怪的字符，如例5.2所示。

【例 5.2】 获取上传文件的 Body 内容。

更改页面 getReaderBody2.jsp 的 form 内容，源代码如下：

```
------------------------getReaderBody2.jsp---------------------------
<%@ page pageEncoding="UTF-8"%>
<!DOCTYPE HTML>
<html>
  <head>
    <title>获取上传文件body内容</title>
  </head>
  <body>
      <form action="<%=request.getContextPath()%>/getReaderBody.do"
```

```
09              method="POST" enctype="multipart/form-data">
10              选择文件：<input  type="file" name="filename"/>
11              <input type="submit" name="file_submit" value="提交" />
12        </form>
13    </body>
14  </html>
```

在上述代码中，第 10 行代码设置 enctype 类别为 multipart/form-data，第 11 行代码设置 input 类型为 file。不选择文件，直接单击“提交”按钮，运行后的效果如图 5.5 所示。

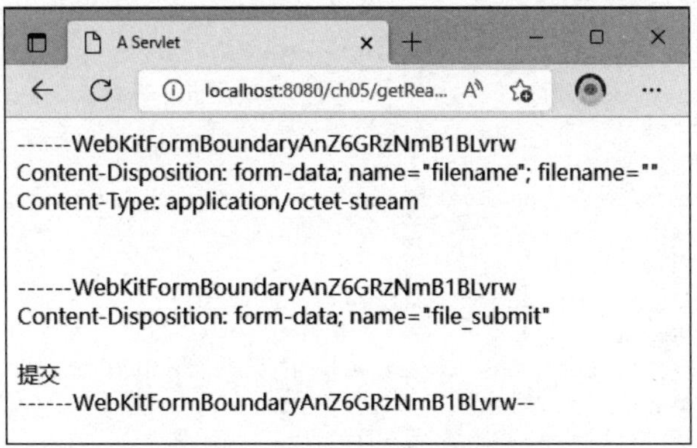

图5.5　上传文件输出Body内容示意图

从图5.5中可以看出，上传文件的内容就是一堆乱码。显然这不是我们想要的结果，可以使用getInputStream()方法来获取内容，如例5.3所示。

【例5.3】　获取上传文件的Body内容。

```
---------------------- GetReaderBody2.java-----------------------
package com.eshore;

import java.io.FileNotFoundException;
import java.io.FileOutputStream;
import java.io.IOException;
import java.io.InputStream;
import java.io.PrintWriter;

import jakarta.servlet.ServletException;
import jakarta.servlet.http.HttpServlet;
import jakarta.servlet.http.HttpServletRequest;
import jakarta.servlet.http.HttpServletResponse;
@WebServlet(name = "getReaderBody2", urlPatterns = { "/upload.do" })
public class GetReaderBody2 extends HttpServlet {

    public void doPost(HttpServletRequest request, HttpServletResponse
response) throws ServletException, IOException {
        //读取请求body
        byte[] body = readBody(request);
        //将内容进行统一编码
        String textBody = new String(body,"UTF-8");
        //获得文件名字
        String filename = getFilename(textBody);
        //写到指定位置
```

```
25            writeToFile(filename, body);
26            //返回页面消息，首先设置响应的类型和编码
27            response.setContentType("text/html;charset=UTF-8");
28            PrintWriter out = response.getWriter();  //取得PrintWriter()对象
29            out.println("<!DOCTYPE HTML>");
30            //输出页面内容
31            out.println("<HTML>");
32            out.println("  <HEAD><TITLE>A Servlet</TITLE></HEAD>");
33            out.println("  <script>alert(\"上传成功\")</script><BODY>");
34            out.println("  </BODY>");
35            out.println("</HTML>");
36            out.flush();
37            out.close();
38        }
39        //读取请求的输入流
40        private byte[] readBody(HttpServletRequest request)
41    throws IOException{
42            //获取请求内容长度
43            int len = request.getContentLength();
44            //获取请求的输入流
45            InputStream is = request.getInputStream();
46            //新建一个字节数组
47            byte b[] = new byte[len];
48            int total = 0;
49            while(total<len){                           //读取字节流
50                int bytes = is.read(b, total, len);
51                total += bytes;
52            }
53            return b;
54        }
55        //获取文件名称
56        private String getFilename(String bodyText){
57            String filename = bodyText.substring(bodyText.indexOf ("filename=\
"")+10);
58            filename = filename.substring(filename.lastIndexOf ("\\")+1,
filename.indexOf("\""));
59            return filename;
60        }
61        //写入指定文件
62        private void writeToFile(String filename,byte[] body)
63    throws FileNotFoundException,IOException{
64            //写到指定的文件地址
65            FileOutputStream fileOutputStream =
66                new FileOutputStream("d:/file/"+filename);
67            fileOutputStream.write(body);
68            fileOutputStream.flush();
69            fileOutputStream.close();                   //注意关闭输出流
70        }
71    }
```

第 27~37 行代码用于输出页面的响应；第 40~54 行代码用于读取请求中的字节流，并用字节数组保存；第 56~60 行代码用于获取文件名称；第 62~70 行代码将字节流输出到指定的位置。页面代码主要使用例 5.2 中的代码，但将 Servlet 更改为 GetReaderBody2，并在@WebServlet 注解中配置请求路径。程序运行结束后会在“d:/file/”文件夹下新建一个上传的文件，打开文件后可发

现多了很多页面中的信息，这是因为没有进行位置的过滤。其实在 Servlet 3.0 及之后的版本中，可以利用 getPart()、getParts()来获取上传文件。

5.2.2 使用getPart()、getParts()取得上传文件

在5.2.1节中，介绍了使用getInputStream()方法获取上传文件的内容，从例5.3可以看出代码比较长，而且得到的结果还不理想。在Servlet 3.0以后新增getPart()和getParts()方法来处理上传文件的内容。getPart()用于处理单文件内容，getParts()用于处理多文件内容。在使用getPart()和 getParts() 方法时，必须使用 MultipartConfig 注解，这样 Servlet 才能获得 Part 对象。MultipartConfig的属性如表5.1所示。

表 5.1　MultipartConfig 属性介绍

属　性	说　明
fileSizeThreshold	数值类型，当上传文件的大小大于该值时，内容将先写入缓存文件，默认值为 0
location	字符串类型，设置存放生成的文件目录地址，也是上传过程中临时文件的保存路径，执行 Part.write()方法之后，临时文件将被清除
maxFileSize	数值类型，设置允许文件上传的最大值，默认值为-1L，表示不限制大小
maxRequestSize	数值类型，限制 multipart/form-data 请求的最大数，默认值为-1L，表示不限制大小

通过改写例 5.3 的 GetReaderBody2 来说明 getPart()的用法，如例 5.4 所示。

【例5.4】　利用getPart()方法获取上传文件的Body内容。

```
----------------------- GetPartBodyContent.java--------------------------
import java.io.PrintWriter;
import jakarta.servlet.ServletException;
import jakarta.servlet.annotation.MultipartConfig;
import jakarta.servlet.annotation.WebServlet;
import jakarta.servlet.http.HttpServlet;
import jakarta.servlet.http.HttpServletRequest;
import jakarta.servlet.http.HttpServletResponse;
import jakarta.servlet.http.Part;
import org.apache.commons.lang3.StringUtils;
@MultipartConfig(location = "D:/tmp/", maxFileSize = 1024 * 1024 * 10)
@WebServlet(name = "getPartBodyContentServlet", urlPatterns =
{ "/upload.do" },loadOnStartup = 0)
public class GetPartBodyContent extends HttpServlet {
    public void doPost(HttpServletRequest request, HttpServletResponse
response) throws ServletException, IOException {
            //设置处理编码
            request.setCharacterEncoding("UTF-8");
            Part part = request.getPart("filename");//取得Part对象
            // 获得文件名字
            String filename = getFilename(part);
            part.write(filename);
            // 返回页面消息，首先设置响应的类型和编码
            response.setContentType("text/html;charset=UTF-8");
            PrintWriter out = response.getWriter();// 取得PrintWriter()对象
            out.println("<!DOCTYPE HTML>");
```

```
26            //输出页面内容
27            out.println("<HTML>");
28            out.println("  <HEAD><TITLE>A Servlet</TITLE></HEAD>");
29            out.println("  <script>alert(\"上传成功\")</script><BODY>");
30            out.println("  </BODY>");
31            out.println("</HTML>");
32            out.flush();
33            out.close();
34        }
35        // 获取文件名称
36        private String getFilename(Part part) {
37            if (part == null)
38                return null;
39            String fileName = part.getHeader("content-disposition");
40            if (StringUtils.isBlank(fileName)) {
41                return null;
42            }
43            return StringUtils.substringBetween(fileName, "filename=\"", "\"");
44        }
45    }
```

从上述代码可以看出，使用 Part 对象来处理上传文件后程序变得简单很多。第 10 行代码用于注入 MultipartConfig，并设置上传路径和最大的上传文件大小；第 11 行代码利用注入的方式配置 Servlet 路径；第 17 行代码用于设置文件编码；第 18 行代码用于取得 Part 对象；第 21 行代码用于直接调用 Part 对象的 write()方法来输出文件。

页面 getReaderBody3.jsp 的源代码如下：

```
------------------------ getReaderBody3.jsp--------------------------
01  <%@ page pageEncoding="UTF-8"%>
02  <!DOCTYPE HTML">
03  <html>
04    <head>
05      <title>getPart()获取上传文件body内容</title>
06    </head>
07
08    <body>
09        <form action="<%=request.getContextPath()%>/upload.do"
10                method="POST" enctype="multipart/form-data">
11                选择文件：<input type="file" name="filename"/>
12                <input type="submit" name="file_submit" value="提交" />
13        </form>
14    </body>
15  </html>
```

上述代码与 getReaderBody2.jsp 的区别在于第 09 行，其余都相同。代码运行之后，在“D:/tmp/”文件夹下新建一个上传文件，打开该文件发现内容与原来的文件相同。

注意：代码中运用了 StringUtils 类，是 commons.lang 包中的 String 工具类。Part 对象必须使用@MultipartConfig 才能取得 Part 对象，否则 getPart()对象会得到 null。

上述代码演示了单个文件的处理过程，如果要处理多个文件，就应使用getParts()方法。getParts()方法返回Collection集合，如例5.5所示。

【例5.5】 利用getParts()方法获取批量上传文件。

```
----------------------- GetPartsBodyContent.java---------------------------
package com.eshore;

import java.io.FileNotFoundException;
import java.io.FileOutputStream;
import java.io.IOException;
import java.io.InputStream;
import java.io.PrintWriter;

import jakarta.servlet.ServletException;
import jakarta.servlet.annotation.MultipartConfig;
import jakarta.servlet.annotation.WebServlet;
import jakarta.servlet.http.HttpServlet;
import jakarta.servlet.http.HttpServletRequest;
import jakarta.servlet.http.HttpServletResponse;
import jakarta.servlet.http.Part;

import org.apache.commons.lang3.StringUtils;

@MultipartConfig(location = "D:/tmp/", maxFileSize = 1024 * 1024 * 10)
@WebServlet(name = "getPartsBodyContentServlet",
urlPatterns = { "/uploads.do" }, loadOnStartup = 0)
public class GetPartsBodyContent extends HttpServlet {

    public void doPost(HttpServletRequest request, HttpServletResponse
response) throws ServletException, IOException {
        //设置处理编码
        request.setCharacterEncoding("UTF-8");
        request.getParts();
        for(Part part:request.getParts()){
            //只处理上传文件，因为“提交”按钮也被当作一个Part对象
            if(part.getName().startsWith("filename")){
                // 获取文件名称
                String filename = getFilename(part);
                part.write(filename);
            }
        }
        // 返回页面消息，首先设置响应的类型和编码
        response.setContentType("text/html;charset=UTF-8");
        PrintWriter out = response.getWriter();   // 取得PrintWriter()对象
        out.println("<!DOCTYPE HTML");
        out.println("<HTML>");
        out.println("  <HEAD><TITLE>A Servlet</TITLE></HEAD>");
        out.println("  <script>alert(\"上传成功\")</script><BODY>");
        out.println("  </BODY>");
        out.println("</HTML>");
        out.flush();
        out.close();
    }
    // 获取文件名称
    private String getFilename(Part part) {
        if (part == null)
            return null;
        String fileName = part.getHeader("content-disposition");
```

```
54              if (StringUtils.isBlank(fileName)) {
55                  return null;
56              }
57              return StringUtils.substringBetween(fileName, "filename=\"", "\"");
58          }
59      }
```

在上述代码中，第29~36行代码用于取得Part集合对象并进行遍历，接着逐个输出单个文件。在这里需要注意的是“提交”按钮也会被当作一个Part对象，所以需要进行判断，只输出上传文件的内容。其余代码与例5.4相同。页面请求代码如下：

```
01  <%@ page pageEncoding="UTF-8"%>
02  <!DOCTYPE HTML>
03  <html>
04    <head>
05      <title>getParts()获取上传文件body内容</title>
06    </head>
07    <body>
08          <form action="<%=request.getContextPath()%>/uploads.do"
09                  method="POST" enctype="multipart/form-data">
10                  选择文件1：<input type="file" name="filename1"/><br/>
11                  选择文件2：<input type="file" name="filename2"/><br/>
12                  选择文件3：<input type="file" name="filename3"/><br/>
13                  <input type="submit" name="file_submit" value="提交" />
14          </form>
15    </body>
16  </html>
```

在上述代码中，第 10~12 行分别定义 3 个 type 为 file 的 input，action 更改为新定义的 Servlet 路径 uploads.do。运行程序后，在“D:/tmp/”路径下新增 3 个上传文件。

5.2.3 使用RequestDispatcher调派请求

在Web程序中，经常由多个Servlet来完成请求。RequestDispatcher接口就是为了多个Servlet之间的调整而实现的。该接口可以由HttpServletRequest的getRequestDispatcher()方法取得。调用时指定跳转的URL地址即可完成跳转动作。RequestDispatcher接口有两种方法实现跳转Servlet：include()和forward()。

1. include()方法

include()方法的含义是可以在当前的Servlet中显示另外一个Servlet的内容，如例5.6所示。

【例5.6】 演示include()方法。

```
---------------------- GetRequestDispatcherDemo.java-------------------------
package com.eshore;

import java.io.IOException;
import java.io.PrintWriter;

import jakarta.servlet.RequestDispatcher;
import jakarta.servlet.ServletException;
import jakarta.servlet.annotation.WebServlet;
import jakarta.servlet.http.HttpServlet;
```

```
10 import jakarta.servlet.http.HttpServletRequest;
11 import jakarta.servlet.http.HttpServletResponse;
12
13 @WebServlet(
14     name = "getDispatcherDemo",
15     urlPatterns = { "/include.do" },
16     loadOnStartup = 0)
17 public class GetRequestDispatcherDemo extends HttpServlet {
18
19     private static final long serialVersionUID = 1L;
20
21     public void doGet(HttpServletRequest request, HttpServletResponse
22 response) throws ServletException, IOException{
23         doPost(request,response);
24     }
25     public void doPost(HttpServletRequest request, HttpServletResponse
26 response) throws ServletException, IOException {// 返回页面消息
27         // 设置响应的类型和编码
28         response.setContentType("text/html;charset=UTF-8");
29         PrintWriter out = response.getWriter();    // 取得PrintWriter()对象
30         out.println("<!DOCTYPE HTML>");
31         //输出页面内容
32         out.println("<HTML>");                      //输出页面信息
33         out.println("  <HEAD><TITLE>A Servlet</TITLE></HEAD>");
34         out.println("  <BODY>");
35         out.println("  The First Servlet<br/>");
36         //获取RequestDispatcher对象
37         RequestDispatcher dispatcher =
38 request.getRequestDispatcher("/includeSeconde.do");
39         dispatcher.include(request, response);
40         out.println("  Including Servlet<br/>");
41         out.println("  </BODY>");
42         out.println("</HTML>");
43         out.close();
44     }
45 }
```

在上述代码中，第13~16行代码利用注入的方式声明一个Servlet；第29行代码取得PrintWriter对象，向浏览器输出内容；第37、38行代码设置包含的Servlet路径。其中包含的Servlet类源代码如下：

```
--------------------GetRequestDispatcherSecondeDemo.java--------------------
package com.eshore;

import java.io.IOException;
import java.io.PrintWriter;

import jakarta.servlet.ServletException;
import jakarta.servlet.annotation.WebServlet;
import jakarta.servlet.http.HttpServlet;
import jakarta.servlet.http.HttpServletRequest;
import jakarta.servlet.http.HttpServletResponse;

@WebServlet(name = "getDispatcherSecondeDemo",
        urlPatterns = { "/includeSeconde.do" },
        loadOnStartup = 0)
```

```
public class GetRequestDispatcherSecondeDemo extends HttpServlet {

    private static final long serialVersionUID = 1L;

    public void doGet(HttpServletRequest request, HttpServletResponse
response) throws ServletException, IOException {
        doPost(request, response);
    }
    public void doPost(HttpServletRequest request, HttpServletResponse
response) throws ServletException, IOException {// 返回页面消息
        // 设置响应的类型和编码
        response.setContentType("text/html;charset=UTF-8");
        PrintWriter out = response.getWriter();    // 取得PrintWriter()对象
        out.println("  The second Servlet<br/>");
    }
}
```

上述代码相对简单，首先声明一个Servlet，然后获得PrintWriter对象并输出页面内容。程序运行效果如图5.6所示。

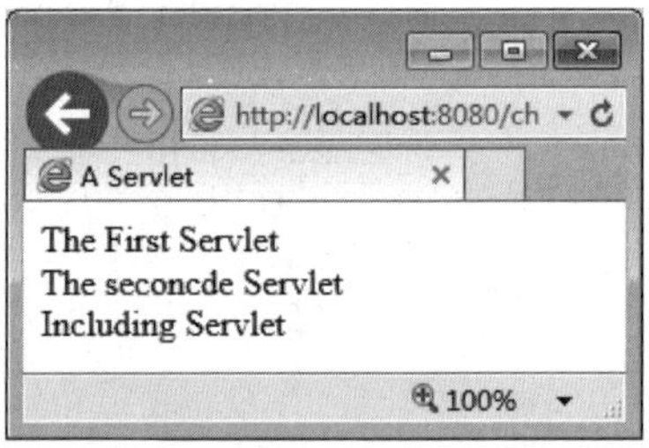

图5.6　例5.6的效果图

2. forward()方法

forward()方法是RequestDispatcher中的另一种Servlet跳转方法，只是它的含义是跳转到其他的Servlet，而不会再返回跳转前的Servlet。forward()方法在Servlet中经常用到，因为很多业务逻辑都是跳转到其他的Servlet中进行处理。下面利用一个实例说明其用法并建立简单的模型架构。

【例5.7】 演示forward()方法。

```
----------------------- HelloServlet.java--------------------------
package com.eshore;

import java.io.IOException;

import jakarta.servlet.ServletException;
import jakarta.servlet.annotation.WebServlet;
import jakarta.servlet.http.HttpServlet;
import jakarta.servlet.http.HttpServletRequest;
import jakarta.servlet.http.HttpServletResponse;

import com.eshore.pojo.HelloUser;

@WebServlet(
    name = "helloServlet",
    urlPatterns = { "/hello.do" })
public class HelloServlet extends HttpServlet{
    private static final long serialVersionUID = 1L;

    private HelloUser user = new HelloUser();
    public void doGet(HttpServletRequest request, HttpServletResponse
response) throws ServletException, IOException{
        doPost(request,response);//调用doPost()方法
    }
    public void doPost(HttpServletRequest request, HttpServletResponse
```

```
25  response) throws ServletException, IOException {
26          //获取参数username的值
27          String userName = request.getParameter("username");
28          String message = user.sayHello(userName);        //获取消息值
29          request.setAttribute("message", message);        //设置参数值
30          request.getRequestDispatcher("/hello.htm").      //跳转到hello.htm
31              forward(request, response);
32      }
33  }
```

在上述代码中，doPost()方法非常简单，第 27 行代码获取参数 username 的值，第 28 行代码通过调用类 HelloUser 中的 sayHello 方法获取欢迎信息，第 30 行代码跳转到以 hello.htm 为 URL 的 Servlet 类中。

类 HelloUser 是普通的 pojo 类，作用是模拟业务上的操作，其源代码如下：

```
----------------------- HelloUser.java-------------------------
01  package com.eshore.pojo;
02
03  import java.util.HashMap;
04  import java.util.Map;
05
06  public class HelloUser {
07      private Map<String,String> helloMessage = new
08  HashMap<String,String>();
09      //构建用户值，模拟从数据库中取出的值
10      public HelloUser(){
11          helloMessage.put("John", "Hello,John");
12          helloMessage.put("Smith", "Welcome,Smith!");
13          helloMessage.put("Rose", "Hi,Rose");
14      }
15      //根据用户返回Map中的消息
16      public String sayHello(String userName){
17          return helloMessage.get(userName);
18      }
19  }
```

在上述代码中，第10~14行代码利用构造方法在Map中存放用户数据，根据用户ID获取欢迎信息。路径为hello.htm的Servlet类的主要作用是显示并打印出欢迎消息，其源代码如下：

```
----------------------- HelloHtml.java-------------------------
package com.eshore;

import java.io.IOException;
import java.io.PrintWriter;

import jakarta.servlet.ServletException;
import jakarta.servlet.annotation.WebServlet;
import jakarta.servlet.http.HttpServlet;
import jakarta.servlet.http.HttpServletRequest;
import jakarta.servlet.http.HttpServletResponse;
@WebServlet(name = "helloHtml", urlPatterns = { "/hello.htm" },
loadOnStartup = 0)
public class HelloHtml extends HttpServlet {
    private static final long serialVersionUID = 1L;

    public void doGet(HttpServletRequest request, HttpServletResponse
```

```
17 response) throws ServletException, IOException {
18         doPost(request, response);
19     }
20
21     public void doPost(HttpServletRequest request, HttpServletResponse
22 response) throws ServletException, IOException { // 返回页面消息
23         String message = (String)request.getAttribute("message");
24         //设置响应的类型和编码
25         response.setContentType("text/html;charset=UTF-8");
26         PrintWriter out = response.getWriter();    //取得PrintWriter()对象
27         out.println("<!DOCTYPE HTML>");
28         //输出页面内容
29         out.println("<HTML>");
30         out.println("  <HEAD><TITLE>A Servlet</TITLE></HEAD>");
31         out.println("  <BODY>");
32         out.println(message);                          //打印出消息到页面
33         out.println("  </BODY>");
34         out.println("</HTML>");
35         out.close();
36     }
37 }
```

在上述代码中，第 23 行代码用于获取上一个 Servlet 中传递的参数 message，第 26 行代码用于获取 PrintWriter()对象的输出页面信息。在浏览器中输入带参数的地址“http://localhost:8080/ch05/hello.do?username=Smith”，运行效果如图 5.7 所示。

在上述代码中，除演示include()方法外，也演示了简单的MVC模型：HelloServlet类可以看作Controller，Controller中不会有HTML代码，它只负责进行业务中的处理；HelloUser是Model，Model负责对数据进行处理，例如在本例中用于获取欢迎信息；HelloHtml是View，用于展示页面结果。当然这只是对MVC的一个简单描述，主要是让读者对MVC有一个初步的概念，在随后章节的实例中会继续介绍MVC模型。

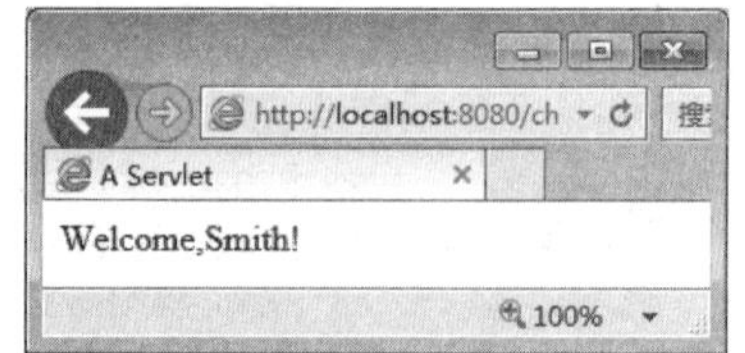

图5.7　例5.7的效果图

5.3 HttpServletResponse 对象

HttpServletResponse是用于对浏览器做出响应的操作对象，可以设置响应类型，也可以直接输出HTML内容。

通常情况下，使用setContentType()设置JSP响应类型，使用getWriter()取得PrintWriter对象或者使用getOutputStream()取得ServletOutputStream流对象，使用setHeader()、addHeader()设置标头，使用sendRedirect()、sendError()对页面进行重定向或者是发送错误消息。

5.3.1 使用getWriter()输出字符

在前面的例子中，经常可以看到使用 getWriter()方法获取 PrintWriter 对象，指定字符串对浏览器输出 HTML 代码，例如：

```
PrintWriter out = response.getWriter();// 取得PrintWriter()对象
out.println("<!DOCTYPE HTML>");
out.println("<HTML>");
out.println("  <HEAD><TITLE>A Servlet</TITLE></HEAD>");
out.println("  <BODY>");
out.println("  </BODY>");
out.println("</HTML>");
```

通常情况下，在对浏览器做出响应的同时会设置字符编码，因为默认的编码是 GBK 或者是 ISO-8859-1，输出中文时会显示乱码，所以需要设置支持中文的字符编码格式。这里有两种设置编码格式的方式：setContentType()或者 setCharacterEncoding()方法。

设置编码语句为：

```
response.setContentType("text/html;charset=UTF-8");
```

或者

```
response.setCharacterEncoding("UTF-8");
```

注意：设置编码需要在获得 PrintWriter 对象或者 ServletOutputStream 流对象之前进行，否则无效。如果请求的参数中有中文，还需要设置在请求中支持中文字符编码，而且要注意与响应的编码保持一致。

在 Servlet 开发中，必须告诉浏览器是以何种方式处理响应（内容类型），所以要设置响应标头。在设置 context-type 中，指定 MIME 类型后，浏览器就能知道标头类型。MIME 类型有 text/html、application/pdf、application/jar、application/x-zip、image/jpeg 等。在应用程序中，可以在 web.xml 中设置 MIME 类型。例如：

```
<mime-mapping>
  <extension>jar</extension>
  <mime-type>application/jar</mime-type>
</mime-mapping>
```

<extension>用于设置文件的后缀，<mime-type>用于设置对应的 MIME 类型名称。在前面的例子中，曾介绍过请求参数是中文的例子。下面通过实例来说明请求参数是中文和响应中也有中文时应该如何发送请求。

【例 5.8】 演示发送中文字符请求。

getResponse.jsp 页面利用一个 form 表单描述喜欢的人物，下拉列表中的值都是中文，最后提交到 Servlet 中进行处理。getResponse.jsp 源代码如下：

```
------------------------ getResponse.jsp---------------------------
<%@ page import="java.util.*" pageEncoding="UTF-8"%>
<!DOCTYPE HTML>
<html>
  <head>
    <title>人物调查</title>
  </head>

  <body>
      <div style="text-align: center;">
                调查问卷
```

```
11    <form action="<%=request.getContextPath()%>/diaocha.do" method="POST">
12                  姓名：<input name="username"/><br/>
13                  邮箱：<input name="email"/><br/>
14                  你喜欢的人物：<br/>
15                  <select name="starname" multiple="multiple">
16                      <option value="袁隆平">袁隆平</option>
17                      <option value="钱学森">钱学森</option>
18                      <option value="杨振宁">杨振宁</option>
19                      <option value="钱钟书">钱钟书</option>
20                      <option value="屠呦呦">屠呦呦</option>
21                      <option value="顾方舟">顾方舟</option>
22                  </select>
23                  <input type="submit" value="提交"/>
24    </form>
25        </div>
26    </body>
27  </html>
```

在上述代码中，第15~22行代码利用下拉列表提供可选项，并且可以多选。diaocha.do中的Servlet代码如下：

```
-----------------------StarSurvey.java--------------------------
package com.eshore;

import java.io.IOException;
import java.io.PrintWriter;

import jakarta.servlet.ServletException;
import jakarta.servlet.annotation.WebServlet;
import jakarta.servlet.http.HttpServlet;
import jakarta.servlet.http.HttpServletRequest;
import jakarta.servlet.http.HttpServletResponse;
@WebServlet(
    name = "starSurvey",
    urlPatterns = { "/diaocha.do" })
public class StarSurvey extends HttpServlet{

    private static final long serialVersionUID = 1L;

    public void doGet(HttpServletRequest request, HttpServletResponse
response) throws ServletException, IOException{
        doPost(request,response);
    }
    public void doPost(HttpServletRequest request, HttpServletResponse
response) throws ServletException, IOException {
        // 设置请求的编码类型
        request.setCharacterEncoding("UTF-8");
        String username = request.getParameter("username");
        String email = request.getParameter("email");
        String[] starname = request.getParameterValues("starname");
        // 设置响应的类型和编码
        response.setContentType("text/html;charset=UTF-8");
        // 取得PrintWriter()对象
        PrintWriter out = response.getWriter();
        out.println("<!DOCTYPE HTML>");
        //输出页面内容
```

```
35          out.println("<HTML>");
36          out.println("  <HEAD><TITLE>感谢您的调查</TITLE></HEAD>");
37          out.println("  <BODY>");
38          out.println("联系人：<a href='"+email+"'>"+username+"</a>");
39          out.println("<br/>喜欢的人物：");
40          String str = "";
41          for(int i=0;i<starname.length;i++){
42              str +=starname[i]+", ";
43          }
44          str = str.substring(0, str.length()-1);
45          out.println(str);
46          out.println("  </BODY>");
47          out.println("</HTML>");
48          out.close();
49      }
50  }
```

在上述代码中，第 25 行代码用于设置请求的编码类型，这一行代码必须在获取参数前设置；第 26~28 行代码用于取得请求中的参数值；第 30 行代码用于设置响应的类型和编码；第 32~47 行代码用于对浏览器进行响应。请求的字符编码与响应的字符编码必须设置一致，这样显示的时候才不会出现乱码。程序的运行效果如图 5.8 所示。

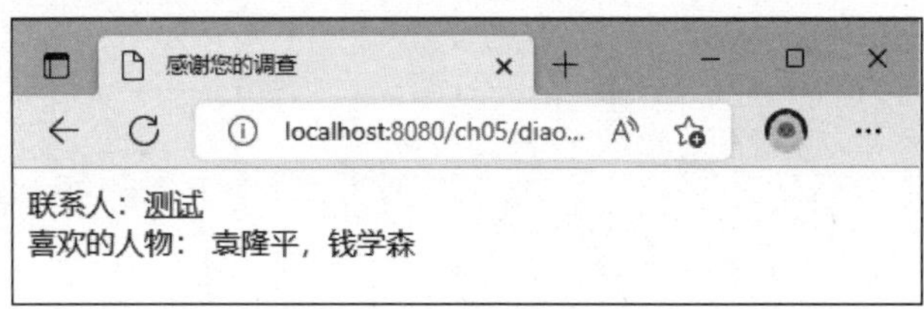

图5.8　例5.8的效果图

5.3.2　使用getOutputStream()输出二进制字符

getOutputStream()是为了取得输出流而设置的，在大多数情况下，PrintWriter对象就能解决问题，但是对于上传文件和下载文件则需要用到字节输出流。通过getOutputStream()方法可以取得ServletOutputStream对象，它是OutputStream的子类。

在例5.2中，介绍了如何上传文件到指定目录。本小节将通过下载文件来说明getOutputStream()的使用方法，如例5.9所示。

【例 5.9】　演示下载文件。

DownloadServlet 类是简单的下载文件类，用于设置响应的内容类型和标头。源代码如下：

```
-----------------------DownloadServlet.java--------------------------
package com.eshore;

import java.io.IOException;
import java.io.InputStream;

import jakarta.servlet.ServletException;
import jakarta.servlet.ServletOutputStream;
import jakarta.servlet.annotation.WebServlet;
import jakarta.servlet.http.HttpServlet;
```

```
10  import jakarta.servlet.http.HttpServletRequest;
11  import jakarta.servlet.http.HttpServletResponse;
12
13  @WebServlet(name = "downloadServlet", urlPatterns = { "/download.do" })
14  public class DownloadServlet extends HttpServlet {
15
16      private static final long serialVersionUID = 1L;
17
18      public void doGet(HttpServletRequest request, HttpServletResponse
19  response) throws ServletException, IOException {
20          doPost(request,response);
21      }
22
23      public void doPost(HttpServletRequest request, HttpServletResponse
24  response) throws ServletException, IOException {
25          //设置响应的内容类型
26          response.setContentType("application/msword");
27          //设置响应的标头内容
28          response.addHeader("Content-disposition","attachment;
filename=test.doc");
29          //获取资源文件
30          InputStream in = getServletContext().getResourceAsStream
("/doc/test.doc");
31          //输出到浏览器中
32          ServletOutputStream os = response.getOutputStream();
33          byte[] bytes = new byte[1024];
34          int len = -1;
35          while((len=in.read(bytes))!=-1){
36              os.write(bytes, 0, len);
37          }
38          //关闭输出输入流
39          in.close();
40          os.close();
41      }
42  }
```

在上述代码中，第 13 行代码利用注入的方式声明 Servlet；第 26 行代码用于设置响应的内容类型为 DOC 文档，所以设置内容类型为 application/msword，若系统安装了 Office 软件，就会直接打开 DOC 文档，还可以进行另存为操作；第 30 行代码使用 HttpServlet 的 getServletContext() 取得 ServletContext 对象，并利用 getResourceAsStream()方法以流形式获取资源文件，指定的路径是相对于 Web 应用程序的根目录；第 32~37 行代码利用 Java IO 流的特性从 DOC 文件中读取字节数据，并利用 ServletOutputStream 输出文件到浏览器中响应。在浏览器中输入 Servlet 地址，即可弹出“新建下载”对话框，如图 5.9 所示。

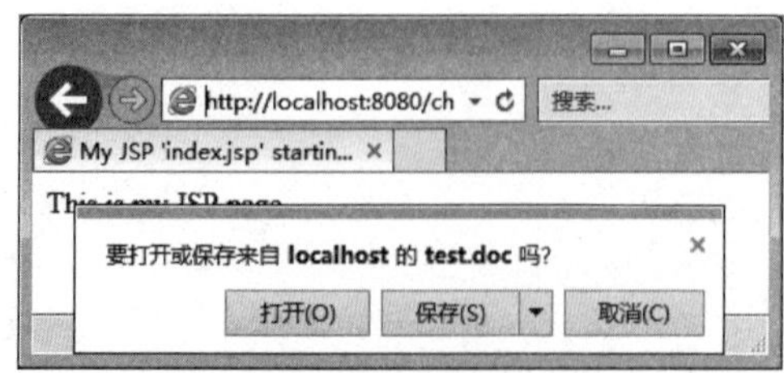

图5.9 “新建下载”对话框

5.3.3 使用sendRedirect()、sendError()方法

1. sendRedirect()方法

在第5.2.3节中，介绍了利用getRequestDispatcher()方法进行Servlet的跳转，forward()方法会将请求转发到指定的URL，该动作是不被浏览器所知的，因此地址栏也不会发生变化。在转发过程中，处于同一Request周期中，因此可以利用setAttribute()方法设置属性对象，利用getAttribute()获取属性对象。本小节将介绍sendRedirect()方法，用于重定向到另外一个URL中。例如“response.sendRedirect("http://localhost:8080/ch05/showview.do")”。

sendRedirect()方法会在响应中设置HTTP状态码和Location标头，当客户端接收到这个标头时，会重新请求指定的URL，所以地址栏上的地址会发生改变。

下面利用实例演示sendRedirect()方法，如例5.10所示。

【例 5.10】 演示 sendRedirect()方法。

ResponseRedirectDemo 类是跳转前的 Servlet 类，用于重定向到 redirSeconde.do 方法中，观察浏览器地址栏是否会发生变化。ResponseRedirectDemo 类的源代码如下：

```
-----------------------ResponseRedirectDemo.java--------------------------
package com.eshore;

import java.io.IOException;
import java.io.PrintWriter;

import jakarta.servlet.RequestDispatcher;
import jakarta.servlet.ServletException;
import jakarta.servlet.annotation.WebServlet;
import jakarta.servlet.http.HttpServlet;
import jakarta.servlet.http.HttpServletRequest;
import jakarta.servlet.http.HttpServletResponse;

@WebServlet(name = "responseRedirectDemo",
        urlPatterns = { "/redirect.do" })
public class ResponseRedirectDemo extends HttpServlet {

    private static final long serialVersionUID = 1L;
    public void doGet(HttpServletRequest request, HttpServletResponse
response) throws ServletException, IOException{
        doPost(request,response);
    }
    public void doPost(HttpServletRequest request, HttpServletResponse
response) throws ServletException, IOException {
        response.setContentType("text/html;charset=UTF-8");  // 设置响应的类型和编码
        PrintWriter out = response.getWriter();   // 取得PrintWriter()对象
        out.println("  Redirect跳转页面的第一个页面<br/>");
        response.sendRedirect(request.getContextPath()+"/redirSeconde.do");
        out.println("  Redirect Servlet<br/>");
        out.close();
    }
}
```

在上述代码中，第24行代码用于设置响应的类型和编码；第25行代码用于取得响应的PrintWriter()对象；第27行代码重定向到redirSeconde.do链接中。链接为redirSeconde.do的Servlet源代码如下：

```
--------------------- ResponseRedirectSecondeDemo.java------------------------
package com.eshore;
import java.io.IOException;
import java.io.PrintWriter;
import jakarta.servlet.RequestDispatcher;
import jakarta.servlet.ServletException;
import jakarta.servlet.annotation.WebServlet;
import jakarta.servlet.http.HttpServlet;
import jakarta.servlet.http.HttpServletRequest;
import jakarta.servlet.http.HttpServletResponse;

@WebServlet(name = "responseRedirectSecodDemo",
        urlPatterns = { "/redirSeconde.do" })
public class ResponseRedirectSecondeDemo extends HttpServlet {

    private static final long serialVersionUID = 1L;
    public void doGet(HttpServletRequest request, HttpServletResponse
response) throws ServletException, IOException{
        doPost(request,response);
    }
    public void doPost(HttpServletRequest request, HttpServletResponse
response) throws ServletException, IOException { // 返回页面消息
        response.setContentType("text/html;charset=UTF-8");// 设置响应的类型
和编码
        PrintWriter out = response.getWriter();    // 取得PrintWriter()对象
        out.println(" Redirect跳转页面的第二个页面<br/>");
    }
}
```

在上述代码中，程序的简单逻辑就是输出页面内容。程序运行效果如图5.10所示。

图5.10 运行结果图

从图中可以看出，跳转到ResponseRedirectSecondeDemo后，只显示其内容，而不显示跳转前的Servlet内容。

2. sendError()方法

如果在处理请求的时候发生错误，那么就可以用sendError()方法传递服务器的状态和错误消息。例如，请求的页面地址不存在，则可以发送如下错误信息：

```
response.sendError(HttpServletResponse.SC_NOT_FOUND);
```

SC_NOT_FOUND表示资源文件不存在，服务器会响应404错误代码，错误代码统一定义在HttpServletResponse接口上。此外还可以自定义错误信息，程序如下：

```
response.sendError(HttpServletResponse.SC_NOT_FOUND, "页面错误");
```

在HttpServlet中的doPost()内就实现了sendError()方法：

```
protected void doPost(HttpServletRequest req, HttpServletResponse resp)
throws ServletException, IOException {

     String protocol = req.getProtocol();
     String msg = lStrings.getString("http.method_post_not_supported");
     if (protocol.endsWith("1.1")) {
         resp.sendError(HttpServletResponse.SC_METHOD_NOT_ALLOWED, msg);
     } else {
         resp.sendError(HttpServletResponse.SC_BAD_REQUEST, msg);
     }
}
```

上述代码用于判断是否存在 sendError() 方法，如果没有，就抛出 SC_METHOD_NOT_ALLOWED 错误，即 405。常见的错误代码如表 5.2 所示。

表 5.2　常见的 HTTP 请求错误代码说明

错误代码	说　明
401	访问被拒绝
401.2	服务器配置导致登录失败
403	禁止访问
403.6	IP 地址被拒绝
403.9	用户数过多
404	没有找到文件或目录
405	用来访问本页面的 HTTP 不被允许（方法不被允许）
406	客户端浏览器不接受所请求页面的 MIME 类型
500	内部服务器错误
504	网关超时
505	HTTP 版本不受支持

5.4　网站注册与登录功能的实现

本节将利用注册与登录两个实例来说明请求与响应之间是如何进行通信的，注册与登录是大多数网站都要实现的功能，只是实现的方法略有不同，但是原理都是一样的。为了说明请求与响应中的方法，注册和登录两个实例都是利用Servlet编写的，包括页面部分。当然读者也可以使用JSP来实现。网站的实现逻辑如图5.11所示。

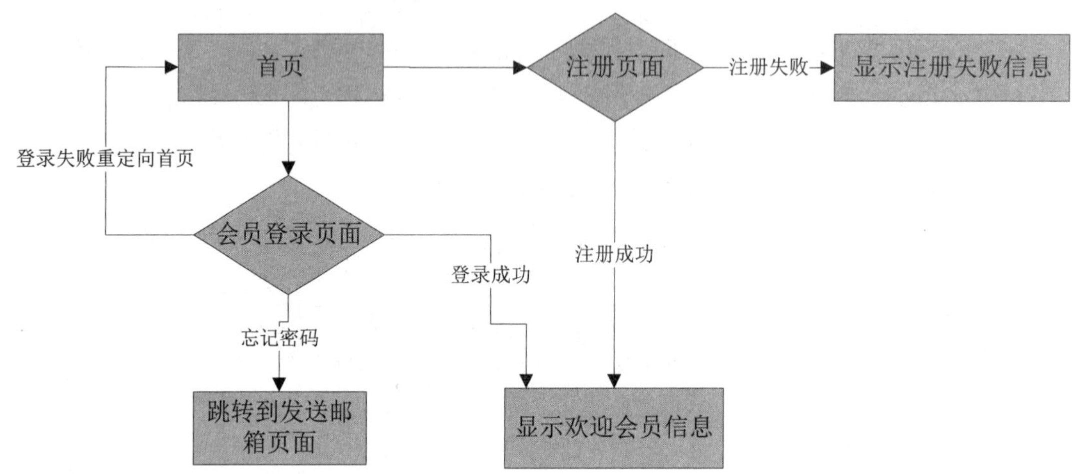

图5.11　注册与登录的实现过程

5.4.1　实现网站注册功能

注册页面register.jsp提供给未注册用户进行注册的功能，由简单的form组成，包含用户名、密码和邮箱。源代码如下：

```
-------------------------register.jsp---------------------------
<%@ page pageEncoding="UTF-8"%>
<!DOCTYPE HTML>
<html>
  <head>
    <title>会员注册页面</title>
  </head>

  <body>
      <div style="text-align: center;">注册会员信息<br/>
              <form action="<%=request.getContextPath()%>/register.htm"
                   method="post">
                  <table border="1">
                      <tr>
                          <td>登录名：</td>
                          <td><input name="username"/></td>
                      </tr>
                      <tr>
                          <td>密码：</td>
                          <td><input name="passwd" type="password"/></td>
                      </tr>
                      <tr>
                          <td>密码确认：</td>
                   <td><input name="confirdPasswd" type="password"/> </td>
                      </tr>
                      <tr>
                          <td>邮箱地址：</td>
                          <td><input name="email"/></td>
                      </tr>
                      <tr align="center">
                   <td colspan="2"><input type="submit" value="提交"/> </td>
```

```
31                    </tr>
32                </table>
33            </form>
34        </div>
35    </body>
36 </html>
```

在上述代码中，form 中的 action 提供注册的 Servlet 地址；第 12~32 行代码提供简单的输入框和“提交”按钮。

注册的 Servlet 类 RegisterServlet 用于接收页面中的参数，并验证用户名、密码和邮箱是否有效。源代码如下：

```
-----------------------------RegisterServlet.java-----------------------------
package com.eshore;
...
import com.eshore.pojo.User;

@WebServlet(name = "registeServlet", urlPatterns = { "/register.htm" })
public class RegisterServlet extends HttpServlet {

    private static final long serialVersionUID = 1L;

    public void doPost(HttpServletRequest request, HttpServletResponse
response) throws ServletException, IOException {
        // 获取参数username的值
        String userName = request.getParameter("username");
        String passwd = request.getParameter("passwd");//获取参数passwd的值
        // 获取参数confirdPasswd的值
        String confirdPasswd = request.getParameter("confirdPasswd");
        String email = request.getParameter("email");  // 获取参数email的值
        List<String> errors = new ArrayList<String>();  // 装载错误信息
        if(!isValidEmail(email)){                       // 验证邮箱
            errors.add("无效的邮箱号码！");
        }
        if(isValidUsername(userName)){                  // 验证用户名
            errors.add("用户名为空或者已经存在！");
        }
        if(isValidPassword(passwd,confirdPasswd)){      // 验证密码
            errors.add("密码为空或者密码不一致！");
        }
        if(!errors.isEmpty()){              //如果List不为空，则跳转到错误页面
            request.setAttribute("errors", errors);
            request.getRequestDispatcher("/error.htm").  // 跳转到错误的页面
            forward(request, response);
        }else{                  // 验证通过，获取User中的Map集合，添加用户
            User user = User.getInstance();;
            Map<String,String> map = user.getUserMap();
            map.put(userName, passwd+"##"+email);
            request.getRequestDispatcher("/success.htm").// 跳转到成功的页面
            forward(request, response);
        }
    }
    //使用正则表达式验证邮箱
    public boolean isValidEmail(String email){
        boolean flag = false;
```

```
43          if(email==null||"".equals(email)) flag = false;
44          if(email!=null&&!"".equals(email)){
45          flag = email.matches("^[\\w-]+(\\.[\\w-]+)*@[\\w-]+(\\.[\\w-]+)+$");
46          }
47          return flag;
48      }
49      //验证用户名是否存在，获取User类中的Map集合
50      //根据key判断Map中是否存在用户
51      public boolean isValidUsername(String userName){
52          User user = User.getInstance();;
53          Map<String,String> map = user.getUserMap();
54          if(userName!=null&&!userName.equals("")){
55              if(map.get(userName)!=null&&!map.get(userName).equals("")){
56                  return true;                    //存在用户
57              }else return false;
58          }
59          if(userName==null&&userName.equals("")){
60              return true;
61          }
62          return true;
63      }
64      //验证密码，如果密码为空或长度小于6或与确认密码不统一
65      //返回true
66      public boolean isValidPassword(String passwd,String confirdPasswd){
67          return passwd==null||confirdPasswd==null
68          ||passwd.length()<6||confirdPasswd.length()<6
69          ||!passwd.equals(confirdPasswd);
70      }
71  }
```

在上述代码中，第13~17行代码用于获取参数值；第19~38行分别验证邮箱、用户名、密码的有效性，如果无效，就将错误写进List中，如果通过验证，就跳转到欢迎界面；第40~48行利用正则表达式验证邮箱的合法性；第51~63行用于验证用户的合法性，如果新注册的用户名已存在，就提示用户名已经存在，否则新增一个用户；第66~70行用于验证密码的合法性，如果密码为空、确认密码为空、密码的长度小于6、确认密码的长度小于6、密码与确认密码不一致，则为不合法。在实际开发中这些验证应该放在JavaScript中，但是这里先在后台进行验证。在用户验证中，用户类本来是要保存到数据库中的，但是这里利用单例模式保存新增用户。User类的源代码如下：

```
----------------------------------- User.java-----------------------------------
package com.eshore.pojo;

import java.util.HashMap;
import java.util.Map;

public class User {

    private Map<String,String> userMap = new HashMap<String,String>();
    private static User user = null;
    private User(){
        userMap.put("zhangsan", "111111##zhangsan@sian.com");
        userMap.put("lisi", "222222##lisi@sian.com");
        userMap.put("wangwu", "333333##wangwu@sian.com");
```

```
14          userMap.put("zhaoliu", "444444##zhaoliu@sian.com");
15      }
16
17      public static User getInstance(){
18          if(user==null){
19              user = new User();
20          }
21          return user;
22      }
23      public Map<String, String> getUserMap() {
24          return userMap;
25      }
26  }
```

在上述代码中，第 10~15 行代码利用私有的构造方法初始化 Map 中的值；第 17~22 行代码用于建立一个静态的方法来获取 User 对象。

如果用户注册失败了，就会跳转到 ErrorServlet 这个类中，源代码如下：

```
----------------------------- ErrorServlet.java-----------------------------------
package com.eshore;

import java.io.IOException;
...
import jakarta.servlet.http.HttpServletResponse;

@WebServlet(name = "errorServlet", urlPatterns = { "/error.htm" })
public class ErrorServlet extends HttpServlet {

    private static final long serialVersionUID = 1L;

    public void doGet(HttpServletRequest request, HttpServletResponse
response) throws ServletException, IOException {
        doPost(request, response);                    // 调用doPost()方法
    }

    public void doPost(HttpServletRequest request, HttpServletResponse
response) throws ServletException, IOException {
        //设置响应的类型和编码
        response.setContentType("text/html;charset=UTF-8");
        PrintWriter out = response.getWriter();   //取得PrintWriter()对象
        out.println("<!DOCTYPE HTML>");
        //输出页面内容
        out.println("<HTML>");
        out.println("  <HEAD><TITLE>新增会员失败</TITLE></HEAD>");
        out.println("  <BODY>");
        out.print("<h2>新增会员失败</h2>");
        //遍历错误信息
        List<String> list = (List<String>)request.getAttribute("errors");
        for(String str:list){
            out.println(str+"<br>");
        }
        out.print("<a href=\""+request.getContextPath()+"/register.jsp\">
                    返回注册首页</a>");
        out.println("  </BODY>");
        out.println("</HTML>");
```

```
37            out.flush();
38            out.close();
39        }
40    }
```

ErrorServlet 类主要用于显示错误信息，也可以利用 JSP 实现，在这里只是为了进一步演示响应的方法，利用 Servlet 来编写 HTML 代码。在上述代码中，第 20 行代码用于设置响应类型和字符编码；第 29~32 行代码用于遍历错误信息；第 33 行代码用于创建超链接并返回注册页面。

如果用户注册成功，就跳转到成功页面 SuccessServlet，其源代码如下：

```
---------------------------- SuccessServlet.java--------------------------------
01  package com.eshore;
02
03  import java.io.IOException;
04  ...
05  @WebServlet(name = "successServlet", urlPatterns = { "/success.htm" })
06  public class SuccessServlet extends HttpServlet {
07
08      private static final long serialVersionUID = 1L;
09
10      public void doGet(HttpServletRequest request, HttpServletResponse
11  response) throws ServletException, IOException {
12          doPost(request, response);          // 调用doPost()方法
13      }
14
15      public void doPost(HttpServletRequest request, HttpServletResponse
16  response) throws ServletException, IOException {
17          // 设置响应的类型和编码
18          response.setContentType("text/html;charset=UTF-8");
19          PrintWriter out = response.getWriter();    // 取得PrintWriter()对象
20          out.println("<!DOCTYPE HTML>");
21          //输出页面内容
22          out.println("<HTML>");
23          out.println("  <HEAD><TITLE>新增会员成功</TITLE></HEAD>");
24          out.println("  <BODY>");
25          out.print("<h2>会员,"+request.getParameter("username")+"注册成功
</h2>");
26          out.print("<a href=\"" + request.getContextPath()
27                  + "/login.jsp\">返回首页登录</a>");
28          out.println("  </BODY>");
29          out.println("</HTML>");
30          out.flush();
31          out.close();
32      }
33  }
```

上述代码的主要作用是显示用户注册成功，例如第 25 行代码；第 26~27 行代码用于创建超链接，以返回到登录页面。程序运行效果依次如图 5.12～图 5.14 所示。

图5.12　注册页面

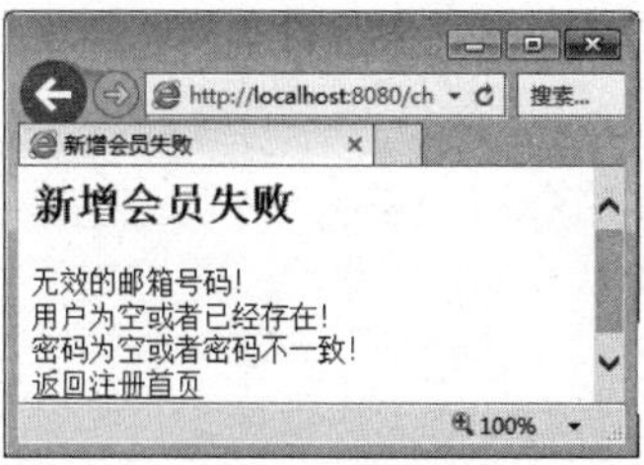

图5.13　错误页面

图5.14　成功页面

5.4.2　实现网站登录功能

登录页面 login.jsp 用于提供给已注册的用户进行登录，由简单的 form 组成，包含用户名和登录密码。源代码如下：

```
------------------------ login.jsp--------------------------
01  <%@ page pageEncoding="UTF-8"%>
02  <!DOCTYPE HTML>
03  <html>
04    <head>
05      <title>会员登录页面</title>
06    </head>
07
08    <body>
09         <div style="text-align: center;">会员登录<br/>
10  <form action="<%=request.getContextPath()%>/login.htm" method="post">
11                   <table border="1">
12                  <tr>
13                      <td>登录名：</td>
14                      <td><input name="username"/></td>
15                  </tr>
16                  <tr>
17                      <td>密码：</td>
18                      <td><input name="passwd" type="password"/></td>
19                  </tr>
20                  <tr align="center">
21                  <td colspan="2"><input type="submit" value="提交"/></td>
22                  </tr>
23                   </table>
24  </form>
25         </div>
26    </body>
27  </html>
```

在上述代码中，form 中的 action 提供登录的 Servlet 地址；第 10~24 行代码提供简单的输入框和“提交”按钮。

登录的 Servlet 类 LoginServlet 用于接收页面中的参数，并验证用户名、密码是否有效。源代码如下：

```
------------------------ LoginServlet.java --------------------------
package com.eshore;
```

```
02
03  import java.io.IOException;
04  ...
05  import com.eshore.pojo.User;
06
07  @WebServlet(name = "loginServlet", urlPatterns = { "/login.htm" })
08  public class LoginServlet extends HttpServlet {
09
10      private static final long serialVersionUID = 1L;
11      public void doPost(HttpServletRequest request, HttpServletResponse
12  response) throws ServletException, IOException {
13          //获取参数username的值
14          String userName = request.getParameter("username");
15          String passwd = request.getParameter("passwd");//获取参数passwd的值
16          if(checkLogin(userName,passwd)){        //用户登录成功，跳转到用户页面
17              request.getRequestDispatcher("/member.htm").
18              forward(request, response);
19          }else{
20              response.sendRedirect("login.jsp");//重定向到登录首页
21          }
22      }
23      //验证登录用户
24      public boolean checkLogin(String username,String passwd){
25          User user = User.getInstance();
26          Map<String,String> map = user.getUserMap();
27          if(username!=null&&!"".equals(username)&&      //用户不为空才判断
28              passwd!=null&&!"".equals(passwd)){
29              String[] arr = map.get(username).split("##"); //分割Map中的值
30              if(arr[0].equals(passwd)) return true;
31              else return false;
32          }else
33              return false;
34      }
35  }
```

在上述代码中，接收用户名与密码，如果经过验证为合法用户，则跳转到会员欢迎页面，否则重定向到登录页面，也可以重定向到首页或者自定义的页面。第 14~15 行代码接收页面参数；第 16~21 行代码判断是否为合法用户，并选择相应的跳转页面；第 24~34 行代码用于获取用户实例，判断用户是否存在，如果存在，就验证密码是否正确。

如果用户登录成功，就跳转到会员欢迎页面，源代码如下：

```
----------------------- MemberServlet .java--------------------------
package com.eshore;

import java.io.IOException;
...
import jakarta.servlet.http.HttpServletResponse;

@WebServlet(name = "memberServlet", urlPatterns = { "/member.htm" })
public class MemberServlet extends HttpServlet {

    private static final long serialVersionUID = 1L;

    public void doGet(HttpServletRequest request, HttpServletResponse
response) throws ServletException, IOException {
```

```
14            doPost(request, response);                    // 调用doPost()方法
15        }
16        public void doPost(HttpServletRequest request, HttpServletResponse
17    response) throws ServletException, IOException {
18            response.setContentType("text/html;charset=UTF-8");
19            PrintWriter out = response.getWriter();   // 取得PrintWriter()对象
20            out.println("<!DOCTYPE HTML>");
21            //输出页面内容
22            out.println("<HTML>");
23            out.println("  <HEAD><TITLE>会员登录成功</TITLE></HEAD>");
24            out.println("  <BODY>");
25            out.print("<h2>会员,"+request.getParameter("username")+",您好!</h2>");
26            out.print("<a href=\"" + request.getContextPath()
27                    + "/register.jsp\">返回首页登录</a>");
28            out.println("  </BODY>");
29            out.println("</HTML>");
30            out.flush();
31            out.close();
32        }
33    }
```

MemberServlet 类主要起到显示会员信息的作用：第 25 行代码用于打印欢迎信息，第 26 行代码用于创建超链接以返回登录页面。程序运行效果如图 5.15、图 5.16 所示。

图 5.15　登录页面

图 5.16　会员欢迎页面

到目前为止，网站的注册与登录功能已实现完毕，但是只能检查用户名与密码是否正确并跳转到指定页面，无法判断用户是否已经登录，有关这方面的内容将在下一章介绍。

5.5　小　　结

本章介绍了HttpServletRequest请求对象及其使用方法，使用getReader()、getInputStream()读取Body内容，利用getPart()和getParts()方法分别获取单个上传文件和多个上传文件，以及使用forward、include这2种的方法跳转页面（读者要深入了解页面跳转原理，因为在Web开发中有些需求一个Servlet是无法实现的，往往需要在多个Servlet之间相互跳转）。本章还介绍了HttpServletResponse响应对象及其使用方法，使用getWriter()输出字符、使用getOutStream()输出二进制流对象，在页面输出时要注意输出类型的设定和中文编码的支持。本章最后，还介绍了网站的注册和登录这两个常用的功能实现。

5.6 习 题

（1）实现一个Web程序，同时上传多个文件到本地文件夹中。

（2）利用HttpServletResponse设置标头、响应内容、编码字符。

（3）实现一个Web程序，下载本地的指定目录文件。

（4）实现一个Web程序，利用JSP实现网站的注册和登录功能中的显示页面部分。

第 6 章 会 话 管 理

在人机交互过程中，会话管理是指保持用户的整个会话活动的交互与计算机系统跟踪的过程。会话管理分为桌面会话管理、浏览器会话管理、Web会话管理。本书讨论的是Web会话管理（通常指的是session以及Cookie），也称为会话跟踪。

本章主要涉及的知识点有：

- 会话管理的基本原理
- HttpSession会话管理
- HttpSession会话管理的实例演示

6.1 会话管理的基本原理

会话管理可以通过以下3种解决方案实现：使用隐藏域、使用Cookie、使用URL重写。下面分别介绍各方案。

6.1.1 使用隐藏域

隐藏域是指在显示页面时隐藏表单中的内容，即不显示数据。在JSP中将input标签的type属性值设定为hidden，即生成一个隐藏表单域。再将会话的唯一标识记录到隐藏域中的value属性中，并设定name属性值。当提交表单时，会话标识也被提交到服务器端，服务器端根据它找到对应的会话对象。

使用隐藏域时，需要在每个页面中都包含会话标识表单，这样才能在许多页面跳转之间保存会话，并对它进行管理。此方法实现起来比较烦琐，安全性较差，不适合隐秘性的数据，在目前的开发中较少使用。

6.1.2 使用Cookie

利用Cookie实现会话管理是最容易的实现方式，也是使用较多的方式。原理是在服务器端保存的会话对象中设定会话的唯一标识，客户端将会话标识保存在Cookie中，当浏览器发起请求时，从Cookie中取得会话标识并发送给服务器端，服务器端在收到请求后，根据发送过来的会话标识查找到对应的会话对象，这样服务器端就清楚当前是哪个客户端在连接，并且可以从会话中获得信息。

利用Cookie实现会话管理是目前开发中采用的主流方法，大多数的动态页面开发技术都实现了这一功能，并且其管理流程是自动完成的，在实现上并不需要多大的开发工作量。使用Cookie实现会话管理的过程如图6.1所示。

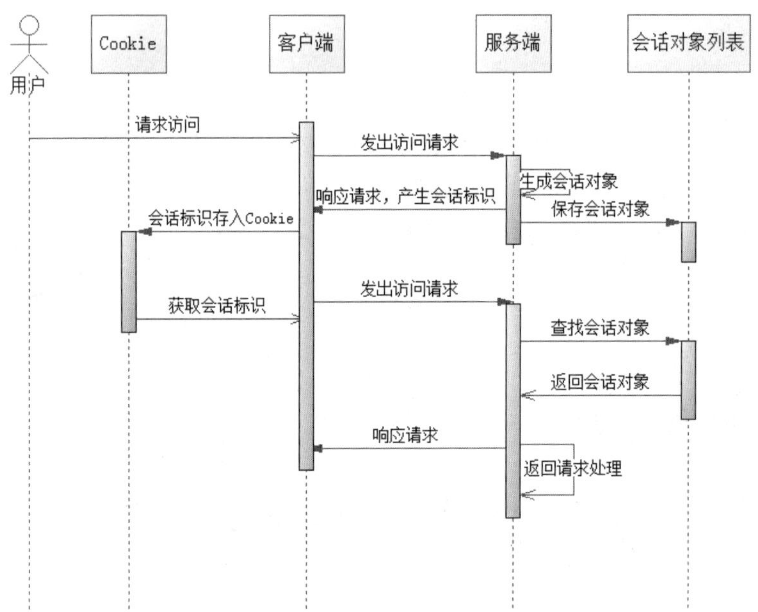

图6.1 使用Cookie实现会话管理的过程

6.1.3 使用URL重写

顾名思义，URL重写是指在URL地址的末尾添加会话标识，改写了原先的URL地址。其本质是用于唯一标识会话的信息以参数的形式添加到URL中，服务器端接收到请求时，解析出会话标识，然后利用会话标识查找出与当前请求对应的会话对象。该方法用在浏览器中的Cookie被禁用的情况下，利用该方法可以很好地实现会话跟踪而不会受到浏览器参数设定的影响。

URL重写方法是在URL后面添加会话标识，因此整个Web应用中的超链接或者脚本中用到的URL都需要添加会话标识，Web应用中的每一个页面都需要动态生成，页面中的每一个超链接或者由客户端生成的跳转指令都必须加上会话标识，这样才能确保是当前会话。当客户端访问静态页面时，会话标识将会丢失，当重回动态页面时将不能继续此前的会话。以上是它的特点，也是它的缺陷。

该方法还有一个问题，即当用户在浏览网页时，可能会将URL复制下来分享给朋友，因为URL中会包含会话标识，那么其他人可能与当前浏览者使用同一会话对象，这样当前浏览者的个人隐私就暴露了，信息不再安全。

无论采用哪种方式实现会话管理，在服务器端都需要完成相关代码才能完整地实现会话管理的任务。这些任务主要是生成唯一的会话标识、存储会话对象、从Web容器中取得当前请求的会话、回收空闲会话。

注意：会话管理会产生会话对象，它存储在服务器的内存中，会占用内存，因此需要注意空闲会话的回收，以提高服务器的性能。如果 Web 应用不需要会话的存在，那么就禁用会话，从而提高服务器的运行速率。

6.2　HttpSession 会话管理

HTTP协议（http://www.w3.org/Protocols/）是无意识的、单向的协议。服务器端不能主动连接客户端，只能等待并答复客户端请求。客户端连接服务器端，发出一个HTTP请求，服务器端处理请求，并返回一个HTTP响应给客户端，至此，本次会话结束。从这一过程可以看出，HTTP协议本身并不支持服务器端保存客户端的状态等信息。于是，Web服务器中引入了session的概念，用来保存客户端的信息。

6.2.1　使用HttpSession管理会话

在 Java 中，使用 jakarta.servlet.http.HttpSession 类来实现 session 会话。每个请求者对应一个 session 对象，客户的所有状态信息都保存在该对象里。当用户第一次请求服务器时，就创建了 session 对象。它以 key-value 的形式进行保存，通过相应的读写方法来保存客户的状态信息，例如 getAttribute(String key)用于获得参数、setAttribute(String key,Object value)用于设定参数和参数值。在 Servlet 中可以通过 request.getSession()方法获取客户的 session 对象，例如：

```
HttpSession session = request.getSession();              //获取session对象
Session.setAttribute("username","John");                 //设置session中的属性
```

在 JSP 中内置了 session 对象，可以直接使用，Servlet 使用上述方法获取 session 对象。若 JSP 页面中设定了<%@page session="false"%>，则该 JSP 页面的 session 对象是不可用的。下面通过例子说明如何利用 session 来保存登录信息。

【例 6.1】　利用 session 来保存登录信息。

首先，新建一个登录页面 login.jsp，源代码如下：

```
-----------------------login.jsp--------------------------
<%@ page pageEncoding="UTF-8"%>
<%
String path = request.getContextPath();
%>
<!DOCTYPE HTML>
```

```
06  <html>
07    <head>
08      <title>用户登录</title>
09    </head>
10
11    <body>
12          <p>用户登录</p>
13          <form action="<%=path%>/servlet/CheckUser" method="post">
14               <table border="1" width="250px;">
15                  <tr>
16                      <td width="75px;">用户名：</td>
17                      <td ><input name="userId"/></td>
18                  </tr>
19                  <tr>
20                      <td width="75px;">密  码：</td>
21                      <td ><input name="passwd" type="password"/></td>
22                  </tr>
23                  <tr>
24                      <td colspan="2">
25                          <input type="submit" value="提交"/>  
26                          <input type="reset" value="重置"/>
27                      </td>
28                  </tr>
29               </table>
30          </form>
31    </body>
32  </html>
```

在上述代码中，表单内容很简单，只有用户名和密码两个输入框，第 13 行代码用于填写提交表单时跳转的 action。

本例是用 Servlet 实现的，所以需要编写跳转的 Servlet 类 CheckUser，其源代码如下：

```
------------------------CheckUser.java---------------------------
public void doPost(HttpServletRequest request, HttpServletResponse
response) throws ServletException, IOException {

     request.setCharacterEncoding("UTF-8");
     String userId = request.getParameter("userId");
     String passwd = request.getParameter("passwd");

     //判断是否是linl用户且密码相符
     if(userId!=null&&"linl".equals(userId)
               &&passwd!=null&&"123456".equals(passwd)){
          //获得session对象
          HttpSession session = request.getSession();
          //设置user参数
          session.setAttribute("user", userId);
          //跳转页面
          RequestDispatcher dispatcher = request.
getRequestDispatcher("/welcome.jsp");
          dispatcher.forward(request, response);
     }else{
          RequestDispatcher dispatcher = request.
          getRequestDispatcher("/login.jsp");
```

```
22              dispatcher.forward(request, response);
23          }
24      }
```

在上述代码中，第 05、06 行用于获取页面请求的用户名与密码参数；第 09~18 行用于判断用户是否存在，如果存在就存放到 session 中并跳转到欢迎页面 welcome.jsp，如果用户不存在，就跳转到登录页面 login.jsp。

欢迎页面显示当前用户的用户名，其实现比较简单，使用 JSP 页面中的 session 对象就可以取出 user 对象，源代码如下：

```
-----------------------welcome.jsp-------------------------
01  <%@ page pageEncoding="UTF-8"%>
02  <!DOCTYPE HTML>
03  <html>
04    <head>
05      <title>欢迎页面</title>
06    </head>
07    <%
08       String user = (String)session.getAttribute("user");
09       if(user==null){
10    %>
11    <jsp:forward page="login.jsp"/>
12    <%} %>
13    <body>
14        欢迎您：<%=user%>。
15    </body>
16  </html>
```

在上述代码中，第 07~10 行是利用 JSP 的 session 获取 user 对象，如果用户为空，就返回登录页面。运行结果如图 6.2 所示。

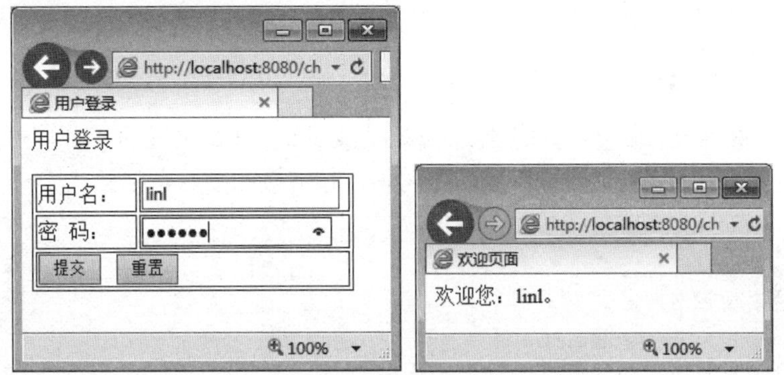

图6.2　保存登录信息效果图

6.2.2　HttpSession管理会话的原理

HttpSession会话管理是利用服务器来管理会话的机制。当程序为某个客户端的请求创建了一个session的时候，服务器会检查客户端的请求是否已经包含了一个session标识，如果已经有了session标识，那么服务器就把该session检索出来使用；如果请求不包含session标识，那就为客户端创建一个该请求的唯一session标识。HttpSession会话管理流程如图6.3所示。

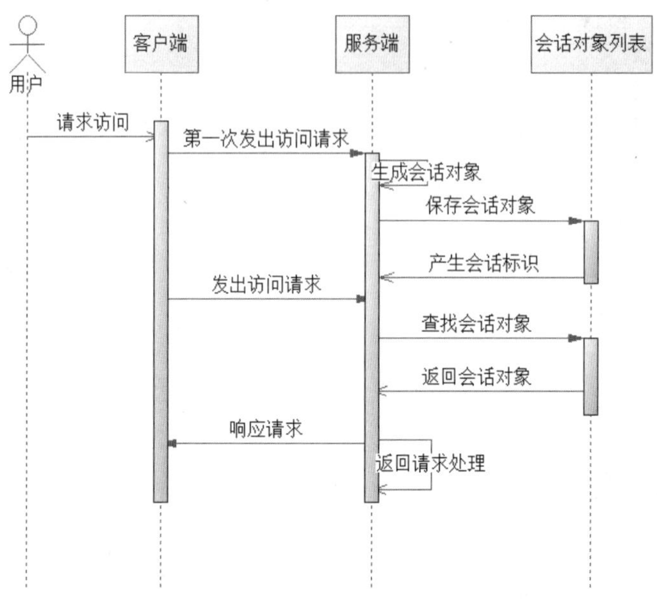

图6.3 HttpSession会话管理流程示意图

6.2.3 HttpSession与URL重写

在第 6.1.2 节中，提到 URL 重写是用在客户端不支持 Cookie 的情况下。它的实现方式是将 session 的标识 ID 添加到 URL 中。服务器通过解析重写后的 URL 获取 session 的标识 ID。这样即使客户端不支持 Cookie，也可以用 session 来记录状态信息。重写的方法如下：

```
response.encodeURL("login.jsp");
```

或者

```
response.sendRedirect(response.encodeRedirectURL("login.jsp"));
```

这两种方法的效果是一样的，如果客户端支持 Cookie，那么生成的 URL 不变；如果不支持，那么生成的 URL 中就会带有 jsessionid 参数字符串，超链接如图 6.4 所示。

文件(F) 编辑(E) 格式(O)

```

<!DOCTYPE HTML>
<html>
  <head>
    <title>欢迎页面</title>
  </head>

   <body>
      <a href="login.jsp;jsessionid=EA27339956F026CA50B1C193DE606243?username=john"></a>
      欢迎您：linl。
   </body>
</html>
```

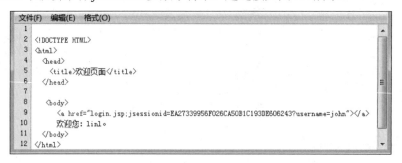

图6.4 URL中带有jsessionid的示意图

从图6.4可以看出，在URL的文件名后面，参数的前面添加了字符串“;jsessionid=标识ID”。用户单击这个超链接时，客户端会把session的ID值通过URL提交到服务器上，服务器通过解析URL地址获得session的ID值。

6.2.4 HttpSession中禁用Cookie

可以通过配置的方式在 Web 项目中禁用 Cookie，禁用方法如下：

- 在Web项目的web目录中的META-INF文件夹下，打开或者创建context.xml文件，编辑如下内容：

```
<?xml version="1.0" encoding="UTF-8"?>
<Context cookies="false" path="/ch06">
</Context>
```

- 打开Tomcat的配置文件context.xml，编辑如下内容：

```
<?xml version="1.0" encoding="UTF-8"?>
<Context cookies="false" >
</Context>
```

上述两种方法都是在 Context 元素中添加属性 cookies="false"，二者的区别在于前者是禁止单个项目使用 Cookie，后者是禁止部署在 Tomcat 服务器里的 Web 项目使用 Cookie。

6.2.5 HttpSession的生命周期

Servlet有生命周期，同样，HttpSession也有生命周期：创建→使用→消亡。

1. HttpSession 对象的创建

当客户端第一次访问服务器时，服务器为每个浏览器创建不同session的ID值。在服务器端使用request.getSession()或者request.getSession(true)方法来获得HttpSession对象。

2. HttpSession 对象的使用

创建 HttpSession 对象后，使用 session 对象进行数据的存取和传输。具体的过程如下：

01 将产生的sessionID存入Cookie中。

02 当客户端再次发送请求时，会将sessionID与request一起传送给服务器端。

03 服务器根据请求过来的sessionID与保存在服务器端的session对应起来，判断是否为同一session。

3. HttpSession 对象的消亡

有如下 3 种方式可以结束 session 对象：

- 关闭浏览器。
- 调用HttpSession的invalidate()方法。
- session超时。

关闭浏览器，这样会使浏览器端的 session 失效，服务器端 session 并不会失效。如果服务器进程终止了，那么 session 会被结束。

在session结束时，服务器会清空当前浏览器的相关数据信息。

以上就是HttpSession的生命周期过程，在请求处理时不断循环第2步（指上面HttpSession对象的使用），直到session对象消亡。

session是保存在服务器内存中的，每个用户都有一个独立的session。session中的内容应该尽量少，这样当有大量客户访问服务器时才不会导致内存溢出。

注意：只有访问 JSP、Servlet 时才会创建 session，访问静态页面时是不会创建 session 对象的。

6.2.6 HttpSession的有效期

通常情况下，会给session设定一个有效期，当某用户访问的session超过这个有效期时，即不是活跃的session，那么session就失效了，服务器会将它从内存中清除。

设定 session 的有效期有以下 3 种方法：

- 在对应的Web服务器配置中设置所有session的有效期。
- 调用session的setMaxInactiveInterval(long interval)进行设定。
- 在web.xml中修改，例如：

```
<session-config>
    <!-- 会话超时时长为30分钟 -->
    <session-timeout>30</session-timeout>
</session-config>
```

注意：调用 session 的 invalidate()可以销毁 session。

6.3 HttpSession 会话管理实例演示

利用HttpSession可以实现Web的基本功能与操作，例如实现简单购物车、在线猜数字游戏等。下面通过在线猜数字游戏来说明HttpSession会话管理的过程。

【例 6.2】 在线猜数字游戏。

首先，新建一个猜数字页面 guessNumber.jsp，源代码如下：

```
------------------------guessNumber.jsp---------------------------
<%@ page pageEncoding="UTF-8"%>
<%
String path = request.getContextPath();
%>
<!DOCTYPE HTML>
<html>
  <head>
    <title>在线猜数字</title>
  </head>
  <%
    String flag = request.getParameter("flag");
    String message="";
```

```
13        if(flag!=null&&"larger".equals(flag)){
14            message="太大了";
15        }else if(flag!=null&&"lessner".equals(flag)){
16            message="太小了";
17        }else if(flag!=null&&"success".equals(flag)){
18            message="您猜对了";
19        }
20    %>
21    <body>
22        <form action="<%=path %>/servlet/Guess" method="post">
23            <span>请输入您所猜数字：</span>
24            <input name="guessNumber" size=""10/>
25            <span style="color: red"><%=message %></span>
26            <input type="submit" value="提交" />
27        </form>
28    </body>
29  </html>
```

在上述代码中，第 10~20 行用于判断返回值的情况，如果所猜值大于随机数的值，就输出“太大了”；如果所猜值小于随机数的值，就输出“太小了”；如果猜对了，就输出“您猜对了”。

其次，编写提交的 Servlet 类 Guess，源代码如下：

```
-----------------------Guess.java---------------------------
01  public void doPost(HttpServletRequest request, HttpServletResponse
02  response) throws ServletException, IOException {
03
04        //获取页面提交的数字
05        String guessNumber = request.getParameter("guessNumber");
06        int number = Integer.parseInt(guessNumber);
07        //产生一个session，并获取存放在session中的currentNumber
08        HttpSession session = request.getSession();
09        Integer currentNumber = (Integer)session.getAttribute ("currentNumber");
10        String context = request.getContextPath();
11        if(currentNumber==null){
12            //产生1~50的随机数
13            currentNumber = 1+(int)(Math.random()*50);
14            session.setAttribute("currentNumber", currentNumber);
15        }
16        //判断所猜数与session中的currentNumber大小
17        if(number>currentNumber){
18            response.sendRedirect(context+"/guessNumber.jsp?flag= larger");
19        }else if(number<currentNumber){
20            response.sendRedirect(context+"/guessNumber.jsp?flag= lessner");
21         }else {
22            currentNumber = 1+(int)(Math.random()*50);
23            session.setAttribute("currentNumber", currentNumber);
24            response.sendRedirect(context+"/guessNumber.jsp?flag= success");
25         }
26  }
```

在上述代码中，第 11~15 行利用 Math.random 方法产生 1~50 的随机数，并把它保存在 session 中，第 17~25 行用于判断提交的数与 session 中的值是否相符，利用重定向的方法跳转到指定页面并携带 flag 标识。代码运行效果如图 6.5 所示。

图6.5　在线猜数字效果图

注意：本例利用重定向跳转到 guessNumber.jsp 页面，读者也可以利用 RequestDispatcher 中的 forward 方法实现跳转。

6.4　小　　结

本章首先简单介绍了会话管理的基本原理，一般有3种实现方式：使用隐藏域、Cookie和URL重写。这3种方式各有其特点，但目前流行的是Cookie的方式。接着介绍了HttpSession会话管理的原理、使用方法、禁用Cookie、HttpSession生命周期及其有效期。最后通过一个实例演示了HttpSession会话管理。熟练掌握会话的功能和特征，并根据需求灵活应用，这是Web开发的必要技能，希望读者能深入研究。

6.5　习　　题

（1）简述会话管理的基本原理。

（2）HttpSession会话管理的原理是什么？如何在HttpSession中禁用Cookie？

（3）编写一个Web应用程序，根据请求者的需要动态设置会话的超时时间。

（4）编写一个简单购物车，实现基本的购物功能。

第 7 章 Servlet 进阶 API、监听器与过滤器

第4章讲解了Servlet的原理、生命周期、部署的方法等，从而让读者了解了如何编写一个Servlet，如何完成一个动作流程。本章将介绍Servlet的一些进阶API、过滤器和监听器。

过滤器和监听器是Servlet规范里的两个高级特性，过滤器通过对request、response的修改实现特定的功能，例如请求数据字符编码、IP地址过滤、异常过滤、用户身份认证等。监听器的作用是监听Web程序中正在执行的程序，根据发生的事件做出特定的响应。合理利用这两个特性，能够轻松解决某些Web开发中的特殊问题。

本章主要涉及的知识点有：

- ServletConfig与GenericServlet的关系以及ServletConfig的用法
- 监听器的作用是什么以及如何编写和部署监听器
- 过滤器的作用是什么以及如何使用过滤器
- 异步处理请求

7.1 Servlet 进阶 API

在编写完一个Servlet类后，通常需要在web.xml中或者通过注解进行相关的配置，这样Web容器才能读取Servlet设置的信息，包括其类地址、初始化等。对于每个Servlet的配置，Web都会生成与之相对应的ServletConfig对象，从ServletConfig对象中可以得到Servlet的初始化参数。

本节将介绍ServletConfig与GenericServlet的关系，以及如何使用ServletConfig和ServletContext对象来获取Servlet初始化参数。

7.1.1 Servlet、ServletConfig与GenericServlet

在Web容器启动后，通过加载web.xml文件读取Servlet的配置信息、实例化Servlet类，并为每个Servlet配置信息产生唯一一个ServletConfig对象。在运行Servlet时，调用Servlet接口的init()方法，将产生的ServletConfig作为参数传入Servlet中，流程如图7.1所示。

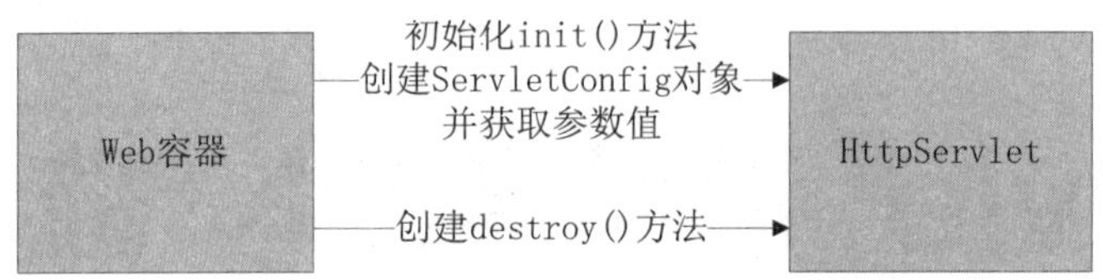

图7.1 创建Servlet与ServletConfig示意图

正如第4章所介绍的，初始化方法只会被调用一次，即容器在启动时实例化Servlet和创建ServletConfig对象，且Servlet与ServletConfig是一一对应关系，之后就直接执行service()方法。

从Jakarta Servlet API中可以得知，GenericServlet类同时实现了Servlet、ServletConfig这两个接口，如图7.2所示。

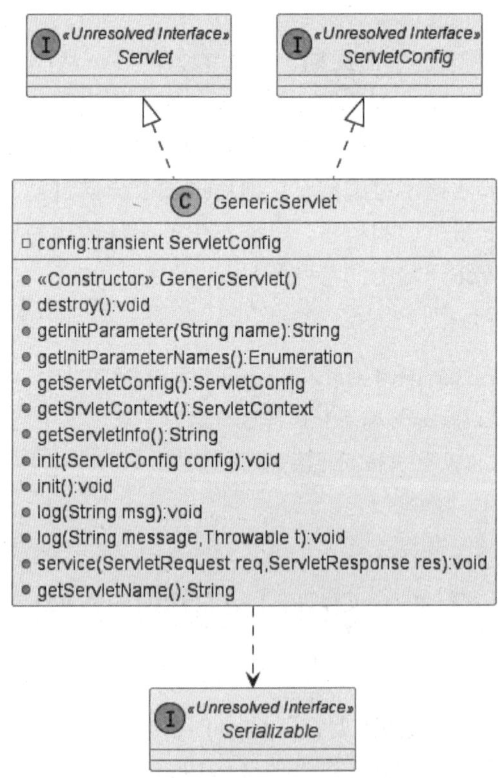

图7.2 GenericServlet类图

GenericServlet类的存在使得编写Servlet更加方便，它提供了一个简单的方法，这个方法用来执行有关Servlet生命周期的方法，以及在初始化时对ServletConfig对象和ServletContext对象进行说明。

GenericServlet 类的源代码如下：

```
-----------------------GenericServlet.java--------------------------
package jakarta.servlet;

import java.io.IOException;
import java.util.Enumeration;
import java.util.ResourceBundle;

public abstract class GenericServlet
implements Servlet, ServletConfig, java.io.Serializable
{
    private static final long serialVersionUID = 1L;
    private transient ServletConfig config;
    //GenericServlet默认的构造方法
    public GenericServlet() { }
    //GenericServlet默认的销毁方法
    public void destroy() {
    }
    //获得初始化参数方法
    public String getInitParameter(String name) {
        return this.getServletConfig().getInitParameter(name);
    }
    //获得参数名称，并返回枚举类型
    public Enumeration<String> getInitParameterNames() {
        return this.getServletConfig().getInitParameterNames();
    }
    //获得ServletConfig对象
    public ServletConfig getServletConfig() {
        return this.config;
    }
    //获得ServletContext对象
    public ServletContext getServletContext() {
        return this.getServletConfig().getServletContext();
    }

    public String getServletInfo() {
        return "";
    }
    //GenericServlet初始化方法
    public void init(ServletConfig config) throws ServletException {
        this.config = config;
        this.init();
    }
    //GenericServlet的空初始化方法
    public void init() throws ServletException {

    }

    public void log(String msg) {
        getServletContext().log(getServletName() + ": "+ msg);
    }

    public void log(String message, Throwable t) {
        getServletContext().log(getServletName() + ": " + message, t);
    }
```

```
54
55      public abstract void service(ServletRequest req, ServletResponse res)
56  throws ServletException, IOException;
57      //GenericServlet获得Servlet名称
58      public String getServletName() {
59          return this.config.getServletName();
60      }
61  }
```

分析上述代码中的第38~40行，可以看到GenericServlet类将ServletConfig封装了。

从代码第21~32行可知，GenericServlet类还定义了获取ServletConfig对象的方法，在编写Servlet类时就可以通过这些方法来获取所要配置的信息，而不用重新创建出ServletConfig对象。

7.1.2 使用ServletConfig

前面已经介绍过，当容器初始化Servlet时，会为Servlet创建唯一的ServletConfig对象。利用Web容器读取web.xml文件，将初始化参数传递给ServletConfig，而ServletConfig作为对象参数传递到init()方法中。

在ServletConfig接口中，各方法的说明如表7.1所示。

表 7.1 ServletConfig 接口中的方法说明

方　法	说　明
getServletName()	该方法返回一个 Servlet 实例的名称
getServletContext()	返回一个 ServletContext 对象的引用
getInitParameter(String name)	返回一个由参数 String name 决定的初始化变量的值。如果该变量不存在，则返回 null
getInitParameterNames()	返回一个存储所有初始化变量的枚举类型。如果 Servlet 没有初始化变量，则返回一个空枚举类型

从Servlet 3.0开始，允许以注入的方式配置Servlet，而不仅仅在web.xml中配置。因此配置Servlet的形式可以有如下形式：

```
@WebServlet(
    urlPatterns = { "/servletConfigDemo.do" },
    loadOnStartup = 1,
    name = "ServletConfigDemo",
    displayName = "demo",
    initParams = {
        @WebInitParam(name = "success", value = "success.html"),
        @WebInitParam(name = "error", value = "error.html")
    }
)
```

它等价于：

```
<servlet>
<display-name>ss</display-name>
<!—Servlet名称-->
<servlet-name>ServletConfigDemo</servlet-name>
<!—Servlet类路径-->
```

```
    <servlet-class>com.eshore.ServletConfigDemo</servlet-class>
<load-on-startup>-1</load-on-startup>
<!—初始化参数-->
    <init-param>
        <param-name>success</param-name>
        <param-value>success.html</param-value>
    </init-param>
    <init-param>
        <param-name>error</param-name>
        <param-value>error.html</param-value>
    </init-param>
</servlet>
<servlet-mapping>
<!—Servlet映射-->
    <servlet-name>ServletConfigDemo</servlet-name>
    <url-pattern>/servletConfigDemo.do</url-pattern>
</servlet-mapping>
```

上述的两个配置过程是等价的，在 Servlet 3.0 及以后版本中可以同时兼容，在 Servlet 2.0 中只能在 web.xml 中配置。@WebServlet 和@WebInitParam 的主要属性分别参见表 7.2 和表 7.3。

表 7.2 @WebServlet 的主要属性列表

属 性 名	描 述
name	指定 Servlet 的 name 属性，等价于<servlet-name>，如果没有指定，则该 Servlet 的取值即为类的全名
value	该属性与 urlPatterns 属性等价，但两个属性不能同时使用
urlPatterns	指定 Servlet 的 URL 匹配模式，等价于<url-pattern>标签
loadOnStartup	指定 Servlet 的加载顺序，等价于<load-on-startup>标签。当值为 0 或者大于 0 时，表示容器在应用启动时就加载并初始化这个 Servlet；当值小于 0 或者没有指定时，则表示容器在该 Servlet 被选择时才会去加载；正数的值越小，该 Servlet 的优先级越高，应用启动时就越先加载；当值相同时，容器就会自己选择顺序来加载
initParams	指定 Servlet 的初始化参数，等价于<init-param>标签
asyncSupported	声明 Servlet 是否支持异步操作模式，等价于<async-supported>标签，该属性在 Servlet 3.0 中才有

表 7.3 @WebInitParam 的主要属性列表

属 性 名	描 述
name	指定参数的名字，等价于 <param-name>
value	指定参数的值，等价于 <param-value>
description	参数的描述，等价于 <description>

下面利用实例说明 ServletConfig 的用法及其自身方法的使用。

【例 7.1】 利用初始化信息设定跳转信息。

编写 ServletConfigDemo 类，输出其初始化参数，当用户成功登录时跳转到成功页面，反之则跳转到错误页面。源代码如下：

```
---------------------- ServletConfigDemo.java--------------------------
@WebServlet(
```

```
    urlPatterns = { "/servletConfigDemo.do" },
    loadOnStartup = 1,
    name = "ServletConfigDemo",
    displayName = "demo",
    initParams = {
        @WebInitParam(name = "success", value = "success.html"),
        @WebInitParam(name = "error", value = "error.html")
    }
)
public class ServletConfigDemo extends HttpServlet {

    public void doPost(HttpServletRequest request, HttpServletResponse
response) throws ServletException, IOException {
        //获取ServletConfig对象
        ServletConfig config = getServletConfig();

        //1.getInitParameter(name)方法
        String success = config.getInitParameter("success");
        String error = config.getInitParameter("error");

        System.out.println("success-----------"+success);
        System.out.println("error-----------"+error);

        //2.getInitParameterNames方法
        Enumeration enumeration = config.getInitParameterNames();
        while(enumeration.hasMoreElements()){
            String name = (String) enumeration.nextElement();
            String value = config.getInitParameter(name);
            System.out.println("name-----------"+name);
            System.out.println("value-----------"+value);
        }
        //3.getServletContext方法
        ServletContext servletContext = config.getServletContext();
        System.out.println("servletContext-----------"+servletContext);
        //4.getServletName方法
        String servletName = config.getServletName();
        System.out.println("servletName-----------"+servletName);

        request.setCharacterEncoding("UTF-8");
        response.setContentType("text/html;charset=UTF-8");
        String userId = request.getParameter("userId");
        String passwd = request.getParameter("passwd");

        //判断是否是linl用户且密码相符
        if(userId!=null&&"linl".equals(userId)
                &&passwd!=null&&"123456".equals(passwd)){
            //获得session对象
            HttpSession session = request.getSession();
            //设置user参数
            session.setAttribute("user", userId);
            //跳转页面
            RequestDispatcher dispatcher = request.
                    getRequestDispatcher(success);
            dispatcher.forward(request, response);
        }else{
            RequestDispatcher dispatcher = request.
```

```
58                    getRequestDispatcher(error);
59            dispatcher.forward(request, response);
60        }
61    }
62    //以下省略init()、destroy()等方法
63 }
```

在上述代码中，第 01~10 行利用注入的方式配置初始化参数，该 Servlet 的名称为 ServletConfigDemo、URL 路径为 servletConfigDemo.do；第 16~38 行代码演示 ServletConfig 类方法的使用；第 40~60 行是具体的执行动作，若验证成功，则跳转到成功页面，否则跳转到失败页面。

注意：使用注入方式配置 Servlet 时，服务器必须支持对应的 Servlet 版本才可以，否则运行 Servlet 时会报错。

7.1.3 使用ServletContext

ServletContext对象是Servlet中的全局存储信息，当服务器启动时，Web容器为Web应用创建唯一的ServletContext对象，应用内的Servlet共享同一个ServletContext。可以认为在ServletContext中存放着共享数据，应用内的Servlet可以通过ServletContext对象提供的方法获取共享数据。ServletContext对象只有在Web应用被关闭的时候才被销毁。

ServletContext接口中定义了运行Servlet应用程序的环境信息，可以用来获取请求资源的URL、设置与存储全局属性和Web应用程序初始化参数。ServletContext中的常见方法如表7.4所示。

表 7.4 ServletContext 中的常用方法

方 法	描 述
getRealPath(String path)	获取给定的虚拟路径所对应的真实路径名
getResource(String uripath)	返回由 path 指定的资源路径对应的一个 URL 对象
getResourceAsStream(String uripath)	返回一个指定位置资源的 InputStream。返回的 InputStream 可以是任意类型和长度。使用时指定路径必须以“/”开头，表示相对于应用程序环境根目录
getRequestDispatcher(String uripath)	返回一个特定 URL 的 RequestDispatcher 对象，否则就返回一个空值
getResourcePaths(String path)	返回一个存储 web-app 中所指资源路径的 Set（集合），如果是一个目录信息，则会以“/”作为结尾
getServerInfo	获取服务器的名字和版本号

下面利用实例说明 ServletContext 的用法。

【例 7.2】 说明 ServletContext 的用法。

编写 ServletContextDemo 类，输出其初始化参数，并利用 getResourceAsStream 方法输出指定文件内容到页面中。源代码如下：

```
----------------------- ServletContextDemo.java---------------------------
@WebServlet(
    urlPatterns = { "/servletContextDemo.do" },
    loadOnStartup = 0,
    name = "ServletContextDemo",
    displayName = "demo",
```

```
06      initParams = {
07          @WebInitParam(name = "dir", value = "/dir"),
08          @WebInitParam(name = "success", value = "success.html"),
09          @WebInitParam(name = "resourcePath", value = "/dir/test.txt")
10      }
11  )
12  public class ServletContextDemo extends HttpServlet {
13
14      public void doPost(HttpServletRequest request, HttpServletResponse
15  response) throws ServletException, IOException {
16          String dir = getInitParameter("dir");
17          String success = getInitParameter("success");
18          String resourcePath = getInitParameter("resourcePath");
19          //获取ServletContext对象
20          ServletContext context =  getServletContext();
21          //getRealPath获得真实路径
22          String path = context.getRealPath(success);
23          System.out.println("path真实路径-----"+path);
24          //getResourcePaths获得指定路径的内容
25          Set set = context.getResourcePaths(dir);
26          for(Object str:set){
27              System.out.println("文件内容-----"+(String)str);
28          }
29          //获得服务器版本
30          String serverInfo = context.getServerInfo();
31          System.out.println("获得服务器版本-----"+serverInfo);
32          //getResourceAsStream获得资源文件内容
33          InputStream in = context.getResourceAsStream(resourcePath);
34          OutputStream out = response.getOutputStream();
35          byte[] buffer = new byte[1024];
36          while(in.read(buffer)!=-1){
37              out.write(buffer);
38          }
39          in.close();
40          out.close();
41      }
42
43      //以下省略init()、destroy()等方法
44  }
```

在上述代码中，第 20 行利用 getServletContext()获得 ServletContext 对象；第 22~31 行代码利用 ServletContext 对象获取指定文件的真实路径、文件夹内容和服务器版本号；第 33~38 行代码利用 getResourceAsStream 方法获取文件流，将文件内容输出到页面中。上述代码的运行结果如图 7.3 和图 7.4 所示。

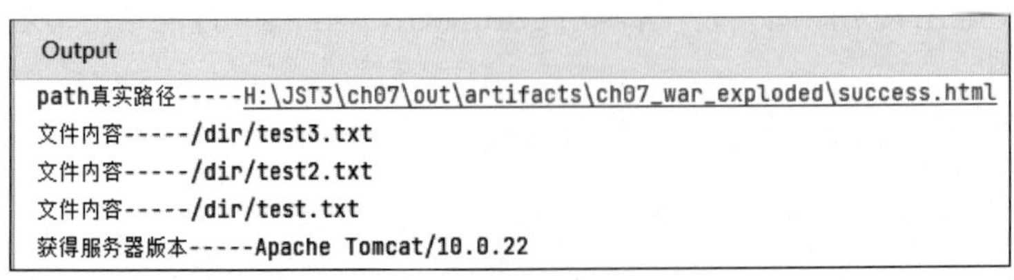

图 7.3　后台输出

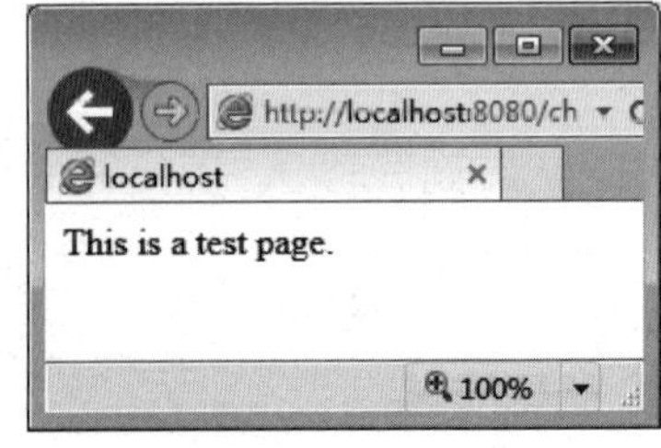

图 7.4　页面输出

注意：以“/”作为开头的路径称为环境相对路径。在 Servlet 中，若是环境相对路径，则直接委托给 ServletContext 的 getRequestDispatcher()。

7.2　应用程序事件、监听器

何谓监听？顾名思义就是监视行为。在Web系统中，所谓的监听器就是为应用监听事件来监听请求中的行为而创建的一组类。HttpServletRequest、HttpSession、ServletContext对象在Web容器中遵循生成、运行、销毁这样的生命周期。当进行相关的监听配置后，Web容器就会调用监听器上的方法，进行对应的事件处理，从而了解运行的情况或者运行其他的程序。各监听接口和事件类如表7.5所示。

表 7.5　监听接口和事件类

<table>
<tr><th>类　别</th><th>监听接口</th><th>监听事件</th></tr>
<tr><td rowspan="2">与 ServletContext 相关</td><td>ServletContextListener</td><td>ServletContexEvent</td></tr>
<tr><td>ServletContexAttributeListener</td><td>ServletContexAttributeEvent</td></tr>
<tr><td rowspan="5">与 HttpSession 相关</td><td>HttpSessionIdListener</td><td rowspan="3">HttpSessionEvent</td></tr>
<tr><td>HttpSessionListener</td></tr>
<tr><td>HttpSessionActivationListener</td></tr>
<tr><td>HttpSessionAttributeListener</td><td rowspan="2">HttpSessionBindingEvent</td></tr>
<tr><td>HttpSessionBindingListener</td></tr>
<tr><td rowspan="2">与 ServletRequest 相关</td><td>ServletRequestListener</td><td>ServletRequestEvent</td></tr>
<tr><td>ServletRequestAttributeListener</td><td>ServletRequestAttributeEvent</td></tr>
</table>

注意：使用监听器需要实现相应的监听接口。在触发监听事件时，应用服务器会自动调用监听方法。开发人员不需要关心应用服务器如何调用，只需要实现这些方法即可。

7.2.1　ServletContext事件、监听器

如上所述，与 ServletContext 有关的监听器有两个，即 ServletContextListener 与 ServletContextAttributeListener。

1. ServletContextListener

ServletContextListener被称为“ServletContext生命周期监听器”，可以用来监听Web程序初始化或者结束时响应的动作事件。

ServletContextListener接口的类是jakarta.servlet.ServletContextListener，该接口提供两个监听方法。

- default void contextInitialized (ServletContextEvent sce)：该方法用于通知监听器，已经加载Web应用和初始化参数。

- default void contextDestroyed (ServletContextEvent sce)：该方法用于通知监听器，Web应用即将关闭。

在 Web 应用程序启动时，会自动开始监听，首先调用的是 contextInitialized()方法，并传入 ServletContextEvent 参数，它封装了 ServletContext 对象，可以通过 ServletContextEvent 的 getServletContext()方法取得 ServletContext 对象，通过 getInitParameter()方法取得初始化参数。在 Web 应用关闭时，会自动调用 contextDestroyed()方法，同样会传入 ServletContextEvent 参数。在 contextInitialized()中可以实现应用程序资源的准备事件，在 contextDestroyed()中可以实现对资源的释放。例如，可以在 contextInitialized()方法中实现 Web 应用的数据库连接、读取应用程序设置等；在 contextDestroyed()中设置数据库资源的释放。

在 Web 应用中，实现 ServletContextListener 的步骤如下：

01 编写一个监听类并实现ServletContextListener接口。

02 进行相关的配置：

```
<listener>
    <listener-class>
        com.eshore.MyServletContextListener
    </listener-class>
</listener>
```

03 或者利用注入的方式注入监听类：

```
@WebListener
public class MyServletContextListener implements ServletContextListener{
}
```

04 若需要初始化参数，则需要在web.xml中进行配置，例如：

```
<context-param>
    <param-name>user_name</param-name>
    <param-value>linlong</param-value>
</context-param>
```

注意：@WebListener 也是 Servlet 3.0 才有的，因为它没有设置初始化参数的属性，所以也需要在 web.xml 中设定。

【例 7.3】 说明 ServletContextListener 的用法。

编写 MyServletContextListener 类，并利用 Log4j 将输出日志写到指定文件中。

```
-----------------------MyServletContextListener.java--------------------------
import jakarta.servlet.ServletContext;
import jakarta.servlet.ServletContextEvent;
import jakarta.servlet.ServletContextListener;
import jakarta.servlet.annotation.WebListener;

import org.apache.logging.log4j.Logger;
import org.apache.logging.log4j.LogManager;

@WebListener
public class MyServletContextListener implements ServletContextListener{

```

```
12        private static Logger log = LogManager.getLogger
13    (MyServletContextListener.class);
14        public void contextInitialized(ServletContextEvent sce) {
15            //通过ServletContextEvent获得ServletContext对象
16            ServletContext context = sce.getServletContext();
17            String name = context.getInitParameter("user_name");
18            log.debug("初始化参数name的值: "+name);
19            log.debug("Tomcat正在启动中......");
20        }
21
22        public void contextDestroyed(ServletContextEvent sce) {
23            log.debug("Tomcat正在关闭中......");
24        }
25    }
```

在上述代码中，第 16 行用于获取 ServletContext 对象，第 17 行用于获取初始化值。部署应用后，在 Tomcat 的 logs 文件夹中的 log.log 中输出初始化参数值和相关信息，如图 7.5 所示。

图7.5　ServletContextListener效果图

注意：本例使用了 Log4j2 日志管理，在 log4j2.xml 中配置输出的日志文件，例如通过 RollingFile 标签设定日志文件的目录以及每个日志文件的容量大小。

2. ServletContextAttributeListener

ServletContextAttributeListener被称为“ServletContext属性监听器”，可以用来监听Application属性的添加、移除或者替换时响应的动作事件。

ServletContextAttributeListener 接口的类是 jakarta.servlet.ServletContextAttributeListener，该接口提供 3 个监听方法。

- default void attributeAdded(ServletContextAttributeEvent scab)：该方法用于通知监听器，有对象或者属性被添加到Application中。
- defaultvoid attributeRemoved(ServletContextAttributeEvent scab)：该方法用于通知监听器，Application中有对象或者属性被移除。
- default void attributeReplaced (ServletContextAttributeEvent scab)：该方法用于通知监听器，有对象或者属性被更改到Application中。

在ServletContext中添加属性、移除属性或者更改属性时，与其相对应的方法就会被调用。同样，在Web应用程序中，实现ServletContextAttributeListener的方法也有两种，形式如下：

- 利用注入的方式注入监听类：

```
    @WebListener
    public class MyServletContextAttributeListener implements
ServletContextAttributeListener{
    }
```

- 在web.xml中配置：

```
<listener>
    <listener-class>
        com.eshore. MyServletContextAttributeListener
    </listener-class>
</listener>
```

7.2.2 HttpSession事件监听器

从表7.5中可以发现，与HttpSession有关的监听器有5个：HttpSessionIdListener、HttpSessionListener、HttpSessionAttributeListener、HttpSessionBindingListener、HttpSession-ActivationListener。从Servlet 3.1版本开始，增加了HttpSessionIdListener。

1. HttpSessionIdListener

HttpSessionIdListener用来监听sessionID的变化。

HttpSessionIdListener 接口的类是 jakarta.servlet.http.HttpSessionIdListener，该接口只提供了一个监听方法。

- public void sessionIdChanged(HttpSessionEvent se, java.lang.String oldSessionId)：该方法用于通知监听器sessionID发生了改变。

请求的 sessionID 发生变化时，会触发 sessionIDChanged()方法，并传入 HttpSessionEvent 和 oldSessionId 参数，可以使用 HttpSessionEvent 中的 getSession().getId()获取新的 sessionID，oldSessionId 代表改变之前的 sessionID。

在 Web 应用程序中，实现 HttpSessionIdListener 的方法同样有两种，形式如下：

- 利用注入的方式注入监听类：

```
@WebListener
public class MyHttpSessionIdListener implements HttpSessionListener{
}
```

- 在web.xml中配置：

```
<listener>
    <listener-class>
        com.eshore. MyHttpSessionIdListener
    </listener-class>
</listener>
```

2. HttpSessionListener

HttpSessionListener是“HttpSession生命周期监听器”，可以用来监听HttpSession对象初始化或者结束时响应的动作事件。

HttpSessionListener 接口的类是 jakarta.servlet.http.HttpSessionListener，该接口提供两个监听方法。

- default void sessionCreated(HttpSessionEventse)：该方法用于通知监听器，产生了新的会话。
- default void sessionDestroyed(HttpSessionEventse)：该方法用于通知监听器，已经消除了一个会话。

在 HttpSession 对象初始化或者结束前，会自动调用 sessionCreated()方法和 sessionDestroyed()方法，并传入 HttpSessionEvent 参数，它封装了 HttpSession 对象，可以通过 HttpSessionEvent 的 getSession()方法取得 HttpSession 对象。

在 Web 应用程序中，实现 HttpSessionListener 的方法同样有两种，形式如下：

- 利用注入的方式注入监听类：

```
@WebListener
public class MyHttpSessionListener implements HttpSessionListener{
}
```

- 在web.xml中配置：

```
<listener>
   <listener-class>
      com.eshore. MyHttpSessionListener
   </listener-class>
</listener>
```

【例 7.4】　说明 HttpSessionListener 的用法，利用 HttpSessionListener 记录在线人数。

首先编写登录的 Servlet，源代码如下：

```
----------------------- Login.java---------------------------
import java.io.IOException;
import java.util.HashMap;
import java.util.Map;

import jakarta.servlet.ServletException;
import jakarta.servlet.annotation.WebInitParam;
import jakarta.servlet.annotation.WebServlet;
import jakarta.servlet.http.HttpServlet;
import jakarta.servlet.http.HttpServletRequest;
import jakarta.servlet.http.HttpServletResponse;
@WebServlet(
    urlPatterns = { "/Login.do" },
    loadOnStartup = 0,
    name = "Login",
    displayName = "demo",
    initParams = {
        @WebInitParam(name = "success", value = "success.jsp")
    }
)
public class Login extends HttpServlet {

    Map<String, String> users;
    //在构造方法中，初始化用户值
```

```
24      public Login() {
25          users = new HashMap<String, String>();
26          users.put("zhangsan", "123456");
27          users.put("lisi", "123456");
28          users.put("wangwu", "123456");
29          users.put("zhaoliu", "123456");
30      }
31
32      public void doGet(HttpServletRequest request, HttpServletResponse
33  response) throws ServletException, IOException {
34          doPost(request, response);
35      }
36      //重写doPost()方法
37      public void doPost(HttpServletRequest request, HttpServletResponse
38  response) throws ServletException, IOException {
39          request.setCharacterEncoding("UTF-8");
40          String userId = request.getParameter("userId");    //获取userId
41          String passwd = request.getParameter("passwd");    //获取passwd
42          //匹配用户名与密码，如果一致则记录数加1
43          if (users.containsKey(userId) && users.get(userId).equals(passwd)) {
44              request.getSession().setAttribute("user", userId);
45              request.getSession().setAttribute("count",
46                      MyHttpSessionListener.getCount());
47          }
48          String success = getInitParameter("success");       //获取初始化参数的
success值
49          response.sendRedirect(success);                      //跳转页面
50      }
51  }
```

第11~19行代码以注入的方式编写Servlet；第37~50行代码从页面获取用户名与密码，如果用户验证通过，则取得HttpSession实例并设置用户属性。如果想在程序中加入统计在线人数的功能，则可以实现HttpSessionListener接口，例如：

```
------------------------ MyHttpSessionListener.java---------------------------
import jakarta.servlet.annotation.WebListener;
import jakarta.servlet.http.HttpSessionEvent;
import jakarta.servlet.http.HttpSessionListener;
@WebListener
public class MyHttpSessionListener implements HttpSessionListener{

    private static int count;          //统计数

    public static int getCount() {
        return count;
    }
    //在session开始时，统计数加1
    public void sessionCreated(HttpSessionEvent se) {
        MyHttpSessionListener.count++;
    }
    //在session销毁时，统计数减1
    public void sessionDestroyed(HttpSessionEvent se) {
        MyHttpSessionListener.count--;
    }

}
```

上述代码以注入的方式注入 HttpSessionListener 中以监听 HttpSession 对象。第 13~19 行代码用于每一次创建 HttpSession 的时候 count 都会递增，销毁 HttpSession 的时候 count 都会递减。

成功显示的页面为 success.jsp，源代码如下：

```
------------------------success.jsp---------------------------
01  <%@ page pageEncoding="UTF-8"%>
02  <%
03  String path = request.getContextPath();
04  %>
05  <!DOCTYPE HTML>
06  <html>
07    <head>
08      <title>登录成功界面</title>
09    </head>
10    <body>
11          <h3>目前在线人数为：${sessionScope.count}</h3>
12          <h4>欢迎您：${sessionScope.user}</h4>
13          <a href="<%=path%>/logout.do?userId=${sessionScope.user}">注销</a>
14    </body>
15  </html>
```

在上述代码中，第 11 行用于显示在线人数，第 13 行用于退出用户登录，运行效果如图 7.6 所示。

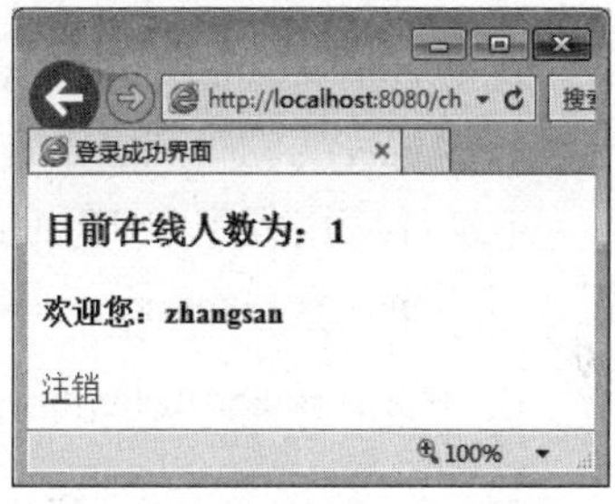

图7.6　HttpSessionListener效果图

3. HttpSessionAttributeListener

HttpSessionAttributeListener是“HttpSession属性改变监听器”，可以用来监听HttpSession对象加入属性、移除属性或者替换属性时响应的动作事件。

HttpSessionAttributeListener 接口的类是 jakarta.servlet.http.HttpSessionAttributeListener，该接口提供 3 个监听方法。

- default void attributeAdded(HttpSessionBindingEvent se)：该方法用于通知监听器，已经在session中添加了一个对象或者变量。
- default void attributeRemoved(HttpSessionBindingEvent se)：该方法用于通知监听器，已经在session中移除了一个对象或者变量。
- default void attributeReplaced(HttpSessionBindingEvent se)：该方法用于通知监听器，已经在session中替换了一个对象或者变量。

当对 session 范围的对象或者变量进行操作时，Web 容器会自动调用与实现接口类相对应的方法。HttpSessionBindingEvent 是一个对象，可以利用其 getName()方法得到操作对象或者变量的名称，利用 getValue()方法得到操作对象或者变量的值。

在 Web 应用程序中，实现 HttpSessionAttributeListener 的方法同样有两种，形式如下：

- 利用注入的方式注入监听类：

```
    @WebListener
    public class MyHttpSessionAttributeListener implements
HttpSessionAttributeListener{
    }
```

- 在web.xml中配置：

```
<listener>
    <listener-class>
        com.eshore. MyHttpSessionAttributeListener
    </listener-class>
</listener>
```

4. HttpSessionBindingListener

HttpSessionBindingListener是“HttpSession对象绑定监听器”，可以用来监听HttpSession中设置成HttpSession属性或者从HttpSession中移除时得到session的通知。

HttpSessionBindingListener 接口的类是 jakarta.servlet.http.HttpSessionBindingListener，该接口提供两个监听方法。

- default void valueBound(HttpSessionBindingEvent event)：该方法用于通知监听器，已经绑定一个session范围的对象或者变量。
- default void valueUnbound(HttpSessionBindingEvent event)：该方法用于通知监听器，已经解绑一个session范围的对象或者变量。

参数 HttpSessionBindingEvent 是一个对象，可以通过 getSession()方法得到当前用户的 session，通过 getName()方法得到操作的对象或者变量名称，通过 getValue()方法得到操作的对象或者变量值。

在Web应用程序中实现HttpSessionBindingListener接口时，不需要注入或者在web.xml中配置，只需将设置成session范围的属性实现HttpSessionBindingListener接口就行。

5. HttpSessionActivationListener

HttpSessionActivationListener是“HttpSession对象转移监听器”，可以用来实现对同一会话在不同JVM中的转移。例如，在负载均衡中，Web的集群服务器中的JVM位于网络中的多台机器中，当session要从一个JVM转移至另一个JVM时，必须先在原来的JVM上序列化所有的属性对象，若属性对象实现HttpSessionActivationListener，就调用sessionWillPassivate()方法，而转移后，就会调用sessionDidActivate()方法。

HttpSessionActivationListener 接口的类是 jakarta.servlet.http. HttpSessionActivationListener，该接口提供两个监听方法。

- default void sessionDidActivate(HttpSessionEvent se)：该方法用于通知监听器，该会话已变为有效状态。
- default void sessionWillPassivate(HttpSessionEvent se)：该方法用于通知监听器，该会话已变为无效状态。

7.2.3 HttpServletRequest事件、监听器

从表7.5中可以发现，与HttpServletRequest有关的监听器有两个：ServletRequestListener、ServletRequestAttributeListener。

1. ServletRequestListener

ServletRequestListener是“Request生命周期监听器”，可以用来监听Reuqest对象初始化或者结束时响应的动作事件。

ServletRequestListener 接口的类是 jakarta.servlet.ServletRequestListener，该接口提供两个监听方法。

- default void requestInitialized(ServletRequestEvent sre)：该方法用于通知监听器，产生了新的request对象。
- default void requestDestroyed(ServletRequestEvent sre)：该方法用于通知监听器，已经消除了一个request对象。

在 request 对象初始化或者结束前，会自动调用 requestInitialized()方法和 requestDestroyed()方法，并传入 ServletRequestEvent 参数，它封装了 ServletRequest 对象，可以通过 ServletRequestEvent 的 getServletContext()方法取得 Servlet 上下文对象，通过 getServletRequest()方法得到请求对象。

在 Web 应用程序中，实现 ServletRequestListener 的方法有两种，形式如下：

- 利用注入的方式注入监听类：

```
@WebListener
public class MyServletRequestListener implements ServletRequestListener{
}
```

- 在web.xml中配置：

```
<listener>
   <listener-class>
      com.eshore. MyServletRequestListener
   </listener-class>
</listener>
```

2. ServletRequestAttributeListener

ServletRequestAttributeListener是“Request属性改变监听器”，可以用来监听Request对象加入属性、移除属性或者替换属性时响应的动作事件。

ServletRequestAttributeListener 接口的类是 jakarta.servlet.http.ServletRequestAttributeListener，该接口提供 3 个监听方法。

- default void attributeAdded(ServletRequestAttributeEvent srae)：该方法用于通知监听器，已经在Request中添加了一个对象或者变量。
- default void attributeRemoved(ServletRequestAttributeEvent srae)：该方法用于通知监听器，已经在Request中移除了一个对象或者变量。
- default void attributeReplaced(ServletRequestAttributeEvent srae)：该方法用于通知监听器，已经在Request中替换了一个对象或者变量。

当对 request 范围的对象或者变量进行操作时，Web 容器会自动调用与实现接口类相对应的方法。ServletRequestAttributeEvent 是一个对象，可以利用其 getName()方法得到操作对象或者变量的名称，利用 getValue()方法得到操作对象或者变量的值。

在 Web 应用程序中，实现 ServletRequestAttributeListener 的方法同样有两种，形式如下：

- 利用注入的方式注入监听类：

```
@WebListener
public class MyServletRequestAttributeListener  implements
HttpSessionAttributeListener{
}
```

- 在web.xml中配置：

```
<listener>
   <listener-class>
      com.eshore. MyServletRequestAttributeListener
   </listener-class>
</listener>
```

【例 7.5】 说明 HttpServletRequest 的监听器示例。

编写 MyRequestListener 监听类，它同时实现 ServletRequestListener 监听器与 ServletRequestAttributeListener 监听器，并用 log4j 将日志输出到指定文件。源代码如下：

```
----------------------- MyRequestListener.java---------------------------
import jakarta.servlet.ServletRequestAttributeEvent;
import jakarta.servlet.ServletRequestAttributeListener;
import jakarta.servlet.ServletRequestEvent;
import jakarta.servlet.ServletRequestListener;
import jakarta.servlet.annotation.WebListener;

import org.apache.logging.log4j2.Logger;
import org.apache.logging.log4j2.LogManager;
@WebListener
public class MyRequestListener implements ServletRequestListener,
        ServletRequestAttributeListener {
    private static Logger log =
LogManager.getLogger(MyRequestListener.class);
    public void requestDestroyed(ServletRequestEvent arg0) {
        log.debug("一个请求消亡");
    }
    //request初始化，日志新增记录
    public void requestInitialized(ServletRequestEvent arg0) {
        log.debug("产生一个新的请求");
    }
    //新增一个request 属性，日志新增属性记录
    public void attributeAdded(ServletRequestAttributeEvent arg0) {
        log.debug("加入一个request范围的属性，名称为："+
           arg0.getName()+",其值为："+arg0.getValue());
    }
    //移除一个request 属性，日志移除属性记录
    public void attributeRemoved(ServletRequestAttributeEvent arg0) {
        log.debug("移除一个request范围的属性，名称为："+arg0.getName());
    }
    //修改一个request 属性，日志修改属性记录
    public void attributeReplaced(ServletRequestAttributeEvent arg0) {
        log.debug("修改一个request范围的属性，名称为："+
           arg0.getName()+",修改前的值为："+arg0.getValue());
    }
}
```

在上述代码中，第09行利用注入的方式注入requestListener监听器，第12行获得Logger对象，第13~33行分配实现监听方法并实现相应的动作。requestListener.jsp页面的源代码如下：

```
----------------------- requestListener.jsp--------------------------
01  <%@ page pageEncoding="UTF-8"%>
02  <%@taglib prefix="c" uri="http://java.sun.com/jsp/jstl/core" %>
03  <%
04  String path = request.getContextPath();
05  %>
06
07  <!DOCTYPE HTML>
08  <html>
09    <head>
10      <title>使用RequestListener监听器</title>
11    </head>
12
13    <body>
14          使用RequestListener监听器<br/>
15          <c:set value="zhangsan" var="username" scope="request"/>
16          姓名为：<c:out value="${requestScope.username}"/>
17        <c:remove var="username" scope="request"/>
18    </body>
19  </html>
```

在上述代码中，第 02 行代码引入 JSTL 标签库，第 15 行代码设定一个 request 参数 username，第 17 行移除一个 request 参数 username。运行上述代码，即可在页面中输入页面路径，在日志文件中打印信息，如图 7.7 所示。

```
Output
10:10:52.971 DEBUG com.eshore.MyRequestListener 21 requestInitialized - 产生一个新的请求
10:10:52.971 DEBUG com.eshore.EncodingFilter 47 doFilter - 请求被encodingFilter过滤
10:10:52.972 DEBUG com.eshore.MyRequestListener 34 attributeReplaced - 修改一个request范围的属性，名称为：org.apache.catalina.ASYNC_SUPPORTED,修改前的值为：true
10:10:52.974 DEBUG com.eshore.MyRequestListener 25 attributeAdded - 加入一个request范围的属性，名称为：username,其值为：zhangsan
10:10:52.974 DEBUG com.eshore.MyRequestListener 30 attributeRemoved - 移除一个request范围的属性，名称为：username
10:10:52.975 DEBUG com.eshore.EncodingFilter 56 doFilter - 响应被encodingFilter过滤
10:10:52.975 DEBUG com.eshore.MyRequestListener 17 requestDestroyed - 一个请求消亡
```

图7.7　HttpServletRequest的监听器效果图

7.3 过 滤 器

上一节介绍了Servlet监听器，使读者了解了什么是监听器以及如何使用监听器。本节将介绍Servlet的另一个高级特性——过滤器，以及如何编写和部署过滤器。

7.3.1 过滤器的概念

何为过滤器？顾名思义，它的作用就是阻挡某些事件的发生。在Web应用程序中，过滤器介于Servlet之间，既可以拦截、过滤浏览器的请求，也可以改变对浏览器的响应。它在服务器端与客户端之间起到了一个中间组件的作用，对二者之间的数据信息进行过滤，其处理过程如图7.8所示。由图7.8可以看出，当客户端浏览器发起一个请求时，服务器端的过滤器将检查请求数据中的内容，它可改变这些内容或者重新设置报头信息，再转发给服务器上被请求的目标资源，处理完毕后再向客户端响应处理结果。

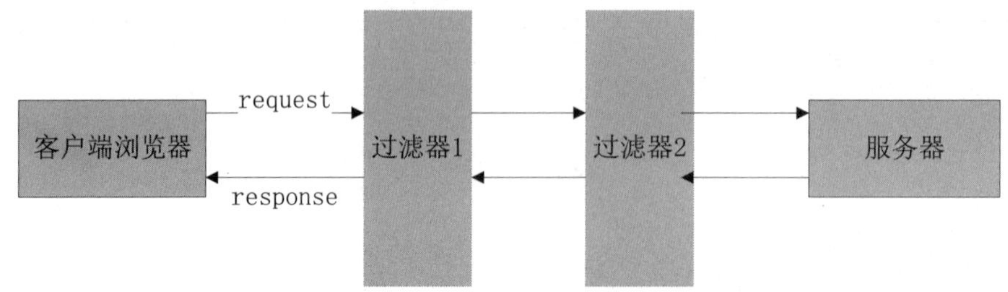

图7.8 处理过程

一个 Web 应用程序，可以有多个过滤器，从而组成一个过滤器链，如经常使用过滤器完成字符编码的设定和验证用户访问的合法性。过滤器链中的每个过滤器都各司其职地处理并转发数据。一般而言，在 Web 开发中，经常利用过滤器来实现如下功能：

- 对用户请求进行身份认证。
- 对用户发送的数据进行过滤或者替换。
- 转换图像的数据格式。
- 数据压缩。
- 数据加密。
- XML数据的转换。
- 修改请求数据的字符集。
- 日志记录和审核。

7.3.2 实现与设置过滤器

在Servlet中要实现过滤器，必须实现Filter接口，并用注入的方式或者在web.xml中定义过滤器，让Web容器知道应该加载哪些过滤器。

1. Filter 接口

Filter 接口的类是 jakarta.servlet.Filter，该接口有 3 个方法。

- default void init(FilterConfig filterConfig)：该方法用来初始化过滤器。filterConfig参数是一个FilterConfig对象，利用该对象可以得到过滤器中初始化的配置参数信息。
- public void doFilter(ServletRequest request, ServletResponse response,FilterChain chain)：该方法是过滤器中主要实现过滤的方法。当客户端请求目标资源时，Web应用程序会调用与此目标资源相关的doFilter()方法，在该方法中，实现对请求和响应的数据处理。参数request表示客户端的请求，response表示对应请求的响应，chain是过滤器链对象。在该方法中的特定操作完成后，可调用FilterChain对象的doFilter(request,response)将请求传递给过滤器链中的下一个过滤器，也可以直接返回响应内容，还可以将目标重定向。
- default void destroy()：该方法用于释放过滤器中使用的资源。

2. FilterConfig 接口

过滤器中还有 FilterConfig 接口，该接口用于在过滤器初始化时由 Web 容器向过滤器传送初始化配置参数，并传入过滤器对象的 init()方法中。FilterConfig 接口中有 4 个方法可以调用。

- public String getFilterName()：用于获得过滤器的名字。
- public String getInitParameter(String name)：用于获得过滤器中初始化的参数值。
- public Enumeration<String> getInitParameterNames()：用于获得过滤器配置中的所有初始化参数名字的枚举类型。
- public ServletContext getServletContext ()：用于获得Servlet上下文文件对象。

3. 设置过滤器

若想实现过滤器，则有两种方法：注入或者在 web.xml 中配置，其形式如下：

- 注入方式：

```
@WebFilter(
    description = "demo",
    filterName = "myfilter",
    servletNames = { "*.do" },
    urlPatterns = { "/*" },
    initParams = {
        @WebInitParam(name = "param", value = "paramvalue")
    },
    dispatcherTypes = { DispatcherType.REQUEST }
)
```

- 在web.xml中配置：

```
<filter>
     <description>demo</description>
     <!—过滤器名称-->
     <filter-name>myfilter</filter-name>
       <!—过滤器类-->
     <filter-class>com.eshore.MyFilter</filter-class>
       <!—过滤器初始化参数-->
     <init-param>
         <param-name>param</param-name>
         <param-value>paramvalue</param-value>
     </init-param>
 </filter>
 <!—过滤器映射配置-->
 <filter-mapping>
     <filter-name>myfilter</filter-name>
     <servlet-name>*.do</servlet-name>
     <url-pattern>/*</url-pattern>
     <dispatcher>REQUEST</dispatcher>
 </filter-mapping>
```

上述的两个配置过程是等价的，在 Servlet 3.0 及以后的版本中可以同时兼容，在 Servlet 2.0 中只能在 web.xml 中配置。@ WebFilter 的主要属性如表 7.6 所示。

表 7.6　@WebFilter 的主要属性列表

属 性 名	描　述
value	该属性与 urlPatterns 属性等价，但两个属性不能同时使用
urlPatterns	指定 Filter 的 URL 匹配模式，等价于<url-pattern>标签
filterName	指定 Filter 的 name 属性，等价于<filter-name>标签
servletNames	指定 Filter 的 Servlet 过滤对象，等价于<servlet-name>标签。当与 urlPatterns 同时存在时，Web 容器先比对 urlPatterns 中的 URL，再比对 servletNames 中的配置
dispatcherTypes	指定 Filter 的过滤时间，等价于<dispatcher>标签，其值有 FORWARD、INCLUDE、REQUEST、ERROR、ASYNC 等，默认值是 REQUEST
asyncSupported	声明 Servlet 是否支持异步操作模式，等价于<async-supported>标签。该属性在 Servlet 3.0 及以后的版本中才有
initParams	设置过滤器的初始参数

7.3.3　请求封装器

请求封装器是指利用HttpServletRequestWrapper类将请求中的内容进行统一修改，例如修改请求字符编码、替换字符、权限验证等。

下面通过例子来说明请求封装器的实现。

【例 7.6】　实现编码过滤器。

通常情况下，在 Web 中实现编码过滤器的形式如下：

```
----------------------- EncodingFilter.java--------------------------
@WebFilter(
    description = "字符编码过滤器",
    filterName = "encodingFilter",
    urlPatterns = { "/*" },
    initParams = {
        @WebInitParam(name = "ENCODING", value = "UTF-8")
    }
)
public class EncodingFilter implements Filter{
    private static Logger log = Logger.getLogger("EncodingFilter");
    private String encoding="";
    private String filterName="";

    public void init(FilterConfig filterConfig) throws ServletException {
        //通过filterConfig获得初始化中的编码值
        encoding = filterConfig.getInitParameter("ENCODING");
        filterName = filterConfig.getFilterName();
        if(encoding==null||"".equals(encoding)){
            encoding="UTF-8";
        }
        log.debug("获得编码值");
    }

    public void doFilter(ServletRequest request, ServletResponse
response, FilterChain chain) throws IOException, ServletException {
        //分别对请求和响应进行编码设置
```

```
27            request.setCharacterEncoding(encoding);
28            response.setCharacterEncoding(encoding);
29            HttpServletRequest req = (HttpServletRequest)request;
30            log.debug("请求被"+filterName+"过滤");
31            //传输给过滤器链过滤
32            chain.doFilter(req, response);
33            log.debug("响应被"+filterName+"过滤");
34        }
35
36        public void destroy() {
37            log.debug("请求销毁");
38        }
39    }
```

上述代码通过注入的方式注入过滤器，并用log4j将日志写到文件中，第16、17行代码取得过滤器中的初始化值，第27~33行代码用于设置编码并传输给过滤器链。第01~08行代码等价于web.xml中的如下配置：

```
<filter>
    <description>字符编码过滤器</description>
    <filter-name>encodingFilter</filter-name>
    <filter-class>com.eshore.EncodingFilter</filter-class>
    <init-param>
        <param-name>ENCODING</param-name>
        <param-value>UTF-8</param-value>
    </init-param>
</filter>
<filter-mapping>
    <filter-name>encodingFilter</filter-name>
    <url-pattern>/*</url-pattern>
</filter-mapping>
```

上述编码过滤器用于处理 post 请求是没有问题的，但是在处理 get 请求获取中文参数时还是会出现乱码问题。这是因为利用 post 方式请求时，参数是在请求数据包的消息体中，而对于 get 请求，参数存放在请求数据包的 URI 字段中；“request.setCharacterEncoding(encoding);”只对消息体中的数据起作用，对 URI 字段中的参数不起作用。基于这种情况，可到请求包装器包装请求，将字符编码转换的工作添加到 getParameter()方法中，这样就可以对请求的参数进行统一转换。

编写请求包装类 RequestEncodingWrapper 并继承 HttpServletRequestWrapper，源代码如下：

```
----------------------- RequestEncodingWrapper.java---------------------------
import java.io.UnsupportedEncodingException;

import jakarta.servlet.http.HttpServletRequest;
import jakarta.servlet.http.HttpServletRequestWrapper;

public class RequestEncodingWrapper extends HttpServletRequestWrapper{
    private String encoding="";
    public RequestEncodingWrapper(HttpServletRequest request) {
        //必须调用父类构造方法
        super(request);
    }
    public RequestEncodingWrapper(HttpServletRequest request,String
encoding) {        //必须调用父类构造方法
        super(request);
```

```
15            this.encoding = encoding;
16        }
17        //重新定义getParameter方法
18        public String getParameter(String name){
19            String value = getRequest().getParameter(name);
20            try {
21                //将参数值进行编码转换
22                if(value!=null&&!"".equals(value))
23                    value = new String(value.trim().getBytes("ISO-8859-1"),
encoding);
24            } catch (UnsupportedEncodingException e) {
25                // TODO Auto-generated catch block
26                e.printStackTrace();
27            }
28            return value;
29        }
30    }
```

在上述代码中，RequestEncodingWrapper类继承了HttpServletRequestWrapper；第12~16行自定义构造方法并实现父类的构造方法，HttpServletRequest将通过此构造方法传入，如果要取得被封装的HttpServletRequest，则可以调用getRequest()方法；第19~23行重构getParameter()方法，在此方法中，从HttpServletRequest对象取得请求参数值并进行编码转换，这时对过滤器EncodingFilter中的doFilter方法更改如下：

```
public void doFilter(ServletRequest request, ServletResponse response,
FilterChain chain) throws IOException, ServletException {
    //分别对请求和响应进行编码设置
    HttpServletRequest req = (HttpServletRequest)request;
    log.debug("请求被"+filterName+"过滤");
    //如果请求的方法是get，则用请求包装器，否则直接设定编码
    if("GET".equals(req.getMethod())){
        req = new RequestEncodingWrapper(req,encoding);
    }else{
        request.setCharacterEncoding(encoding);
    }
    response.setCharacterEncoding(encoding);
    //传输给过滤器链过滤
    chain.doFilter(req, response);
    log.debug("响应被"+filterName+"过滤");
}
```

请求参数的编码设置是通过过滤器的初始化参数设置的，并在过滤器中的 init()中设置，过滤器仅在 get 方法请求时创建 RequestEncodingWrapper 实例，post 方法则通过直接设定编码方式实现，最后调用 FilterChain 的 doFilter()方法传入实例。

7.3.4 响应封装器

响应封装器是指利用HttpServletResponseWrapper类将响应中的内容进行统一修改，例如压缩输出内容、替换输出内容等。有些时候需要对网站的输出内容进行控制，一般有两种方法：一是在保存数据库前对不合法的内容进行替换，二是在输出端进行替换。若是对每一个Servlet都进行输出控制，则任务量将非常大而且烦琐。可利用过滤器对Servlet进行统一处理，但是因

为HttpServletResponse不能缓存输出内容，所以需要自定义一个具备缓存功能的response。下面通过两个例子说明响应封装器的实现。

【例 7.7】 实现内容替换过滤器。

首先，编写一个响应的封装器 ResponseReplaceWrapper，用它来缓存 response 中的内容，源代码如下：

```
----------------------- ResponseReplaceWrapper.java--------------------------
01  import java.io.CharArrayWriter;
02  import java.io.IOException;
03  import java.io.PrintWriter;
04  import jakarta.servlet.http.HttpServletResponse;
05  import jakarta.servlet.http.HttpServletResponseWrapper;
06
07  public class ResponseReplaceWrapper extends HttpServletResponseWrapper{
08
09      private CharArrayWriter charWriter = new CharArrayWriter();
10      public ResponseReplaceWrapper(HttpServletResponse response) {
11          //必须调用父类构造方法
12          super(response);
13      }
14
15      public PrintWriter getWriter() throws IOException{
16          //返回字符数组Writer，缓存内容
17          return new PrintWriter(charWriter);
18      }
19
20      public CharArrayWriter getCharWriter() {
21          return charWriter;
22      }
23  }
```

在上述代码中，第 12 行代码用于调用父类的构造方法，第 15~18 行代码用于重构 getWriter() 方法并用字符数组缓存输出内容。

然后，编写内容过滤器 ReplaceFilter，源代码如下：

```
----------------------- ReplaceFilter.java--------------------------
...
@WebFilter(
    description = "内容替换过滤器",
    filterName = "replaceFilter",
    urlPatterns = { "/*" },
    initParams = {
        @WebInitParam(name = "filePath", value = "replace_ZH.properties")
    }
)
public class ReplaceFilter implements Filter{

    private Properties propert = new Properties();
    public void init(FilterConfig filterConfig) throws ServletException {
        //通过filterConfig获得初始化文件名
        String filePath = filterConfig.getInitParameter("filePath");
        try {
            //导入资源文件
```

```
18                propert.load(ReplaceFilter.class.getClassLoader()
19                    .getResourceAsStream(filePath));
20            } catch (FileNotFoundException e) {
21                e.printStackTrace();
22            } catch (IOException e) {
23                e.printStackTrace();
24            }
25        }
26        public void doFilter(ServletRequest request, ServletResponse
27    response, FilterChain chain) throws IOException, ServletException {
28            HttpServletResponse res = (HttpServletResponse)response;
29            //实例化响应包装类
30            ResponseReplaceWrapper resp = new ResponseReplaceWrapper(res);
31            chain.doFilter(request, resp);
32            //缓存输出字符
33            String outString = resp.getCharWriter().toString();
34            //循环替换不合法的字符
35            for(Object o:propert.keySet()){
36                String key = (String) o;
37                outString = outString.replace(key, propert.getProperty(key));
38            }
39            //利用原先的HttpServletResponse输出字符
40            PrintWriter out = res.getWriter();
41            out.write(outString);
42        }
43        public void destroy() {
44
45        }
46    }
```

在上述代码中，第 02~09 行代码利用注入的方式注入内容过滤器并设定文件名称；第 13~25 行代码在 init 初始化过滤器方法中，利用类反射加载内容、过滤文件内容；第 30~33 行代码实例化响应包装类 ResponseReplaceWrapper 并缓存输出内容；第 35~38 行代码替换不合法的字符；第 41 行代码利用原先的 HttpServletResponse 将替换后的内容输出到浏览器中。

本例中ResponseReplaceWrapper只是一个假的response，它不负责输出到客户端，只负责将输出内容缓存起来。输出到客户端还是通过原来的response完成。

replace_ZH.properties 配置文件的内容如下：

```
\u8272\u60c5=****
\u60c5\u8272=****
\u8d4c\u535a=****
```

注意：replace_ZH.properties 文件是经过 native2ascii 编码得到的，实现命令为："native2ascii-encoding utf-8 源文件 转换文件"。

最后，编写一个测试的 Servlet，源代码如下：

```
---------------------- TestServlet.java--------------------------
@WebServlet(
    urlPatterns = { "/Test.do" },
    loadOnStartup = 0,
    name = "testServlet"
)
```

```
06  public class TestServlet extends HttpServlet {
07      public void doGet(HttpServletRequest request, HttpServletResponse
08  response) throws ServletException, IOException {
09          doPost(request, response);
10      }
11
12      public void doPost(HttpServletRequest request, HttpServletResponse
13  response) throws ServletException, IOException {
14          response.setContentType("text/html;charset=utf-8");
15          PrintWriter out = response.getWriter();
16          out .println("<!DOCTYPE HTML>");
17          //输出页面内容
18          out.println("<HTML>");
19          out.println("  <HEAD><TITLE>测试内容输出过滤</TITLE></HEAD>");
20          out.println("  <BODY>");
21          out.println("dfdasf <br/>消极  <br/>悲观  <br/>恐怖");
22          out.println("  </BODY>");
23          out.println("</HTML>");
24          out.flush();
25          out.close();
26      }
27  }
```

上述测试的 Servlet 很简单，第 01~05 行代码利用注入的方式声明一个 Servlet，第 14~25 行代码直接利用 response 输出内容。上述代码的运行效果如图 7.9 所示。

（a）过滤前

（b）过滤后

图7.9 内容过滤效果图

上面实例介绍了内容替换的过滤器，使读者了解了如何编写一个自定义的响应封装器以及如何使用。下面介绍页面内容压缩过滤器，网站经常使用GZIP压缩算法对网页内容进行压缩，然后传给浏览器，这样可以减少数据传输量，提高响应速度。当浏览器接收到GZIP压缩数据后会自动解压并正确显示。

【例 7.8】 实现 GZIP 压缩过滤器。

首先，编写一个自定义的 ServletOutputStream 类使它具有压缩功能，这里的压缩功能采用 GZIP 算法实现，这是现在主流浏览器都可以接受的压缩格式，应用 JDK 自带的 GZIPOutputStream 类来完成，其源代码如下：

```
---------------------- GZIPResponseStream.java--------------------------
import java.io.*;
import java.util.zip.GZIPOutputStream;
```

```
import jakarta.servlet.*;
import jakarta.servlet.http.*;

public class GZIPResponseStream extends ServletOutputStream {
    //将压缩后的数据存放在ByteArrayOutputStream对象中
    protected ByteArrayOutputStream bArrayOutputStream = null;
    //JDK中自带的GZIP压缩类
    protected GZIPOutputStream gzipOutputStream = null;
    protected boolean closed = false;
    protected HttpServletResponse response = null;     //原先的response
    protected ServletOutputStream outputStream = null; //response中的输出流
    //构造方法，初始化定义值
    public GZIPResponseStream(HttpServletResponse response) throws IOException {
        super();
        closed = false;
        this.response = response;
        this.outputStream = response.getOutputStream();
        bArrayOutputStream = new ByteArrayOutputStream();
        gzipOutputStream = new GZIPOutputStream(bArrayOutputStream);
    }
    //执行压缩，并将数据输出到浏览器
    public void close() throws IOException {
        if (closed) {
            throw new IOException("This output stream has already been closed");
        }
        //执行压缩，必须调用这个方法
        gzipOutputStream.finish();
        //将压缩后的数据输出到浏览器中
        byte[] bytes = bArrayOutputStream.toByteArray();
        //设置压缩算法为GZIP，浏览器会自动解压数据
        response.addHeader("Content-Length",Integer.toString(bytes.length));
        response.addHeader("Content-Encoding", "gzip");
        //输出到浏览器
        outputStream.write(bytes);
        outputStream.flush();
        outputStream.close();
        closed = true;
    }

    public void flush() throws IOException {
        if (closed) {
            throw new IOException("不能刷新关闭的流！");
        }
        gzipOutputStream.flush();
    }
    //重写write方法，如果流关闭，则抛出异常
    public void write(int b) throws IOException {
        if (closed) {
            throw new IOException("输出流关闭中！");
        }
        gzipOutputStream.write((byte)b);
    }
    //重写write方法
    public void write(byte b[]) throws IOException {
```

```
57            write(b, 0, b.length);
58        }
59        //重写write方法，如果流关闭，则抛出异常
60        public void write(byte b[], int off, int len) throws IOException {
61            if (closed) {
62                throw new IOException("输出流关闭中！");
63            }
64            gzipOutputStream.write(b, off, len);
65        }
66
67        public boolean closed() {
68            return (this.closed);
69        }
70    }
```

在上述代码中，GZIPResponseStream 继承自 ServletOutputStream，使用时传入原始的响应对象，它的主要功能是将 response 中的数据用 ByteArrayOutputStream 缓存起来，然后用 GZIPOutputStream 类将数据压缩并输出到客户端浏览器中。第 15~22 行代码用于初始化自定义参数；第 24~40 行代码是执行压缩方法，并将缓存中的数据输出到浏览器中；第 49~65 行代码都是重构输出方法，即利用 GZIPOutputStream 输出。

然后，自定义 response 包装类 GZIPResponseWrapper，它只对输出的内容进行压缩，不进行将内容输出到客户端的操作。因为 response 要处理的不单单是字符内容，还有压缩的内容，即二进制内容，所以它需要重写 getOutputStream()和 getWriter()方法，源代码如下：

```
----------------------- GZIPResponseWrapper.java--------------------------
import java.io.*;
import java.util.*;
import jakarta.servlet.*;
import jakarta.servlet.http.*;

public class GZIPResponseWrapper extends HttpServletResponseWrapper {
    //原始的response
    private HttpServletResponse response = null;
    //自定义的outputStream，对数据进行压缩并输出
    private ServletOutputStream outputStream = null;
    //自定义PrintWriter，将内容输出到ServletOutputStream
    private PrintWriter printWriter = null;

    public GZIPResponseWrapper(HttpServletResponse response) {
        super(response);
        this.response = response;
    }

    //利用GZIPResponseStream创建输出流
    public ServletOutputStream createOutputStream() throws IOException {
        return (new GZIPResponseStream(response));
    }
    //执行这个方法对数据进行GZIP压缩，并输出到浏览器中
    public void finishResponse() {
        try {
            if (printWriter != null) {
                printWriter.close();
            } else {
```

```
29                 if (outputStream != null) {
30                     outputStream.close();
31                 }
32             }
33         } catch (IOException e) {}
34     }
35     //刷新ServletOutputStream输出流
36     public void flushBuffer() throws IOException {
37        outputStream.flush();
38     }
39     //覆盖getOutputStream方法，处理二进制内容
40     public ServletOutputStream getOutputStream() throws IOException {
41         if (printWriter != null) {
42             throw new IllegalStateException("getWriter() has already been called!");
43         }
44         //如果outputStream为空，则创建流
45         if (outputStream == null)
46             outputStream = createOutputStream();
47             return (outputStream);
48     }
49     //处理getWriter方法，处理字符内容
50     public PrintWriter getWriter() throws IOException {
51         if (printWriter != null) {
52             return (printWriter);
53         }
54
55         if (outputStream != null) {
56             throw new IllegalStateException("getOutputStream() has already been called!");
57         }
58         outputStream = createOutputStream();
59         //通过outputStream获得printWriter方法
60         printWriter = new PrintWriter(new OutputStreamWriter(outputStream, "UTF-8"));
61         return (printWriter);
62      }
63      //压缩后数据长度有变化，所以不用重写该方法
64      public void setContentLength(int length) {}
65 }
```

在上述代码中，第 24~34 行执行数据压缩的方法，getOutputStream()方法和 getWriter()方法都能实现 GZIPResponseStream 实例，这样就使得它们都具有压缩功能。在同一个 Servlet 请求中，getWriter()和 getOutputStream()只能调用一个，若两个同时调用则必将抛出 IllegalStateException 异常，如第 42 行和第 56 行代码所示。

接着，编写压缩过滤器类 GZIPFilter，在过滤器类中通过检查 Accept-Encoding 标头是否包含 gzip 字符来判断浏览器是否支持 GZIP 压缩算法，如果支持，则进行 GZIP 压缩数据，否则直接输出，其源代码如下：

```
------------------------ GZIPFilter.java---------------------------
01  import java.io.*;
02  import jakarta.servlet.*;
03  import jakarta.servlet.annotation.WebFilter;
04  import jakarta.servlet.annotation.WebInitParam;
05  import jakarta.servlet.http.*;
06
07  @WebFilter(
08      description = "内容替换过滤器",
09      filterName = "gzipFilter",
10      urlPatterns = { "/*" }
11  )
12  public class GZIPFilter implements Filter {
13
14      public void doFilter(ServletRequest req, ServletResponse res,
15              FilterChain chain) throws IOException, ServletException {
16          if (req instanceof HttpServletRequest) {
17              HttpServletRequest request = (HttpServletRequest) req;
18              HttpServletResponse response = (HttpServletResponse) res;
19              //依据浏览器的Header信息，判断支持的编码方式
20              String ae = request.getHeader("Accept-Encoding");
21              //如果浏览器支持GZIP格式，则使用GZIP压缩数据
22              if (ae != null && ae.toLowerCase().indexOf("gzip") != -1) {
23                  GZIPResponseWrapper wrappedResponse =
24              new GZIPResponseWrapper(response);
25                  chain.doFilter(req, wrappedResponse);
26                  //输出压缩数据
27                  wrappedResponse.finishResponse();
28                  return;
29              }
30              chain.doFilter(req, res);
31          }
32      }
33  }
```

在上述代码中，第 22 行代码用于判断浏览器标头是否包含 gzip 字符，第 23~24 行代码传入压缩响应对象 GZIPResponseWrapper，第 27 行代码输出压缩后的数据到浏览器中。

最后，编写一个测试的 GzipServlet 测试压缩结果，源代码如下：

```
-----------------------GzipServlet.java--------------------------
@WebServlet(
    urlPatterns = { "/gzip.action" },
    loadOnStartup = 0,
    name = "gzipServlet"
)
public class GzipServlet extends HttpServlet{

    public void doGet(HttpServletRequest request, HttpServletResponse
response) throws ServletException, IOException {
        doPost(request, response);
    }
    public void doPost(HttpServletRequest request, HttpServletResponse
response) throws ServletException, IOException {
        response.setContentType("text/html;charset=utf-8");
        response.setCharacterEncoding("UTF-8");
```

```
16          PrintWriter out = response.getWriter();
17          String[] urls = {          //设置一组URL地址
18              "http://localhost:8080/ch07/11.png",
19              "http://code.jquery.com/ui/1.10.3/jquery-ui.js",
20              "http://localhost:8080/ch07/login.jsp"
21          };
22          out.println("<!DOCTYPE HTML>");
23          //输出页面内容
24          out.println("<HTML>");
25          out.println("  <HEAD><TITLE>测试内容压缩输出过滤</TITLE></HEAD>");
26          out.println("  <BODY>");
27
28          for(String url:urls){
29              //模拟一个浏览器
30              URLConnection connGzip = new URL(url).openConnection();
31              //模拟实质浏览器的表头信息支持GZIP压缩格式
32              connGzip.setRequestProperty("Accept-Encoding", "gzip");
33              int lengthGzip = connGzip.getContentLength();  //获取压缩后的长度
34              //模拟另一个浏览器
35              URLConnection connCommon = new URL(url).openConnection();
36              int lengthCommon =connCommon.getContentLength();//获取压缩前的长度
37              double rate = new Double(lengthGzip)/lengthCommon;//计算压缩比率
38              out.println("<table border=\"1\" cellpadding=\"2\"
cellspacing=\"1\">");
39              out.println("<tr>");
40              out.println("<td colspan=\"3\">网址: "+url+"</td>");
41              out.println("</tr>");
42              out.println("<tr>");
43              out.println("<td>压缩后数据: "+lengthGzip+"byte</td>");
44              out.println("<td>压缩前数据: "+lengthCommon+"byte</td>");
45              out.println("<td>压缩率  :
46             "+NumberFormat.getPercentInstance().format(1-rate)+"</td>");
47              out.println("</tr>");
48              out.println("</table>");
49          }
50          out.println("  </BODY>");
51          out.println("</HTML>");
52          out.flush();
53          out.close();
54      }
55  }
```

在上述代码中，利用注入的方式声明Servlet，第 17~21 行代码存放一组 URL 地址；第 29~33 行代码模拟两个浏览器，一个支持 GZIP 压缩算法，另一个不支持；第 38~48 行代码用于输出结果到浏览器中。运行效果如图 7.10 所示。

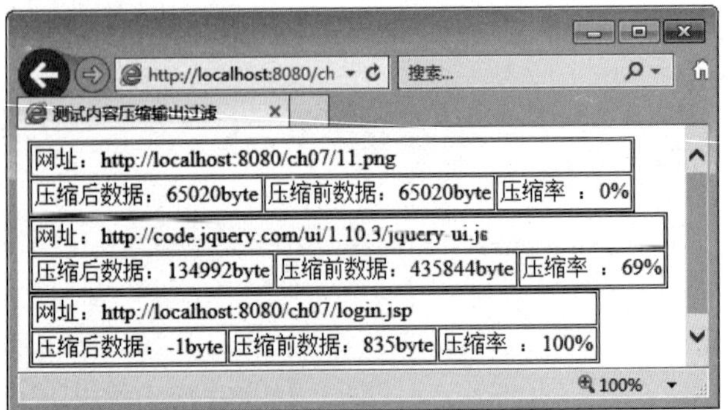

图7.10 压缩过滤器效果图

7.4 异步处理

在Servlet 2.0中，一个普通的Servlet工作流程大致如下：首先，Servlet接收请求，对数据进行处理；然后，调用业务接口方法，完成业务处理；最后，将结果返回到客户端。在Servlet中最耗时的是第二步的业务处理，因为它会执行一些数据库操作或者其他的跨网络调用等。在处理业务的过程中，该线程占用的资源不会被释放，这有可能造成性能的瓶颈。

异步处理是Servlet 3.0以后新增的一个特性，它可以先释放容器被占用的资源，将请求交给另一个异步线程来执行，业务方法执行完成后再生成响应数据。

本节将讲述异步处理接口AsyncContext的使用和一个异步处理应用实例的实现。

7.4.1 AsyncContext简介

在 Servlet 3.0 之后的版本中，ServletRequest 提供了两个方法来启动 AsyncContext：

- AsyncContext startAsync()
- AsyncContext startAsync(ServletRequest servletRequest,ServletResponse servletResponse)

上述两个方法都能得到 AsyncContext 接口的实现对象。当一个 Servlet 调用了 startAsync()方法之后，该 Servlet 的响应就会被延迟，并释放容器分配的线程。AsyncContext 接口的主要方法如表 7.7 所示。

表 7.7 AsyncContext 接口的主要方法

方 法	描 述
void addListener(AsyncListener listener)	添加 AsyncListener 监听器
complete()	响应完成
dispatch()	指定 URL 进行响应完成
getRequest()	获取 Servlet 请求对象
getResponse ()	获取 Servlet 响应对象
setTimeout(long timeout)	设置超时时间
start(java.lang.Runnable run)	异步启动线程

在Servlet 3.0及之后的版本中，有两种方式支持异步处理：注入声明和在web.xml中配置。其形式分别如下所示。

- 注入声明：

```
@WebServlet(
   asyncSupported=true,
   urlPatterns={"/asyncdemo.do"},
   name="myAsyncServlet"
)
public class MyAsyncServlet extends HttpServlet{
   ...
}
```

- 在web.xml中配置：

```
<servlet>
    <servlet-name>myAsyncServlet</servlet-name>
    <!-异步Servlet类路径->
    <servlet-class>com.eshore.MyAsyncServlet</servlet-class>
        <!-异步支持属性-->
    <async-supported>true</async-supported>
</servlet>
```

注意： 如果支持异步处理的 Servlet 前面有 Filter，则 Filter 也需要支持异步处理。

【例 7.9】 演示异步处理 Servlet。

编写一个支持异步通信的 Servlet 类，对于每个请求，该 Servlet 会取得异步信息 AsyncContext，同时释放占用的内存，延迟响应，然后启动 AsyncRequest 对象定义的线程，在该线程中进行业务处理，等业务处理完成后，输出页面信息并调用 AsyncContext 的 complete()方法表示异步完成。异步处理 Servlet 的源代码如下：

```
----------------------- MyAsyncServlet.java--------------------------
@WebServlet(
    asyncSupported = true,
    urlPatterns = { "/asyncdemo.do" },
    name = "myAsyncServlet"
)
public class MyAsyncServlet extends HttpServlet {
    SimpleDateFormat sdf = new SimpleDateFormat("yyyy-MM-dd HH:mm:ss");
    public void doGet(HttpServletRequest request, HttpServletResponse
response) throws ServletException, IOException {
        doPost(request, response);
    }

    public void doPost(HttpServletRequest request, HttpServletResponse
response) throws ServletException, IOException {
        response.setContentType("text/html;charset=UTF-8");
        PrintWriter out = response.getWriter();
        out.println("开始时间: " + sdf.format(new Date()) + "  ");
        out.flush();
        //在子线程中执行业务调用，并由它负责输出响应，主线程退出
        AsyncContext asyncContext = request.startAsync(request,response);
        asyncContext.setTimeout(900000000);//设置最大的超时时间
        new Thread(new Executor(asyncContext)).start();
        out.println("结束时间: " + sdf.format(new Date())+ "  ");
        out.flush();
    }
    //内部类
    public class Executor implements Runnable {
        private AsyncContext ctx = null;
        public Executor(AsyncContext ctx){
            this.ctx = ctx;
        }
        public void run(){
            try {
                //等待20秒，以模拟业务方法的执行
                Thread.sleep(20000);
```

```
36                  PrintWriter out = ctx.getResponse().getWriter();
37                  out.println("业务处理完毕的时间：" + sdf.format(new Date())+ ".");
38                  out.flush();
39                  ctx.complete();
40              } catch (Exception e) {
41                  e.printStackTrace();
42              }
43          }
44      }
45  }
```

在上述代码中，第 02 行表示该 Servlet 支持异步处理；第 20 行用于获得 AsyncContext 对象；第 21 行设置 AsyncContext 超时时间，默认的超时时间是 10 秒；第 27 行启动一个线程 Executor 类；第 35 行等待 20 秒，用于模拟业务处理；第 36 行输出响应结果；第 39 行表示 AsyncContext 类处理完成。运行效果如图 7.11 所示。

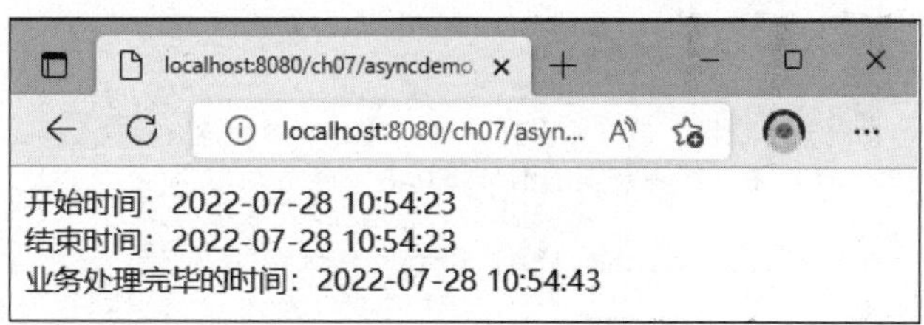

图7.11　异步处理效果图

注意：运行时如果报“is not surpported”错误，可能是因为没有将所有的过滤器或者经过的 Servlet 都设置成支持异步处理。

7.4.2　模拟服务器推送

模拟服务器推送是指模拟由服务器端向客户端推送消息。在HTTP协议中，服务器是无法直接对客户端传送消息的，必须得有一个请求服务器端才能够响应。可以利用Servlet 3.0以后的异步处理技术，主动推送消息到客户端。下面以一个例子说明这种技术的实现过程。

【例 7.10】　模拟服务器推送。

首先，编写一个负责存储消息的队列类 ClientService，该类的作用是利用 Queue 添加异步所有的 AsyncContext 对象，利用 BlockingQueue 阻塞队列存储页面请求的消息队列，当 Queue 队列中有数据时，就启动一个线程，将 BlockingQueue 阻塞的内容输出到页面中。ClientService 的源代码如下：

```
----------------------- ClientService.java-------------------------
public class ClientService {

    private final Queue<AsyncContext> ASYNC_QUEUE =
        new ConcurrentLinkedQueue<AsyncContext>();  //异步Servlet上下文队列

    private final BlockingQueue<String> INFO_QUEUE =
        new LinkedBlockingQueue<String>();         //消息队列

    private static ClientService instance  = new ClientService();
```

```java

    private ClientService() {                         //构造方法，启动线程
        new ClientThread().start();
    }

    public static ClientService getInstance() {   //获得ClientService单例
        return instance;
    }

    public void addAsyncContext(final AsyncContext asyncContext) {
        ASYNC_QUEUE.add(asyncContext);                //添加异步Servlet上下文
    }

    public void removeAsyncContext(final AsyncContext asyncContext) {
        ASYNC_QUEUE.remove(asyncContext);             //删除异步Servlet上下文
    }

    /**
     *
     * 发送消息到异步线程，最终输出到http response 流 <br>
     * @param str 发送给客户端的消息.<br>
     */
    public void callClient(final String str) {
        try {
            INFO_QUEUE.put(str);
        } catch (Exception ex) {
            throw new RuntimeException(ex);
        }
    }
    /**
     *
     * 将数据发送到response流上
     */
    protected class ClientThread extends Thread {
        public void run() {
            boolean done = false;
            while (!done) {
                try {
                    final String script = INFO_QUEUE.take();//当消息队列中有数据时，就调用take()方法
                    for (AsyncContext ac : ASYNC_QUEUE) {
                        try {
                            //调用响应中的getWriter方法
                            PrintWriter writer = ac.getResponse().getWriter();
                            writer.println(escapeHTML(script));
                            writer.flush();
                        } catch (IOException ex) {
                            ASYNC_QUEUE.remove(ac);
                            throw new RuntimeException(ex);
                        }
                    }
                }catch(InterruptedException e) {
                    //抛出异常，开关为true
                    done = true;
                    e.printStackTrace();
                }
```

```
65                  }
66              }
67          };
68
69          /**
70           * 删除多余的回车符和换行符
71           */
72          private String escapeHTML(String str) {
73              return "<script type='text/javascript'>\n"
74              + str.replaceAll("\n", "").replaceAll("\r", "")//替换回车符或换行符
75              + "</script>\n";
76          }
77      }
```

在上述代码中，第 12 行代码表示启动一个线程，该线程的作用是输出内容到页面中；第 49~66 行代码用于遍历所有的异步队列，将内容输出到页面中；第 72~76 行代码用于过滤回车符、换行符。

其次，编写异步的 Servlet 类，该类的作用是将客户端注册到发送消息的监听队列中，当产生超时、错误等事件时，将异步上下文对象从队列中移除。同时当访问该 Servlet 的客户端时，在 ASYNC _QUEUE 中注册一个 AsyncContext 对象，这样当服务器端需要调用客户端时，就会输出 AsyncContext 内容到客户端。源代码如下：

```
---------------------- AsyncContextServlet.java-------------------------
@WebServlet(urlPatterns = { "/AsyncContextServlet" }, asyncSupported = true)
public class AsyncContextServlet extends HttpServlet {

    private static final long serialVersionUID = 1L;

    @Override
    protected void doGet(HttpServletRequest request, HttpServletResponse
response) throws ServletException, IOException {

        request.setCharacterEncoding("UTF-8");                    //设置字符集
        response.setContentType("text/html;charset=UTF-8");    //设置响应类型和字符集
        final AsyncContext asyncContext = request.startAsync();
        //注册AsyncContext对象超时时间
        asyncContext.setTimeout(1000000);
        asyncContext.addListener(new AsyncListener() {     //添加异步监听器
            public void onComplete(AsyncEvent event) throws IOException {
                ClientService.getInstance().removeAsyncContext(asyncContext);
            }
            public void onTimeout(AsyncEvent event) throws IOException {
                ClientService.getInstance().removeAsyncContext(asyncContext);
            }
            public void onError(AsyncEvent event) throws IOException {
                ClientService.getInstance().removeAsyncContext(asyncContext);
            }
            public void onStartAsync(AsyncEvent event) throws IOException {
            }
        });
        //添加异步asyncContext对象
```

```
29            ClientService.getInstance().addAsyncContext(asyncContext);
30        }
31    }
```

在上述代码中，第 14 行注册 AsyncContext 对象的超时时间，第 15~27 行为异步对象添加监听器。

为了显示这个异步 Servlet 发出的信息，通过一个隐藏的 frame 去读取，源代码如下：

```
01  <html>
02  <head>
03  <script type="text/javascript" src="${pageContext.request.contextPath}
04  /js/jquery-3.6.0.js"> </script>
05  <style>
06  . textareaStyle {
07      width: 100%;
08      height: 100%;
09      border: 0;
10  }
11  </style>
12  <script type="text/javascript">
13      function update(data) {
14          var result = $('#result')[0];
15          result.value = result.value + data + '\n';
16      }
17  </script>
18  </head>
19      <body style="margin: 0; overflow: hidden">
20          <table width="100%" height="100%" border="0" >
21              <tr>
22                  <td colspan="2">
23                      <textarea name="result" id="result" readonly="true"
24  wrap="off" style="padding: 10; overflow: auto" rows="50"
25                  class=" textareaStyle"></textarea>
26                  </td>
27              </tr>
28          </table>
29          <iframe id="autoFrame" style="display:
30              none;" src="/ch07/AsyncContextServlet"></iframe>
31      </body>
32  </html>
```

在上述代码中，第 30 行 iframe 中的 src 值为支持异步的 Servlet 路径。

最后，编写一个测试的 Servlet，每隔 2 秒调用一次客户端方法。

```
------------------------ TestServlet.java---------------------------
@WebServlet("/test.action")
public class TestServlet extends HttpServlet {

    public void doGet(HttpServletRequest request,
HttpServletResponse response)throws ServletException, IOException {
        try {
            //隔2秒调用一次客户端方法
            for (int i = 0; i < 20; i++) {
                final String str = "window.parent.update(\""
                                        + String.valueOf(i) + "\");";
                ClientService.getInstance().callClient(str);
                Thread.sleep(2 * 1000);  //线程暂停2秒，模拟业务执行方法
```

```
13                  if (i == 10) {              //执行到第11次时，跳出循环
14                      break;
15                  }
16              }
17          } catch (InterruptedException e) {
18              e.printStackTrace();
19          }
20      }
21  }
```

在上述代码中，第 08~16 行用于循环调用客户端方法、推送信息。运行结果如图 7.12 所示。

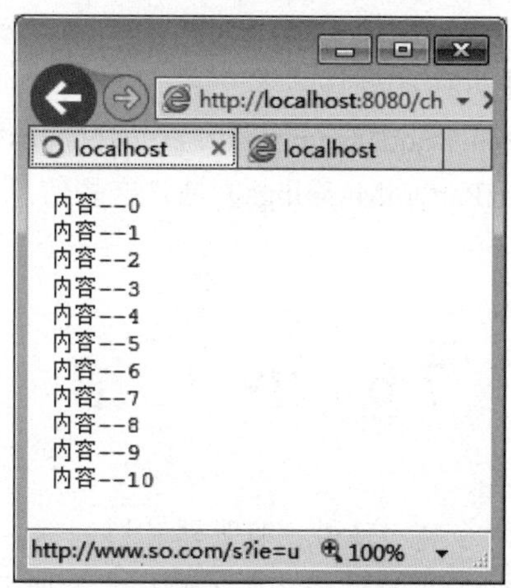

图7.12 模拟服务器推送效果图

7.5 Registration 动态注入的基础

前面的章节中介绍了Servlet的两种配置方式，一种是通过注解进行注入，另一种是通过web.xml进行配置。其中，注入方式是Servlet 3.0之后新增的特性。实现动态注入的基础是Registration接口。该接口是Servlet 3.0后引入的接口，主要用于向ServletContext中动态注册Servlet、Filter的实例，从而减轻web.xml繁重的配置。

Registration接口定义了如表7.8所示的一些方法。

表 7.8 Registration 接口的方法

返 回 值	方 法	说 明
java.lang.String	getClassName()	返回类名
java.lang.String	getInitParameter(java.lang.String name)	根据参数 name 获取启动时的初始化参数
java.util.Map<String>	getInitParameters()	获取所有的初始化参数和值，封装到 map 中
java.lang.String	getName()	返回对应的 Servlet 或 Filter 对应的 name
boolean	setInitParameter(java.lang.Stringname, java.lang.String value)	设置单个初始化参数

（续表）

返 回 值	方　法	说　明
java.util.Set<String>	setInitParameters(java.util.Map<java.lang.String,java.lang.String> initParameters)	批量设置初始化参数

ServletRegistration 接口和 FilterRegistration 接口继承了 Registration 接口，并且添加了各自的内容。

ServletRegistration 接口中添加了 addMapping、getMappings、getRunAsRole 方法，在 ServletRegistration.Dynamic接口中，添加了setLoadOnStartup、setMultipartConfig等接口，这些信息与之前@WebServlet注解中介绍的属性内容一致。

FilterRegistration接口中添加了addMappingForServletNames、addMappingForUrlPatterns、getServletNameMappings、getUrlPatternMappings方法，这些信息与之前@WebFilter注解中介绍的属性内容一致。

7.6 小　结

本章详细介绍了Servlet的两个高级特性：监听器和过滤器。监听器用来监听Web容器中正在执行的程序情况，并根据发生的事情做出相应的响应。监听范围可以是整个ServletContext，也可以是Session或者Request。要实现相对应的接口，可注入声明或者在web.xml中配置。

过滤器在服务器端和客户端之间起到一个中间件的作用，对二者之间的交互信息进行过滤。一个Web系统可以部署一个或者多个过滤器，从而组成一个过滤器链，对整个系统起到整体的过滤作用。

7.7 习　题

（1）对于一个Web应用程序而言，与HttpSession有关的监听器有哪些？在哪个期间有效？

（2）利用监听器监听整个应用程序中的共享数据，应该选用哪个监听器？

（3）配置监听器有几种方法？

（4）监听器的开发步骤有哪些？

（5）过滤器的作用是什么？在Web中，过滤器常被用来做什么？

（6）过滤器的开发步骤有哪些？

（7）编写一个监听器，实现记录在线用户数的功能。

（8）编写一个过滤器，过滤IP地址，使得某些IP地址不能访问。

第 8 章 MySQL 8 数据库开发

在Web应用技术中，数据库的操作是必不可少的，包括对数据库表的增加、删除、修改、查询等。现今，数据库可以分为关系型数据库和非关系型数据库。常用的关系型数据库主要有MySQL、Oracle、DB2、Informix、SQL Server等。常用的非关系型数据库主要有Memcached、Redis、MongoDB等。本章将主要介绍MySQL数据库的开发及其在Web开发中的操作和应用。

注意：本书若无特别说明，则数据库操作都是在 MySQL 数据库环境下进行的。

本章主要涉及的知识点有：

- MySQL数据库的安装和配置
- MySQL数据库的操作（包括增加、删除、修改等操作）
- MySQL数据的管理
- MySQL中的图形化界面管理

8.1 MySQL 数据库入门

MySQL数据库是一款小型的关系型数据库，以体积小、速度快、成本低等特点独树一帜。MySQL数据库是目前最受欢迎的开源数据库，已经被Oracle公司收购。

本节主要介绍MySQL的版本特点、安装和配置。

8.1.1 MySQL的版本特点

MySQL 被 Oracle 收购后，针对不同的用户群体提供了不同的版本。

- MySQL Community Server：社区版，完全免费，但是官方不提供技术支持。

- MySQL Enterprise Edition：企业版，能够为企业提供高性能的数据库应用，以及高稳定性的数据库系统，提供完整的数据库提交、回滚以及锁机制等功能，但是该版本收费，官方只提供电话支持。

注意：MySQL Cluster 主要用于建立数据库集群服务器，需要在上述两个版本的基础上使用。

MySQL 的命名机制由 3 个数字组成，例如 MySQL 8.0.11。

- 第1个数字8是主版本号，用于描述文件格式，表示所有版本8的发行版都有相同的文件格式。
- 第2个数字0是发行级别，与主版本号组合在一起构成发行序列号。
- 第3个数字11是此发行系列的版本号。

注意：旧版本的 MySQL，例如 MySQL 4.1、4.0 以及 3.13 版本，官方将不再提供技术支持。而所有发布的 MySQL 版本都经过严格的测试，可以保证能正常使用。针对不同的系统，读者可以从 MySQL 官方网址下载相应的安装文件。

8.1.2 MySQL 8的安装和配置

MySQL支持不同的操作系统平台，虽然在不同平台下的安装和配置都不相同，但是差别也不是很大。在Windows平台下可以使用二进制的安装包或者免安装版的软件包进行安装，安装包提供图形化的安装向导过程，免安装版则直接解压就能用。在Linux平台下使用命令安装MySQL，但由于Linux有很多的版本，因此不同的Linux平台需要下载相应的MySQL安装包。本小节主要讲解 Windows平台下的MySQL安装和配置过程。

在Windows平台下提供两种安装方式：MySQL二进制版和免安装版。一般来说，应该采用二进制版，因为该版本使用起来比较简单，不用第三方工具启动就可以运行MySQL。这里采用二进制的安装方式。

1. 下载 MySQL 8 安装文件

具体的下载操作步骤如下：

01 打开常用浏览器，输入网址“http://dev.mysql.com/downloads/mysql”，页面自动跳转到MySQL Community Server 8.0.11下载页面，选择Generally Available(GA) Release选项卡，在选择平台的下拉列表框中选择Microsoft Windows平台，如图8.1所示。

02 可以选择网络安装二进制文件或者直接下载二进制文件，在这里选择直接下载二进制文件，单击 Download 按钮进行下载，如图8.2所示。

2. 安装和配置 MySQL 数据库

MySQL二进制文件下载完成后，找到下载文件（例如 d:\mysql-installer-community-8.0.11.0.msi），双击文件进行安装，具体操作步骤如下：

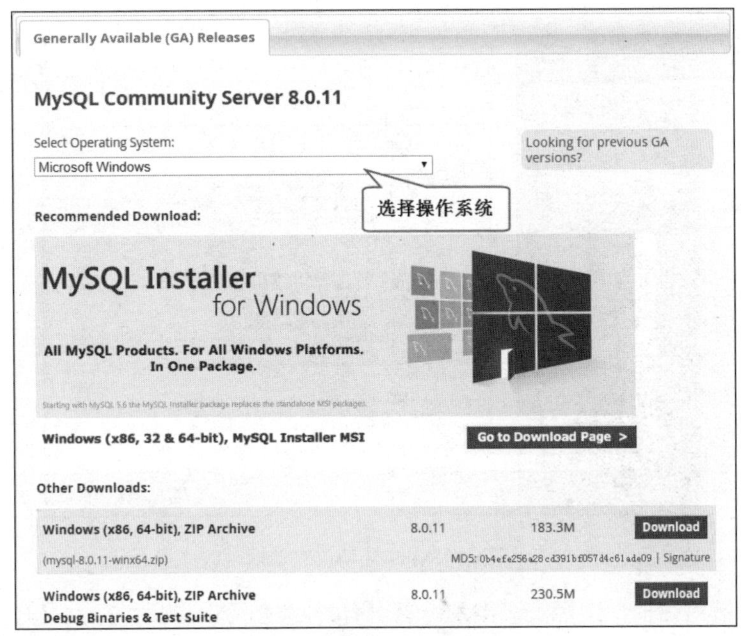

图8.1　MySQL下载页面

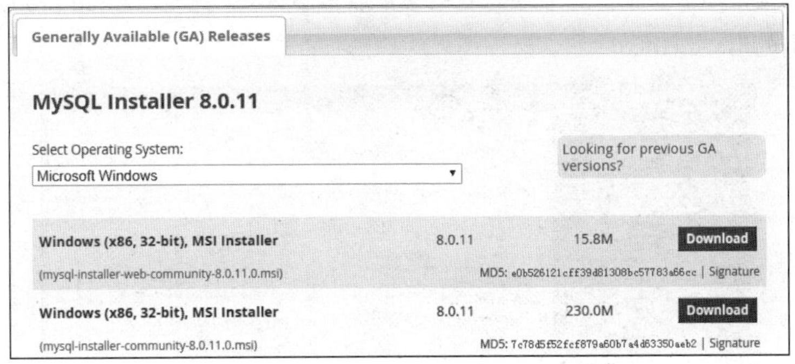

图8.2　单击下载MySQL二进制文件

01 双击下载的mysql-installer-community-8.0.11.0.msi文件，如图8.3所示。

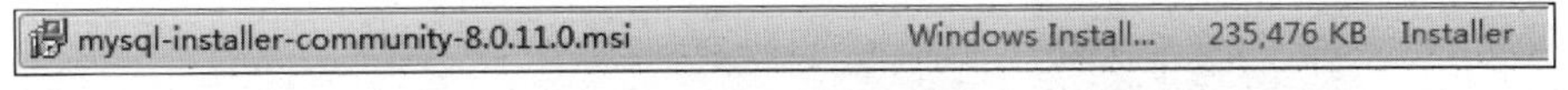

图8.3　MySQL安装文件

02 等待Windows系统检测MySQL安装环境，如图8.4所示。

图8.4　Windows系统检测MySQL安装环境

03 弹出安装对话框，勾选同意协议复选框，单击Next按钮，如图8.5所示。

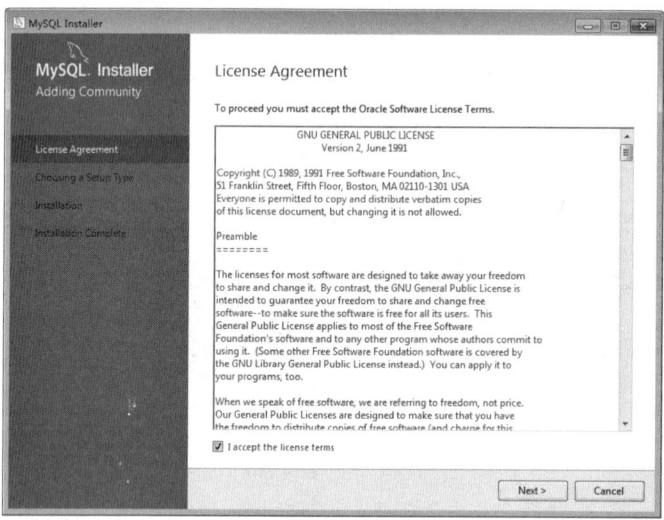

图8.5　MySQL协议对话框

04 弹出安装对话框，选择安装类别。这里有5种类别可以选择，分别是开发版（Developer Default）、服务器版（Server Only）、客户端（Client Only）、全部MySQL（Full）、定制版（Custom）。在此，本书选择开发版并默认勾选右侧Install all products单选按钮，单击Next按钮，如图8.6所示。

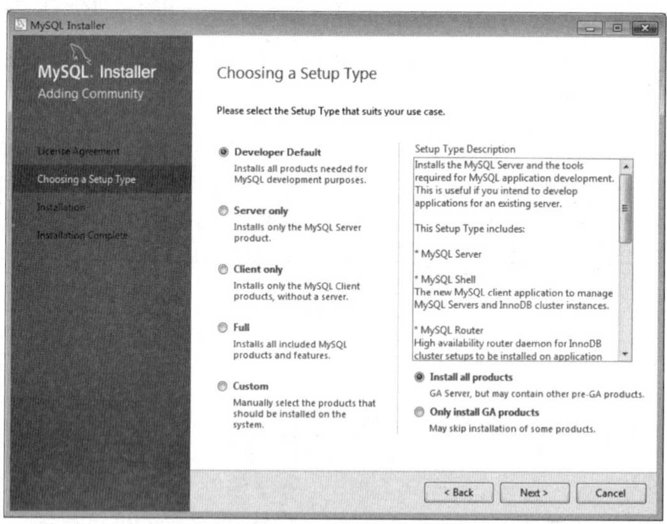

图8.6　选择安装类型

注意： MySQL 8.0.11 默认的安装路径为"C:\Program Files\MySQL"，使用 msi 安装方式首次安装时无法修改安装路径，如果之前曾安装过 MySQL 产品，那么在发生冲突时会提示修改安装路径，并弹出修改输入框。

05 检查安装的必需产品，验证Python依赖是否安装，如果未安装，单击弹出框给出的下载地址http://www.python.org/download/，下载对应的Python版本，安装完成后单击页面中的Check按钮，验证通过后，继续执行，如图8.7所示。

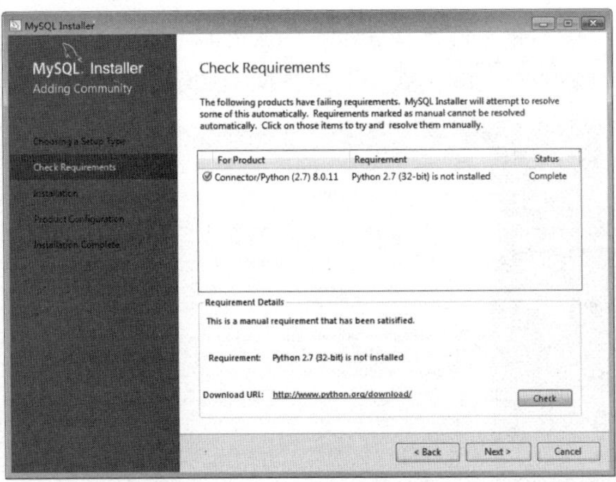

图8.7　验证Python的安装

06 开始安装，单击下方的Execute按钮，如图8.8所示。

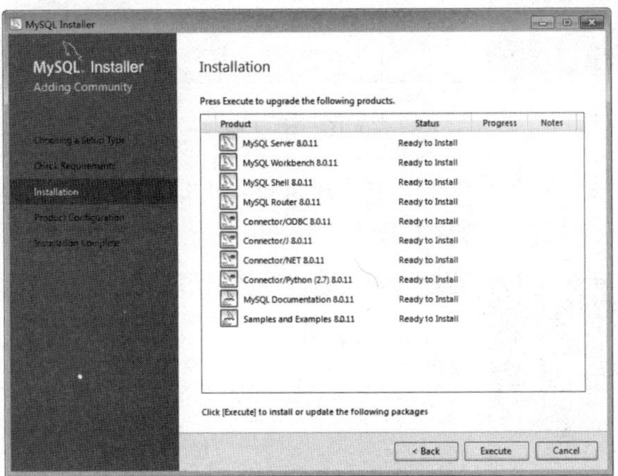

图8.8　安装MySQL产品

07 安装完成后，所有产品名右侧的Status栏都显示complete，单击Next按钮进入产品配置页面，如图8.9所示。

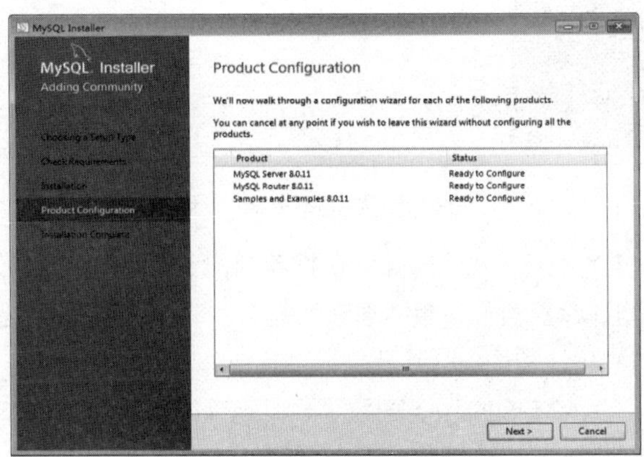

图8.9　产品配置

08 单击Next按钮，在界面左侧选择Group Replication选项（实现数据最终一致性的MySQL插件），界面右侧的选项默认即可，单击Next按钮，如图8.10所示。

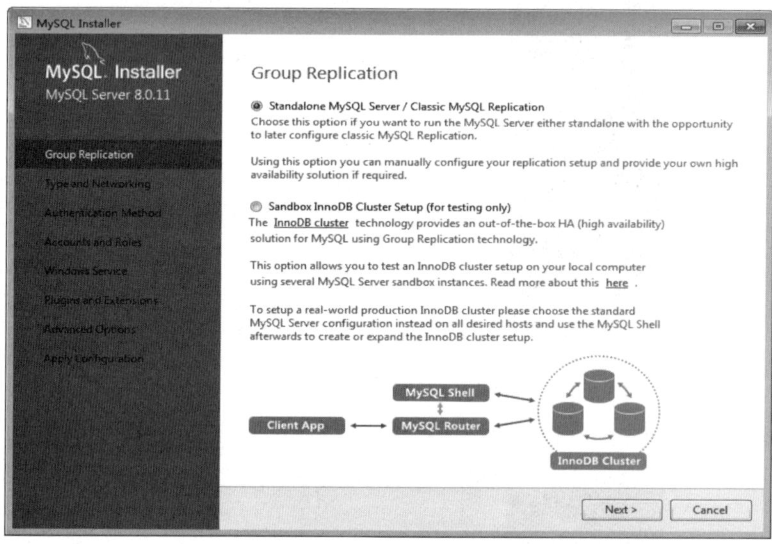

图8.10　组复制

09 选择类型和网络化内容，类型选择Development Computer（开发计算机），端口默认为3306，单击Next按钮，如图8.11所示。

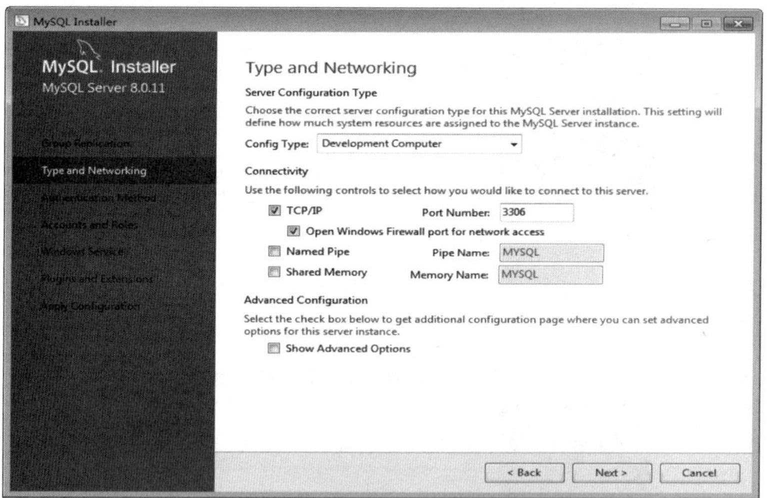

图8.11　类型和网络化参数

10 配置登录MySQL的验证方式，选择默认推荐项，使用密码登录，如图8.12所示，单击Next按钮，跳转到密码设置页面，输入管理员密码，输入完成后会提示密码强度，最好设置为strong类型的密码，如图8.13所示。

11 单击Next按钮，进入MySQL服务配置页面，选择默认项。单击Next按钮，进入插件和扩展页面。不做任何更改，直接单击Next按钮，进入配置执行页面。单击Execute按钮，配置生效，如图8.14所示。

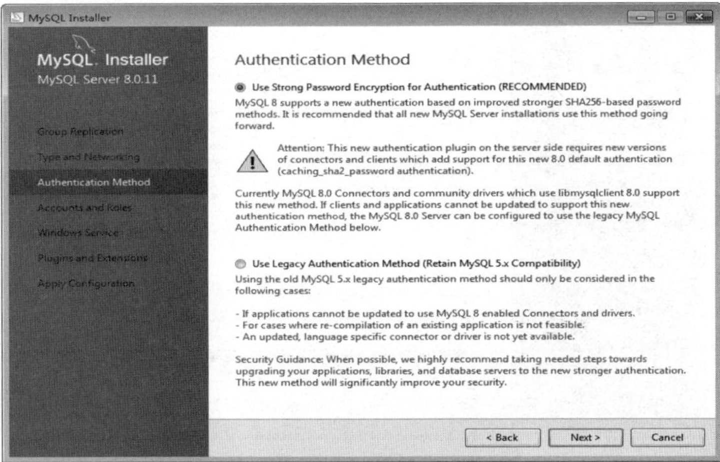

图8.12 选择验证方式

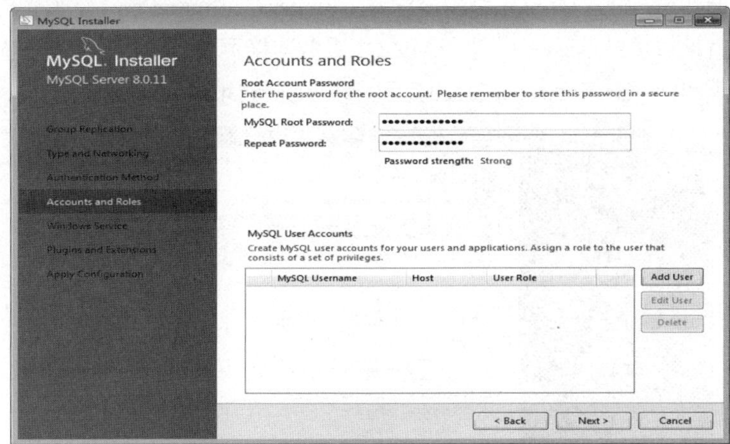

图8.13 设置MySQL管理员密码

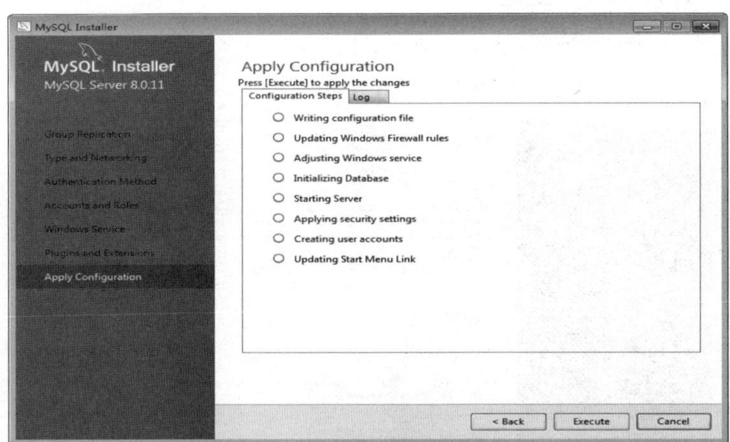

图8.14 MySQL应用配置

12 设置完后，单击Finish按钮，进入其他附带产品配置页面，也可取消，用到时再进行配置，配置界面如图8.15所示。

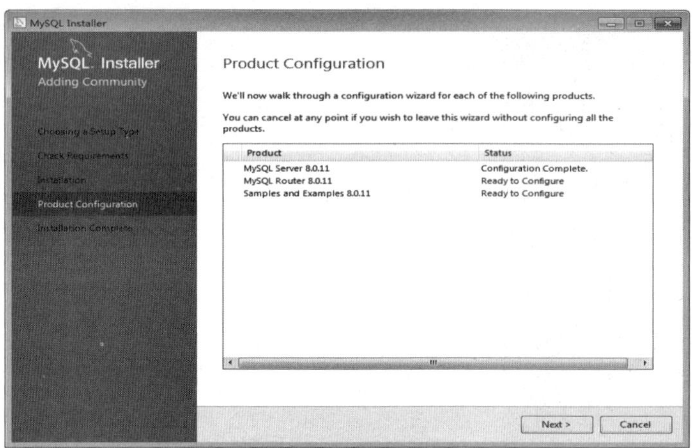

图8.15　产品配置

13 单击Next按钮，进入MySQL路由器配置（Router Configuration），如图8.16所示，不做任何更改直接单击Finish按钮，进入MySQL服务连接页面，输入设置的密码，单击Check按钮，出现All connections succeeded提示，表示连接成功，如图8.17所示。

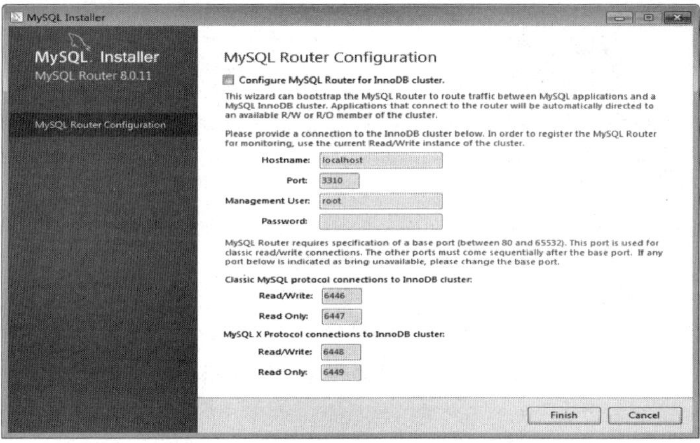

图8.16　MySQL路由器配置

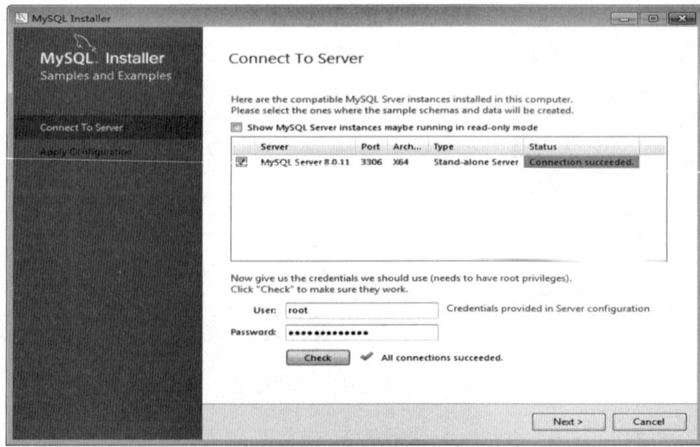

图8.17　MySQL服务连接

14 单击Next按钮，继续进入配置生效页面，单击Execute按钮，应用配置生效后进入安装完成页面，如图8.18所示。

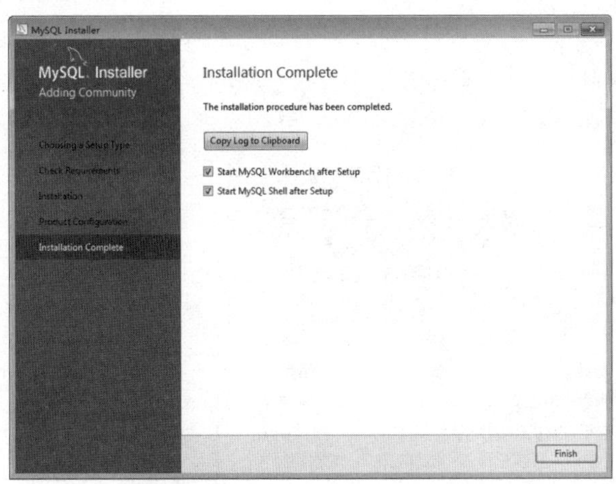

图8.18　MySQL安装完成界面

15 单击Finish按钮，系统出现MySQL Workbench图形化界面，如图8.19所示。

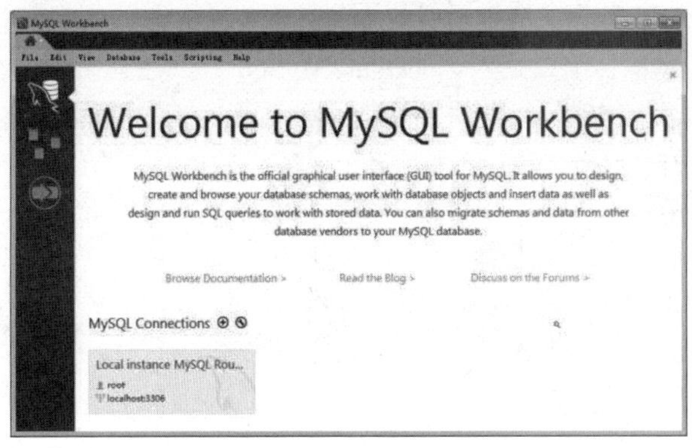

图8.19　MySQL Workbench 操作界面

8.2　启动 MySQL 服务并登录数据库

安装完MySQL数据库之后，启动服务进程，相应的客户端就可以连接数据库，客户端可以通过命令行或者图形界面工具登录数据库。本节将介绍如何启动MySQL服务和登录MySQL数据库。

8.2.1　启动MySQL服务

在默认的配置中，已经将MySQL设置为Windows服务，当系统启动或停止时，MySQL服务会自动启动或者关闭，但是用户还可以通过图形服务工具来控制MySQL服务器或者从命令行使用命令启动。

通过 Windows 的服务进行管理，具体的操作步骤如下：

01 单击“开始” | “运行”菜单，在弹出的对话框中输入“services.msc”命令，打开Windows的“服务管理器”窗口，在其中可以看到服务名为“MySQL”的服务项，右边状态为“正在运行”，表明该服务已经启动，如图8.20所示。

02 从图8.21可以看出MySQL的启动类型为自动，且该服务已经启动。如果状态为空白，说明服务未启动。启动方法为：双击MySQL服务名，打开“MySQL的属性”对话框，在其中通过单击“启动”或者“停止”按钮来改变服务状态，具体如图8.21所示。

MySQL		正在运行	自动	本地系统
Net.Tcp Port Sharing Ser...	提供...		禁用	本地服务
Netlogon	为用...		手动	本地系统
Network Connected Devi...	网络...	正在运行	手动(触发...	本地服务
Network Connection Bro...	允许 ...	正在运行	手动(触发...	本地系统

图 8.20 “服务管理器”窗口

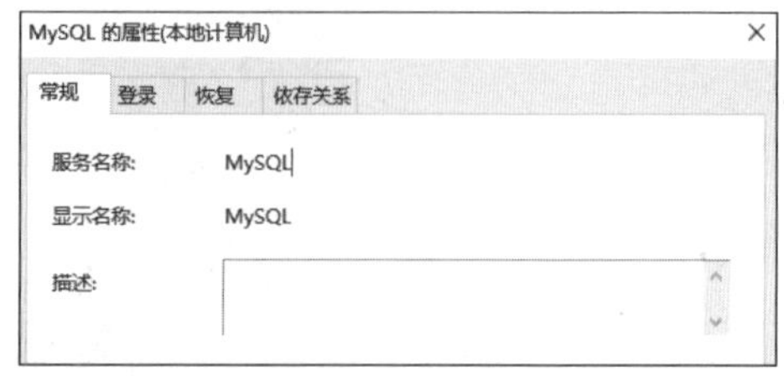

图 8.21 “MySQL 的属性”对话框

03 也可以通过命令行启动，启动方法为：单击“开始”菜单，在搜索框中输入“cmd”，按回车键弹出Windows命令提示符界面，如图8.22所示。

04 进入“命令提示符”窗口输入“net start MySQL”，按回车键，就可以启动MySQL服务了，停止MySQL服务的命令为“net stop MySQL”，如图8.23所示。

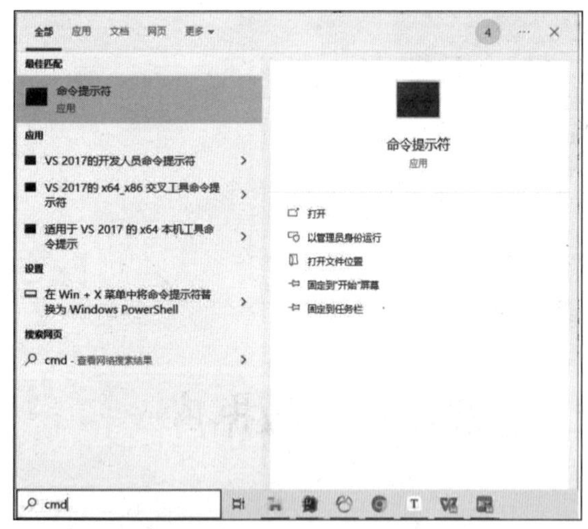

图 8.22 Windows 命令操作界面

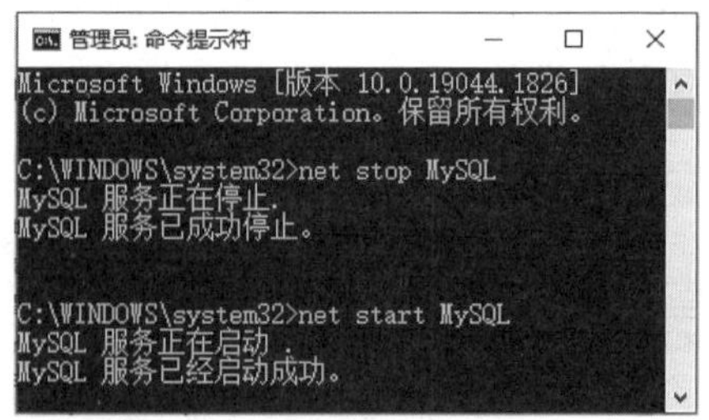

图 8.23 在命令行中启动和停止 MySQL

注意：“net start MySQL”中的“MySQL”是 MySQL 服务的名字。如果 MySQL 服务的名字是其他名字，应该输入“net start XX”。

8.2.2 登录MySQL数据库

当MySQL服务启动后，可以通过客户端来登录MySQL数据库。在Windows系统中，有以下两种登录MySQL数据库的方式。

1. 利用 Windows 命令行登录

具体的操作步骤如下：

01 单击“开始” | “运行”菜单，在弹出的对话框中输入命令“cmd”。在DOS窗口中通过登录命令连接MySQL数据库。连接MySQL的命令为：

```
MySQL -h hostname  -u username -p
```

其中，MySQL为命令，-h后面是服务器主机地址，-u后面是登录数据库的用户名，-p后面是用户登录密码。在这里由于MySQL客户端和服务器是同一台机器，所以输入命令如下：

```
MySQL -h localhost -u root -p
```

02 按回车键，系统会提示“Enter password”（输入密码），如图8.24所示。输入前面配置中的密码，验证正确后，即可登录到MySQL数据库，如图8.25所示。

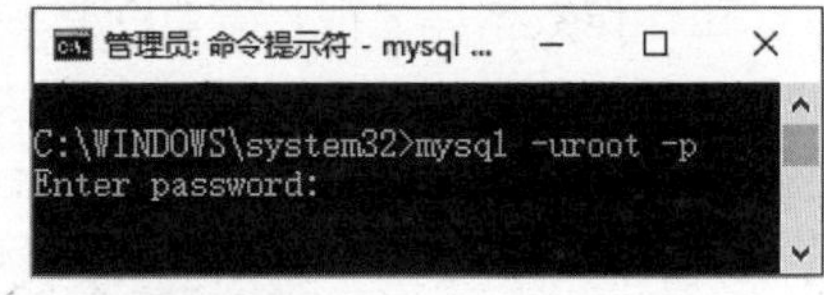

图8.24 按照命令提示输入密码

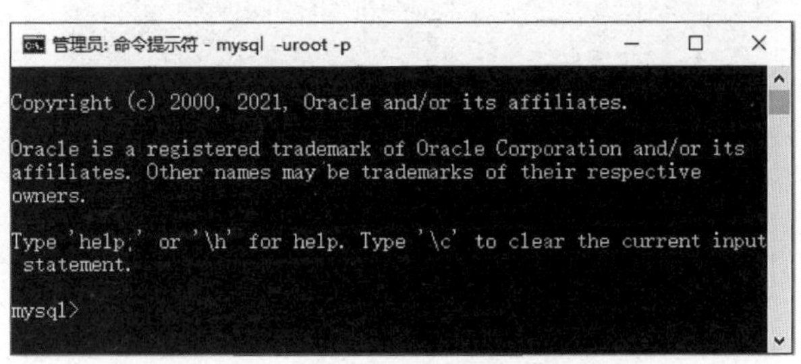

图8.25 Windows命令行登录窗口

注意：

① 当窗口中出现如图 8.25 所示的描述信息且命令提示符变为“mysql>”时，表明已成功登录 MySQL 服务器了。

② 如果在输入正确的密码后，出现 ERROR 1045 拒绝连接的错误，请先检查 Path 变量中是否配置了 MySQL 的 bin 目录，如果没有配置，就需要在 Path 中加上，例如 C:\Program Files\MySQL\MySQL Server 8.0\bin。

2. 利用 MySQL 命令行登录

依次单击“开始” | “MySQL” | “MySQL Server 8.0” | “MySQL 8.0 Command Line Client”菜单命令，进入密码输入窗口，如图8.26所示。

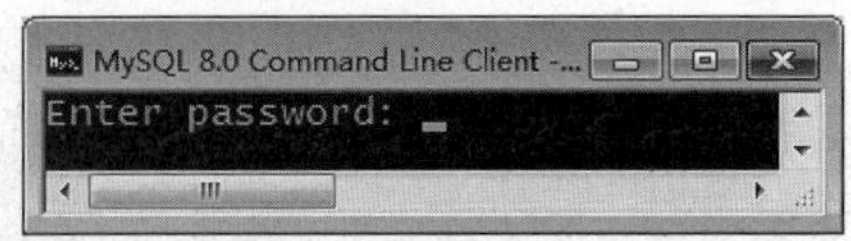

图8.26 密码输入窗口

输入正确的密码后，就可以登录MySQL数据库了。显示的结果与Windows命令行登录的结果是一样的。

8.3 MySQL 数据库的基本操作

MySQL数据库安装完成并启动后，就需要对数据库进行简单的操作了，其内容涉及创建数据库、修改数据库、删除数据库等。本节将介绍如何在MySQL数据库中进行简单的操作。

8.3.1 创建数据库

在操作数据库之前，需要创建数据库。MySQL安装完成之后，将会在data目录下自动创建几个必需的数据库，可以通过命令“show databases;”查看当前的所有数据库，注意命令以分号结尾，如图8.27所示。

```
MySQL 8.0 Command Line Client - Unicode
Type 'help;' or '\h' for help. Type '\c' to clear

mysql> show databases;
+--------------------+
| Database           |
+--------------------+
| information_schema |
| mysql              |
| performance_schema |
| sakila             |
| sys                |
| world              |
+--------------------+
6 rows in set (0.00 sec)

mysql>
```

图8.27　显示必需的数据库界面

可以看出，数据库列表中包含了6个数据库。其中，mysql是必需的，用于描述用户访问权限；sys数据库是MySQL 5.7版本之后新增的数据库，用于查看系统信息。

MySQL 创建数据库的基本语法如下：

```
create database database_name;
```

“database_name”为要创建的数据库名称。

【例 8.1】 创建测试数据库 test_database。

输入如下命令：

```
create database test_database;
```

效果如图 8.28 所示。

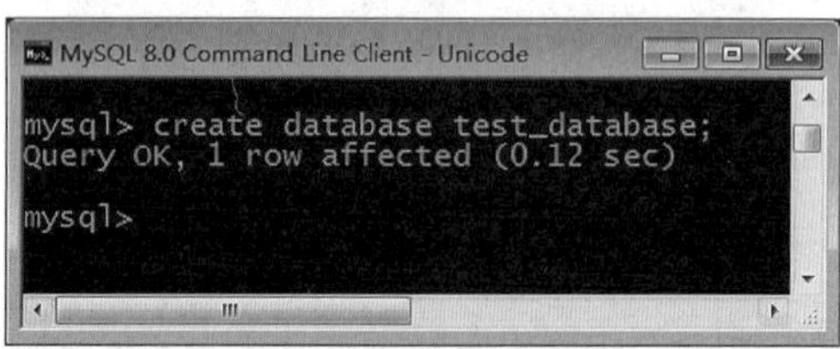

图8.28　创建数据库

数据库创建好之后，可以使用如下命令查看数据库的定义：

```
show create database test_database;
```

效果如图 8.29 所示。

```
mysql> show create database test_database;
+---------------+------------------------------------------------------------------
-----------------------------------+
| Database      | Create Database
                                   |
+---------------+------------------------------------------------------------------
-----------------------------------+
| test_database | CREATE DATABASE `test_database` /*!40100 DEFAULT CHARACTER SET
 utf8mb4 COLLATE utf8mb4_0900_ai_ci */ |
+---------------+------------------------------------------------------------------
-----------------------------------+
1 row in set (0.01 sec)

mysql>
```

图8.29　查看数据库的定义

可以看出，如果数据库创建成功，就将显示数据库的创建信息等内容。输入命令“show databases;”再次查看当前存在的数据库，如图8.30所示。

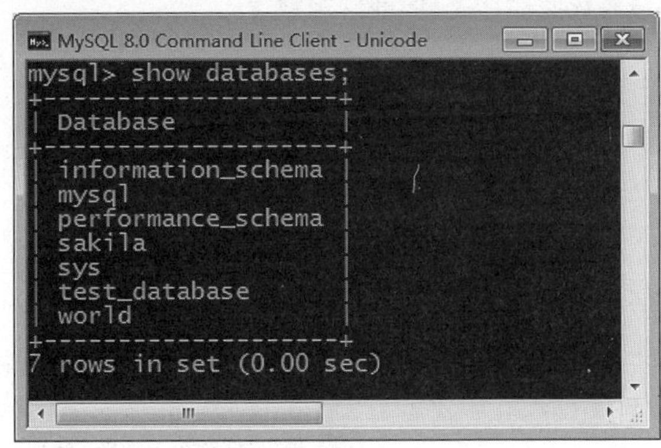

图8.30　查看当前存在的数据库

可以看到，数据库列表中包含了刚刚创建的数据库test_database。

8.3.2　删除数据库

删除数据库是将已经创建的数据库和数据库中的数据一并从磁盘空间中删除。删除数据库的基本语法格式如下：

```
drop database database_name;
```

“database_name”是数据库名称，如果数据库不存在，那么删除数据库后会出现错误。

【例 8.2】　删除测试数据库 test_database。

输入如下命令：

```
drop database test_database;
```

效果如图 8.31 所示。

```
mysql> drop database test_database;
Query OK, 0 rows affected (0.15 sec)

mysql>
```

图8.31　MySQL删除数据库

执行完删除命令后，数据库将被删除，再次删除“test_database”数据库时，系统会提示不存在数据库，删除出错，如图8.32所示。

```
mysql> drop database test_database;
ERROR 1008 (HY000): Can't drop database 'test_database'; database doesn't exist
mysql>
```

图8.32　MySQL再次执行删除数据库命令

注意：使用删除数据库命令要十分谨慎，因为在执行删除命令时，MySQL 不会给出提醒确认信息。数据库中存储的数据都将被删除且不可恢复。

8.3.3　创建数据库表

在创建数据库表之前，需要使用语句“use database_name”指定在哪个库中进行创建。创建数据库表的基本语法如下：

```
CREATE TABLE <TABLE_NAME>（字段名,数据类型［默认值］）;
```

【例 8.3】　创建数据库表 tb_temp1。

首先选择创建表的数据库，SQL 命令如下：

```
use test_database;
```

然后创建tb_temp1表，SQL命令如下：

```
CREATE TABLE tb_temp1(
  id int(10),
  name varchar(20)
);
```

语句执行后，创建了一个名为 tb_temp1 的表。可以使用命令“show tables;”查看是否创建成功，如图 8.33 所示。

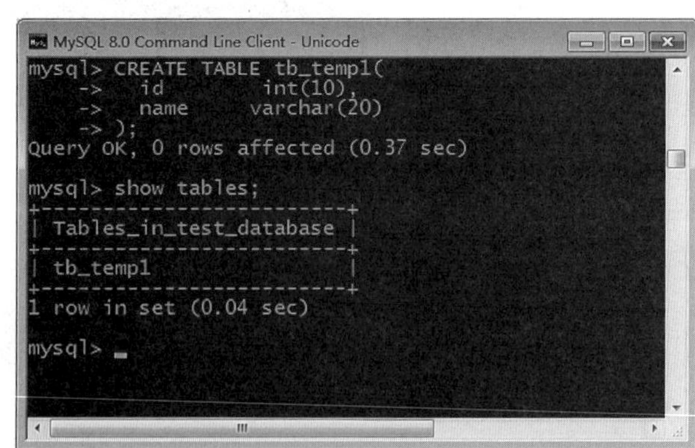

图8.33　创建表tb_temp1

可以看到，tb_temp1数据库表已经在test_database中创建成功了。

8.3.4　修改数据库表

MySQL 修改数据库表的基本语法如下：

```
alter table <旧表名> rename [to] <新表名>
```

在修改表之前，也要选定在哪个库中执行修改操作。

【例 8.4】　修改数据库表 tb_temp1 为 tb_temp2。

首先选择修改表的数据库，SQL 命令如下：

```
use test_database;
```

然后修改tb_temp1表，SQL命令如下：

```
alter table tb_temp1 rename to tb_temp2;
```

语句执行后，即修改tb_temp1的表名为tb_temp2。使用命令“show tables;”查看是否修改成功，如图8.34所示。

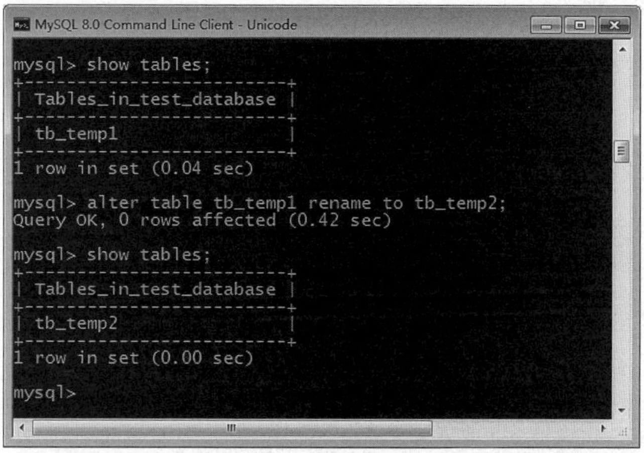

```
MySQL 8.0 Command Line Client - Unicode
mysql> show tables;
+-------------------------+
| Tables_in_test_database |
+-------------------------+
| tb_temp1                |
+-------------------------+
1 row in set (0.04 sec)

mysql> alter table tb_temp1 rename to tb_temp2;
Query OK, 0 rows affected (0.42 sec)

mysql> show tables;
+-------------------------+
| Tables_in_test_database |
+-------------------------+
| tb_temp2                |
+-------------------------+
1 row in set (0.00 sec)

mysql>
```

图8.34　修改表tb_temp1

经过比较可以看到，表tb_temp1已经修改为tb_temp2了。

注意：可以利用命令 desc 查看修改完的表结构是否与修改前的表结构相同。

8.3.5　修改数据库表的字段名

在 MySQL 中修改表字段名的基本语法如下：

```
alter table <表名> change <旧字段名> <新字段名> <新数据类型>;
```

【例 8.5】　修改数据库表 tb_temp2 中的 id 字段名称为 temp_id、数据类型为 varchar(15)。

语句如下：

```
alter table tb_temp2 change id temp_id  varchar(15);
```

使用命令“desc tb_temp2”查看 tb_temp2 表的结构，结果如图 8.35 所示。

经过比较可以发现，字段id已经修改成功，且类型被修改为varchar（15）。

注意：由于不同类型的数据在 MySQL 中的存储方式并不一定相同，修改数据类型可能会影响到表中的数据，因此当表中存在数据时最好不要修改数据类型。

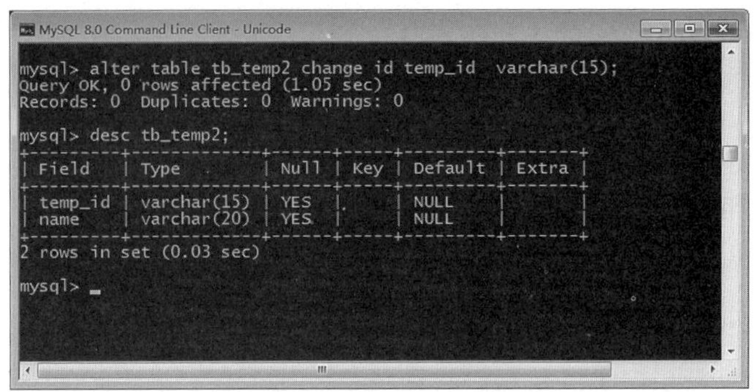

图8.35 修改表tb_temp2字段

8.3.6 删除数据表

删除表是将库中存在的表删除。MySQL 删除表的基本语法如下：

```
drop table [if exists]table_name1,table_name2,…;
```

【例 8.6】 删除表 tb_temp2。

语句如下：

```
drop  table tb_temp2;
```

可以使用命令“show tables;”查看表tb_temp2是否删除成功，结果如图8.36所示。从执行结果可以看到已经不存在tb_temp2表，说明删除成功。

```
MySQL 8.0 Command Line Client - Unicode
mysql> desc tb_temp2;
+---------+-------------+------+-----+---------+-------+
| Field   | Type        | Null | Key | Default | Extra |
+---------+-------------+------+-----+---------+-------+
| temp_id | varchar(15) | YES  |     | NULL    |       |
| name    | varchar(20) | YES  |     | NULL    |       |
+---------+-------------+------+-----+---------+-------+
2 rows in set (0.03 sec)

mysql> drop  table tb_temp2;
Query OK, 0 rows affected (0.25 sec)

mysql> show tables;
Empty set (0.00 sec)

mysql>
```

图8.36 删除表tb_temp2

8.4 MySQL 数据库的数据管理

上节介绍了MySQL数据库的基本操作，使读者对数据库的操作有了基本印象。在MySQL数据库中，对数据的管理是数据库中最重要又是最基本的操作。数据在结构化数据库中是按照行、列的格式进行存储的，行代表一条记录，列代表记录中的域。

数据的基本操作主要涉及插入数据、修改数据、删除数据等。本节将逐一介绍数据的具体操作。

8.4.1　插入数据

在 MySQL 数据库中插入数据的语法结构也遵循标准的 SQL 语法结构，其基本语法如下：

```
insert into table_name(column1,column2,…)  value(value1,value2,…);
```

“table_name”为要插入的数据表，column1 是表中的列，value1 为插入的列值，如果不指明列，则按照表中默认的顺序插入值。

【例 8.7】　向数据表 tb_temp1 插入数据。

输入如下命令：

```
nsert into tb_temp1(id,name,address,phone) value(1,'TOM','Guangzhou',
'1890220XXXX');
```

执行命令：

```
select * from tb_temp1;
```

查看是否添加成功，结果如图 8.37 所示。

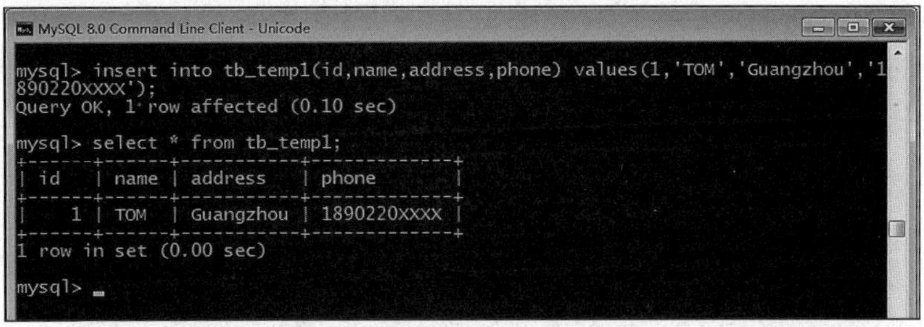

```
MySQL 8.0 Command Line Client - Unicode
mysql> insert into tb_temp1(id,name,address,phone) values(1,'TOM','Guangzhou','1
890220xxxx');
Query OK, 1 row affected (0.10 sec)

mysql> select * from tb_temp1;
+------+------+-----------+-------------+
| id   | name | address   | phone       |
+------+------+-----------+-------------+
|    1 | TOM  | Guangzhou | 1890220xxxx |
+------+------+-----------+-------------+
1 row in set (0.00 sec)

mysql> _
```

图8.37　向表tb_temp1插入数据

从结果可以看出，数据添加成功。

8.4.2　修改数据

在 MySQL 数据库中修改数据是修改表中已经存在的数据，其基本语法如下：

```
update table_name set column1='value1', column2='value2' [condition...]
```

其中，table_name 为要更新的数据表，column1、column2 是更新的列，value1、value2 为更新的列值，condition 为更新的条件。如果不指明条件，就将表中所有的 column1、column2 列值都更新为 value1 和 value2。

【例 8.8】　更新数据表 tb_temp1 中的数据，将表中 name 为“Smith”的数据更新为“John”。

输入如下命令：

```
update tb_temp1 set name='John' where name='Smith';
```

执行命令：

```
select * from tb_temp1;
```

查看是否修改成功，结果如图 8.38 所示。从结果可以看出，数据添加成功。

```
mysql> select * from tb_temp1;
+------+-------+-----------+-------------+
| id   | name  | address   | phone       |
+------+-------+-----------+-------------+
|    1 | Smith | Guangzhou | 1890220xxxx |
|    2 | TOM   | Guangzhou | 1890220xxxx |
+------+-------+-----------+-------------+
2 rows in set (0.00 sec)

mysql> update tb_temp1 set name='John' where name='Smith';
Query OK, 1 row affected (0.08 sec)
Rows matched: 1  Changed: 1  Warnings: 0

mysql> select * from tb_temp1;
+------+------+-----------+-------------+
| id   | name | address   | phone       |
+------+------+-----------+-------------+
|    1 | John | Guangzhou | 1890220xxxx |
|    2 | TOM  | Guangzhou | 1890220xxxx |
+------+------+-----------+-------------+
2 rows in set (0.00 sec)

mysql> _
```

图8.38　更新表tb_temp1中指定的数据

如果更新的语句为：

```
update  tb_temp1 set name='John';
```

那么运行结果如图 8.39 所示。

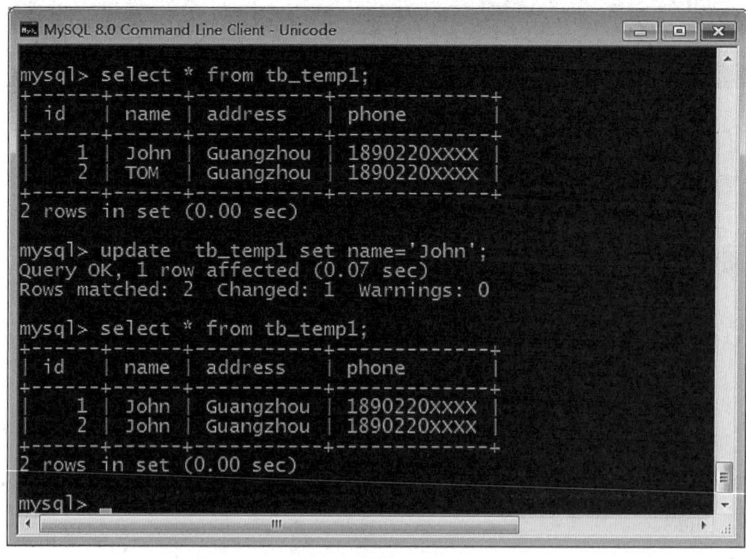

```
mysql> select * from tb_temp1;
+------+------+-----------+-------------+
| id   | name | address   | phone       |
+------+------+-----------+-------------+
|    1 | John | Guangzhou | 1890220xxxx |
|    2 | TOM  | Guangzhou | 1890220xxxx |
+------+------+-----------+-------------+
2 rows in set (0.00 sec)

mysql> update  tb_temp1 set name='John';
Query OK, 1 row affected (0.07 sec)
Rows matched: 2  Changed: 1  Warnings: 0

mysql> select * from tb_temp1;
+------+------+-----------+-------------+
| id   | name | address   | phone       |
+------+------+-----------+-------------+
|    1 | John | Guangzhou | 1890220xxxx |
|    2 | John | Guangzhou | 1890220xxxx |
+------+------+-----------+-------------+
2 rows in set (0.00 sec)

mysql> _
```

图8.39　更新表中列的所有数据

8.4.3　删除数据

在 MySQL 数据库中删除数据时，可以指定删除行数据或者全部数据，基本语法如下：

```
delete from table_name  [condition…]
```

table_name 为要删除的数据表，condition 是删除的条件。如果不指明条件，就删除表中的所有数据。

注意：删除表中的数据时注意备份数据，以便在误删数据后可以及时恢复。

【例 8.9】　删除数据表 tb_temp1 中的数据。

输入如下命令：

```
delete from tb_temp1 where id=1;
```

执行命令：

```
select * from tb_temp1;
```

查看是否删除成功，结果如图 8.40 所示。从结果可以看出，数据删除成功。

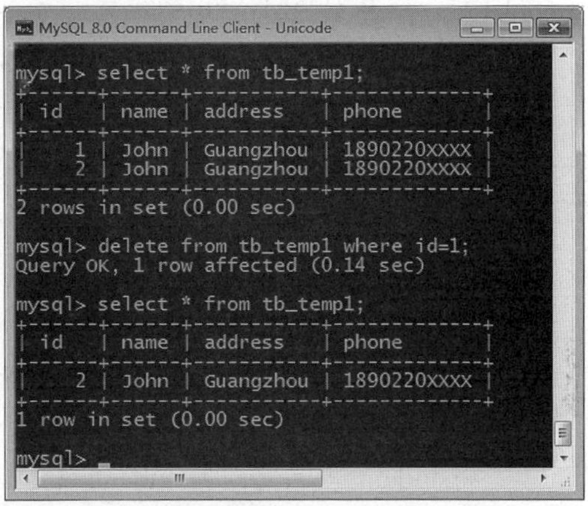

```
mysql> select * from tb_temp1;
+------+------+-----------+-------------+
| id   | name | address   | phone       |
+------+------+-----------+-------------+
|    1 | John | Guangzhou | 1890220XXXX |
|    2 | John | Guangzhou | 1890220XXXX |
+------+------+-----------+-------------+
2 rows in set (0.00 sec)

mysql> delete from tb_temp1 where id=1;
Query OK, 1 row affected (0.14 sec)

mysql> select * from tb_temp1;
+------+------+-----------+-------------+
| id   | name | address   | phone       |
+------+------+-----------+-------------+
|    2 | John | Guangzhou | 1890220XXXX |
+------+------+-----------+-------------+
1 row in set (0.00 sec)

mysql>
```

图8.40　删除表tb_temp1中的指定数据

执行命令：

```
delete from tb_temp1;
```

将表 tb_temp1 中所有的数据都删除，执行结果如图 8.41 所示。

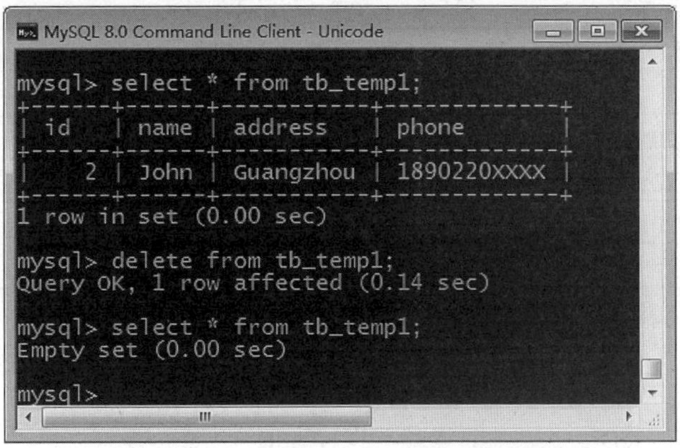

```
mysql> select * from tb_temp1;
+------+------+-----------+-------------+
| id   | name | address   | phone       |
+------+------+-----------+-------------+
|    2 | John | Guangzhou | 1890220XXXX |
+------+------+-----------+-------------+
1 row in set (0.00 sec)

mysql> delete from tb_temp1;
Query OK, 1 row affected (0.14 sec)

mysql> select * from tb_temp1;
Empty set (0.00 sec)

mysql>
```

图8.41　删除表tb_temp1中的所有数据

8.5　小　　结

本章主要介绍MySQL数据库的基本知识，包括如何安装、配置一个MySQL数据库实例，如何创建、删除数据库，如何为数据库添加表数据、更改表数据以及删除表数据。掌握操作MySQL的基本技能，是Web开发技术的一个必备要求。

8.6 习 题

（1）下载并安装MySQL数据库。

（2）配置MySQL为系统服务，在系统服务中，手动启动或者关闭MySQL服务。

（3）创建、删除一个自定义的数据库，例如Notice。

（4）创建数据库School，并在School库中创建数据表students。students的表结构如表8.1所示。按下面的要求进行操作：

① 创建数据库 School。

② 创建数据表students，在s_id字段上添加主键约束和自增约束。

③ 将s_name字段的数据类型改为varchar(30)。

④ 将s_contact字段名更改为s_phone。

⑤ 将students表名修改为students_info。

⑥ 删除字段s_city。

表 8.1 students 表结构

字段名	数据类型	主键	外键	非空	唯一	自增
s_id	int(11)	是	否	是	是	是
s_name	varchar(20)	否	否	否	否	否
s_contact	varchar(20)	否	否	否	否	否
s_city	varchar(20)	否	否	否	否	否
s_birth	datatime	否	否	否	否	否

第 9 章 JSP 与 Java Bean

开发出来的应用代码应该具有较高的可维护性，以方便后续的代码维护，而实现高可维护性的有效途径是要实现软件的低耦合、高内聚。软件设计分层的概念主要就是将软件各部分进行解耦合设计，对于JSP动态开发技术而言，Java Bean是最基础的分层技术。Bean是一种软件组件，在JSP开发中经常用来封装事务逻辑、数据库操作等。本章将介绍JSP中Bean的使用。

本章主要涉及的知识点有：

- Bean的基本概念
- 在JSP中如何使用Bean
- Bean的属性以及应用
- Bean的作用域

9.1 Java Bean 的基本概念

在开发软件过程中，应尽量将业务逻辑和表现层分开，从而达到完全解耦，这是软件分层设计的基本理念。在JSP中，经常利用Java Bean实现核心的业务逻辑，而JSP页面用于表现层。

Java Bean完全符合Java语言编码规范的要求和特性，形式上就是纯Java代码，它是可以重复使用的一个组件。在这种设计模式下，JSP页面只用于接收用户的输入以及显示处理之后的结果，因此不再需要在JSP页面中嵌入大量的Java代码，不但提高了系统的可维护性，而且方便开发人员的分工。

根据 Java 规范，Java Bean 具有以下特性：

- 支持反射机制：利用反射机制可以分析出Java Bean是如何运行的。
- 支持事件：事件是一种简单的通信机制，利用它可以将相应的信息通知给Java Bean。
- 支持属性：可以自定义属性，利用标准标签与JSP页面交互数据。
- 支持持久性：持久性是指可以将Java Bean进行保存，在需要的时候又可以重新载入。

下面以一个例子来说明Java Bean的创建以及要遵循的规范：

```
import java.io.Serializable;

public class User implements Serializable{
    private String username;                //用户名
    private String passwd;                  //密码
    private String sex;                     //性别
    private String address;                 //地址

    public User() {                         //无参数的构造方法
        super();

    }

    //以下是属性的get和set方法，但必须是public
    public String getUsername() {
        return username;
    }
    public void setUsername(String username) {
        this.username = username;
    }
    ...
}
```

以上就是新建一个Java Bean的基本结构，其要遵循的规范大致如下：

- Java Bean类必须是public类。
- 提供给JSP页面调用的方法必须赋予public访问权限。
- Java Bean类中的属性提供给JSP页面调用时必须提供public的get和set方法。
- 必须拥有不带参数的构造方法。

9.2 JSP中使用Bean

在JSP页面中，要正确使用Bean，应注意以下3个问题：

- 按照规范定义Bean类，并给出类属性的相应get和set方法。
- 在页面中要导入相应的Bean类。
- 在JSP页面中利用<jsp:useBean>标签使用Bean类。

1. 按照规范定义Bean类

定义Bean类与定义普通的Java类相同，如果在IntelliJ IDEA中，则在包中新增一个类就可以，软件会自动编译。如果是记事本或者其他编辑器，那么编辑完后需要手动进行编译。Bean类文件有两种部署方法：一种是将Bean的Class文件部署在Web服务器的公共目录中；另一种是将Bean的Class文件部署在Web服务器的特定项目的目录中，即将Bean类的Class文件部署在Web项目的Web-INF文件夹的classes目录内的指定文件夹下，例如ch09\Web-INF\classess\com\eshore\pojo\User.class。在实际项目开发中，后一种情况比较常见。

2. 在页面中要导入相应的 Bean 类，并用<jsp:useBean>标签获取 Bean 对象

在页面中使用标签 useBean，以便 JSP 页面访问，其语句形式如下：

```
<jsp:useBean
   id="beanInstanceName"
   scope="page | request | session | application"
   {
      class="package.class" |
      type="package.class" |
      class="package.class"  type="package.class" |
      beanName="{package.class | <%= expression %>}"
      type="package.class"
   }
</jsp:useBean>
```

语法中的属性说明如表 9.1 所示。

表 9.1 <jsp:useBean>标签属性说明

属 性 值	说 明
id	Bean 的变量名，可以在指定的范围中使用该变量名
class	Bean 的类路径，必须是严格的 package.class，不能指定其父类
scope	Bean 的有效范围，取值有 page、request、session、application
beanName	实例化的类名称或序列化模板的名称，指定的名称可以为其接口、父类
type	指定其父类或接口的类型，例如想实例化 HashMap，type 可以填为 Map

在使用<jsp:useBean>标签时需要注意以下几点：

- class或beanName不能同时存在；若Java Bean对象已存在，则class和beanName属性可以不指定，只需指定type属性。
- class可省去type独立使用，但beanName必须和type一起使用。
- class指定的类必须包含public、无参数的构造方法。
- class或beanName指定的类必须包括包名，而type可以省去包名，通过<%@page import=""%>指定所属包。

注意：class 通过 new 创建 Java Bean 对象，beanName 由 java.beans.Beans.instantiate 初始化 Java Bean 对象。

<jsp:useBean>标签中Bean的作用域有4个：page、request、session、application，其中page是默认作用域，可以省略。JSP引擎会根据作用域给用户分配不同的Bean，每个用户都拥有这些Bean，但不同用户间的Bean是相互独立的。

- 当指定的范围为request时，针对同一用户的不同请求，JSP引擎都会给用户分配不同的Bean对象。当响应结束时，取消分配的Bean。Bean的生命周期就是从客户请求开始到响应结束这段时间。

- 当指定的范围为page时，针对同一用户访问不同的页面，JSP引擎都会给用户分配不同的Bean对象。当用户进入当前页时，JSP引擎给用户分配一个Bean对象，当用户离开当前页时，取消分配的Bean对象。因此，该Bean的生命周期就位于当前页。
- 当指定的范围为session时，针对同一用户访问同一Web项目下的不同页面，JSP引擎都会给用户分配不同的Bean对象。当用户访问Web项目中某一目录下的页面时，JSP引擎给用户分配一个Bean对象，当用户离开Web目录时，取消分配的Bean对象。因此，该Bean的生命周期是Web项目的一个session时间。
- 当指定的范围为application时，JSP引擎为访问用户分配同一个Bean对象。它的生命周期就是该Web应用的存在时间，即Web应用在服务器中存在的时间。

9.3 访问Bean属性

上一节已经介绍过在JSP中使用Bean的方法以及规范，本节将介绍如何运用JSP标签设置和取得Bean的属性值。

9.3.1 设置属性：<jsp:setProperty>

使用标签<jsp:setProperty>可以设置Bean的属性值，其语法形式如下：

```
<jsp:setProperty
name="beanInstanceName"
{
    property= "*" |
    property="propertyName" [ param="parameterName" ] |
    property="propertyName" value="{string | <%= expression %>}"
}
/>
```

在使用该标签时，注意必须使用<jsp:useBean>标签创建一个Bean。<jsp:setProperty>标签的属性说明如表9.2所示。

表9.2 <jsp:setProperty>属性说明

属 性	说 明
name	Bean的名字，用来指定被使用的Bean，它的值必须是<jsp:useBean>标签中的id值
property	Bean的属性名，也就是Bean类中的属性，将值赋给Bean类的属性
param	表单参数名称
value	要设定的属性值

从语法形式中可以看出，<jsp:setProperty>标签有以下3种使用方式：

- 使用字符串或者表达式给Bean属性赋值，其要求是表达式的值与Bean属性的值类型相同。
- 使用表单参数形式给Bean属性赋值，其要求是表单中提供的参数名字与Bean属性的名字相同。这种形式不用具体指定每个Bean属性的名字，系统会自动根据表单的参数名字与Bean属性名字进行一一对应。该形式不用value属性，因此其语法形式为：

```
<jsp:setProperty property="" name="" />
```

- 使用表单参数值给Bean属性赋值，其要求是表单中提供的参数名字与<jsp:setProperty>标签中的param属性值名字相同，其语法形式为：

```
<jsp:setProperty property="Bean类中的属性名" param="表单参数名" name="Bean的变量名" />
```

下面以例子来分别说明3种设置Bean属性的方法。

【例 9.1】 直接利用表达式设置 Bean 属性值。

product1.jsp 页面直接利用表达式设置 Bean 属性值，源代码如下：

```
-----------------------product1.jsp--------------------------
01  <%@ page pageEncoding="UTF-8"%>
02  <!-- 导入引用的bean -->
03  <jsp:useBean id="product" class="com.eshore.pojo.Product"/>
04  <!DOCTYPE HTML>
05  <html>
06    <head>
07      <title>设置Bean属性</title>
08    </head>
09
10    <body>
11        <!-- 设置产品名称 -->
12        <jsp:setProperty name="product" property="product_name"
13            value="Struts开发教程"/>
14        <br/>产品名称是:
15        <!-- 获取产品名称 -->
16        <%=product.getProduct_name() %>
17        <!-- 设置产品编号 -->
18        <jsp:setProperty name="product" property="product_id"
value="111100123689"/>
19        <br/>产品编号是:
20        <!-- 获取产品编号 -->
21        <%=product.getProduct_id() %>
22        <!-- 设置产品价格 -->
23        <%
24            double price = 68.23;
25         %>
26        <jsp:setProperty name="product" property="price" value=
"<%=price+23.67 %>"/>
27        <br/>产品价格是:
28        <!-- 获取产品价格 -->
29        <%=product.getPrice() %>
30        <!-- 设置产品信息 -->
31        <jsp:setProperty name="product" property="info"
32            value="Struts开发教程是一本介绍如何使用Struts的专业图书......"/>
33        <br/>产品信息是:
34        <!-- 获取产品信息 -->
35        <%=product.getInfo() %>
36    </body>
37  </html>
```

在上述代码中只是简单设置一下Bean值，第03行代码是导入引用的Bean类，第11~35行代码分别设定Bean属性值和获取Bean属性值。Product类的源代码如下：

```
----------------------- Product.java--------------------------
package com.eshore.pojo;
import java.io.Serializable;

public class Product implements Serializable{

    private static final long serialVersionUID = 1L;
    private String product_id;          //产品号
    private String product_name;        //产品名称
    private double price;               //产品价格
    private String info;                //产品信息
    public Product() {                  //无参数的构造方法
        super();
    }
    ...//省略get和set方法
}
```

在上述代码中，分别设定 Bean 类中的属性成员及其 get 方法和 set 方法。成员变量最好为 private，这是编码的良好习惯。页面运行结果如图 9.1 所示。

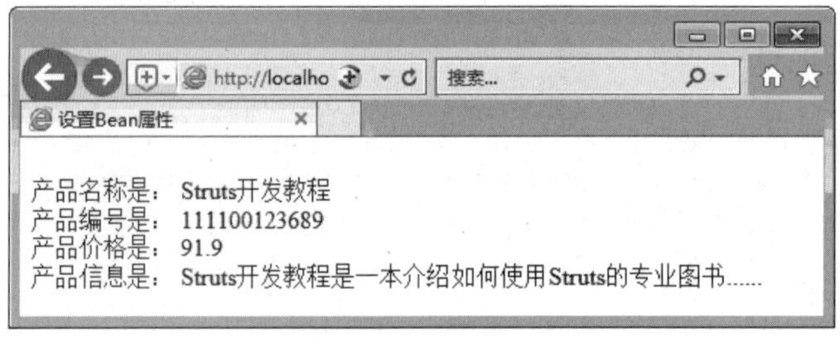

图9.1 product1.jsp运行效果图

【例 9.2】 通过表单参数名设置 Bean 属性值。

通过表单参数设置Bean属性值，不用设置value值，JSP引擎会自动根据表单中的参数名与Bean属性名对应，并转换为对应的数据类型。

product2.jsp 页面通过表单参数设置 Bean 属性值，源代码如下：

```
-----------------------product2.jsp--------------------------
<%@ page pageEncoding="UTF-8"%>

<!-- 设定参数编码 -->
<%
    request.setCharacterEncoding("UTF-8");
 %>
<!-- 导入引用的Bean -->
<jsp:useBean id="product" class="com.eshore.pojo.Product"/>
<!DOCTYPE HTML>
<html>
  <head>
    <title>通过表单参数设置Bean属性值</title>
    <meta http-equiv="Content-Type" content="text/html;charset=utf-8">
  </head>

  <body>
     <form action="" method="post">
```

```
18            <br>
19            输入产品名称：<input name="product_name"/><br/>
20            输入产品编号：<input name="product_id"/><br/>
21            输入产品价格：<input name="price"/><br/>
22            输入产品信息：<input name="info"/><br/>
23            <input type="submit" value="提交"/>
24        </form>
25        <!-- 设定product的属性值 -->
26        <jsp:setProperty property="*" name="product"/>
27        <br/>产品名称是：
28        <!-- 获取产品名称 -->
29        <%=product.getProduct_name() %>
30        <br/>产品编号是：
31        <!-- 获取产品编号 -->
32        <%=product.getProduct_id() %>
33        <br/>产品价格是：
34        <!-- 获取产品价格 -->
35        <%=product.getPrice() %>
36        <br/>产品信息是：
37        <!-- 获取产品信息 -->
38        <%=product.getInfo() %>
39    </body>
40 </html>
```

在上述代码中，第 04~06 行代码设定参数编码；第 17~24 行代码设定 form 架构；第 26 行代码设置 Bean 属性值，但是没有设定具体的值；第 27~38 行代码获取 form 中设定的 Bean 值，Bean 类不变，继续使用例 9.1 中的 Product 类。页面运行效果如图 9.2 所示。

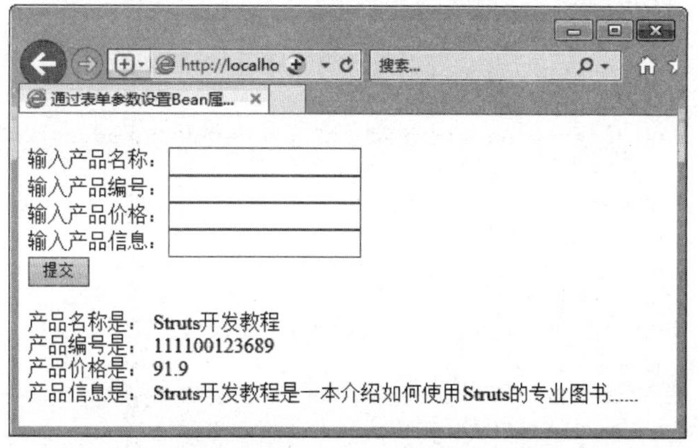

图9.2 product2.jsp运行效果图

注意：如果在页面中输入中文，那么提交后会显示为乱码，因此应该设定中文编码，有关参数编码问题请参见第 3.1.3 节的说明。

【例 9.3】 通过表单参数值设置 Bean 属性值。

通过表单参数值和使用param属性设置Bean属性值，与第2种方式相似。在第2种方式中，表单中的name名字就是Bean名字；在第3种方式中，表单中的name名字与Bean名字不同，在标签设置时，可利用param属性引用。

product3.jsp 页面通过表单参数值设置 Bean 属性值，源代码如下：

```
-----------------------product3.jsp--------------------------
01  <%@ page pageEncoding="UTF-8"%>
02
03  <!-- 设定参数编码 -->
04  <%
05     request.setCharacterEncoding("UTF-8");
06   %>
07  <!-- 导入引用的Bean -->
08  <jsp:useBean id="product" class="com.eshore.pojo.Product"/>
09  <!DOCTYPE HTML>
10  <html>
11    <head>
12      <title>通过表单参数值设置Bean属性值</title>
13      <meta http-equiv="Content-Type" content="text/html;charset=utf-8">
14    </head>
15
16    <body>
17        <form action="" method="post">
18            <br>
19            输入产品名称：<input name="product_name1"/><br/>
20            输入产品编号：<input name="product_id1"/><br/>
21            输入产品价格：<input name="price1"/><br/>
22            输入产品信息：<input name="info1"/><br/>
23            <input type="submit" value="提交"/>
24        </form>
25        <!-- 设置产品名称 -->
26        <jsp:setProperty name="product" property="product_name"
27            param="product_name1"/>
28        <br/>产品名称是：
29        <!-- 获取产品名称 -->
30        <%=product.getProduct_name() %>
31        <!-- 设置产品编号 -->
32        <jsp:setProperty name="product" property="product_id" param=
"product_id1"/>
33        <br/>产品编号是：
34        <!-- 获取产品编号 -->
35        <%=product.getProduct_id() %>
36        <!-- 设置产品价格 -->
37        <jsp:setProperty name="product" property="price" param="price1"/>
38        <br/>产品价格是：
39        <!-- 获取产品价格 -->
40        <%=product.getPrice() %>
41        <!-- 设置产品信息 -->
42        <jsp:setProperty name="product" property="info" param="info1"/>
43        <br/>产品信息是：
44        <!-- 获取产品信息 -->
45        <%=product.getInfo() %>
46    </body>
47  </html>
```

上述代码与例 9.2 的代码的主要区别在于第 25~45 行，即应用 param 参数引用表单中的参数名。代码运行效果如图 9.3 所示。

从图9.3可以看出，效果与图9.2所示是一样的。

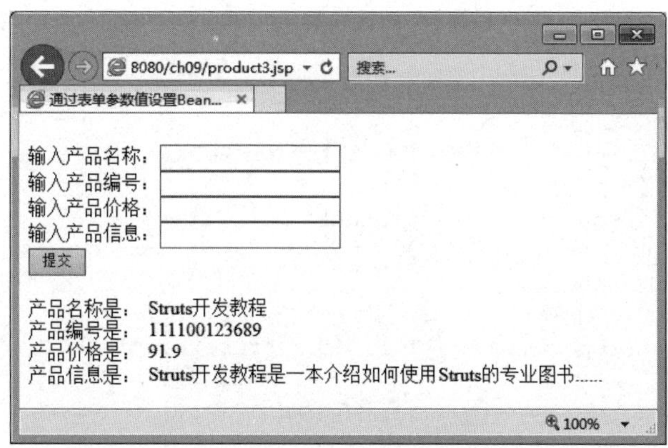

图9.3 product3.jsp运行效果图

9.3.2 取得属性：<jsp:getProperty>

<jsp:getProperty>标签用来获得 Bean 属性值，并且可以显示在浏览器中。该标签必须和<jsp:useBean>标签一起使用。其语法形式如下：

```
<jsp:getProperty  name="beanInstanceName" property="propertyName"/>
```

在上述语法中，name值是指Bean的名字，用来指定被使用的Bean，它的值必须是<jsp:useBean>标签中的id值。property是指Bean的属性名，也就是Bean类中的属性，将值赋给Bean类的属性。应用<jsp:getProperty>标签获取Bean属性值，将例9.1的代码修改如下：

```
<%@ page pageEncoding="UTF-8"%>
<!-- 导入引用的Bean -->
<jsp:useBean id="product" class="com.eshore.pojo.Product"/>
<!DOCTYPE HTML>
<html>
  <head>
    <title>设置Bean属性</title>
  </head>

  <body>
      <!-- 设置产品名称 -->
      <jsp:setProperty name="product" property="product_name"
          value="Struts开发教程"/>
      <br/>产品名称是：
      <!-- 获取产品名称 -->
      <jsp:getProperty  name="product" property="product_name"/>
      <%=product.getProduct_name() %>
      <!-- 设置产品编号 -->
      <jsp:setProperty name="product" property="product_id"
value="111100123689"/>
      <br/>产品编号是：
      <!-- 获取产品编号 -->
      <jsp:getProperty  name="product" property="product_id"/>
      <!-- 设置产品价格 -->
      <%
         double price = 68.23;
       %>
```

```
27        <jsp:setProperty name="product" property="price"  value=
"<%=price+23.67 %>"/>
28        <br/>产品价格是：
29        <!-- 获取产品价格 -->
30        <jsp:getProperty  name="product" property="price"/>
31        <!-- 设置产品信息 -->
32        <jsp:setProperty name="product" property="info"
33           value="Struts开发教程是一本介绍如何使用Struts的专业图书......"/>
34        <br/>产品信息是：
35        <!-- 获取产品信息 -->
36        <jsp:getProperty  name="product" property="info"/>
37     </body>
38   </html>
```

上述代码与例 9.1 的代码的主要区别在于第 16、27、30、36 行，其余均相同。运行结果如图 9.1 所示。

9.4 Bean的作用域

前面讲过Bean的作用域有4个，分别是page、request、session和application。Bean的作用范围是由标签中的scope属性指定的，默认是page范围，即该Bean在当前页有效。设置为request时，表示该Bean对当前用户的当前请求有效。设置为session时，表示该Bean在当前用户的session范围内有效。设置为application时，表示该Bean对该系统的所有页面有效。scope属性决定了在使用<jsp:useBean>标签时是否要重新创建新的对象。如果某个Bean在其有效的范围内又出现一个id和scope都相同的Bean，那么就可以重用已经被实例化的Bean，而不是重新创建。

【例 9.4】 演示 Bean 的 request 生命周期。

在 requestScope.jsp 页面中，先提交圆半径，然后用 reqest.getParameter()获取半径，并计算圆周长和圆面积，默认的圆半径为 1。源代码如下：

```
------------------------ requestScope.jsp--------------------------
<%@ page pageEncoding="UTF-8"%>
<!-- 导入引用的Bean -->
<jsp:useBean id="circle" class="com.eshore.pojo.Circle" scope="request"/>
<!DOCTYPE HTML>
<html>
  <head>
    <title>测试scope为request</title>
  </head>
  <%
    //获取圆半径，如果没有，则默认为1
     String radius = request.getParameter("radius");
     if(radius==null||radius.equals("")){
          radius = "1";
     }
     double rad = Double.parseDouble(radius);
     //设置圆半径
     circle.setRadius(rad);
   %>
```

```
19    <body>
20        <form action="" method="post">
21            请输入圆的半径：<input name="radius"/><br/>
22            <input type="submit" value="提交"/><br/>
23            该Bean类对象为：<%=circle.toString()%><br/>
24            <br/>圆的半径为：<jsp:getProperty property="radius" name="circle"/>
25            <br/>圆的周长为：<jsp:getProperty property="circumference" name="circle"/>
26            <br/>圆的面积为：<jsp:getProperty property="circleArea" name="circle"/>
27        </form>
28    </body>
29 </html>
```

在上述代码中，第09~18行用于获取页面获得的半径值，并转换为Double格式；第20~27行利用form提交表单数据和获取Bean的相关值，其中第23行输出Circle类的对象值。Circle类的源代码如下：

```
-----------------------Circle.java--------------------------
package com.eshore.pojo;
import java.io.Serializable;
public class Circle implements Serializable{
    private static final long serialVersionUID = 1L;
    private double radius = 1.0d;                          //半径
    private double circleArea = 0.0d;                      //圆面积
    private double circumference=0.0d;                     //圆周长
    public Circle() {                                      //无参数的构造方法
        super();

    }
    //属性的get和set方法
    public double getRadius() {
        return radius;
    }
    public void setRadius(double radius) {
        this.radius = radius;
    }
    public double getCircleArea() {
        circleArea = Math.PI*radius*radius;                //设置圆面积
        return circleArea;
    }
    public void setCircleArea(double circleArea) {
        this.circleArea = circleArea;
    }
    public double getCircumference() {
        circumference = 2*Math.PI*radius;                  //设置圆周长
        return circumference;
    }
    public void setCircumference(double circumference) {
        this.circumference = circumference;
    }
}
```

页面运行效果如图 9.4 所示。

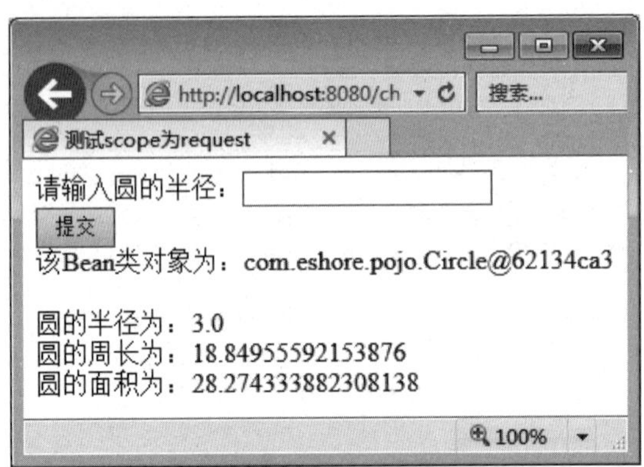

图9.4 requestScope.jsp页面运行效果图

从图9.4中可以看出，每次提交显示出的Circle类都是不同的。

【例 9.5】 演示 Bean 的 page 生命周期。

在 pageScope.jsp 页面中，先设置圆的半径并计算圆周长和圆面积，然后跳转到 pageScope2.jsp 页面，在 pageScope2.jsp 中设定 Circle 类的范围与 pageScope.jsp 的范围相同。可以发现 Bean 的有效范围就是在当前页有效，在 pageScope2.jsp 中获取的是默认值。它们的源代码分别为：

```
------------------------pageScope.jsp---------------------------
01  <%@ page pageEncoding="UTF-8"%>
02  <!-- 导入引用的Bean -->
03  <jsp:useBean id="circle2" class="com.eshore.pojo.Circle"/>
04  <!DOCTYPE HTML>
05  <html>
06    <head>
07      <title>测试scope为page</title>
08    </head>
09    <body>
10          pageScope.jsp页面信息：<br/>
11            该Bean类对象为：<%=circle2.toString()%><br/>
12            设置该Bean的半径为20：
13          <%
14            circle2.setRadius(20);
15          %>
16          <!-- 获取Bean的属性值 -->
17         <br/>圆的半径为：<jsp:getProperty property="radius" name="circle2"/>
18         <br/>圆的周长为：<jsp:getProperty property="circumference" name="circle2"/>
19         <br/>圆的面积为：<jsp:getProperty property="circleArea" name="circle2"/>
20         <!-- 跳转pageScope2.jsp页面 -->
21         <form action="pageScope2.jsp" method="get">
22                <input type="submit" value="跳转pageScope2.jsp页面"/><br/>
23         </form>
24    </body>
25  </html>
```

在上述代码中，第13~15行代码设置Bean的半径值，第16~19行代码获得Bean的各属性值，第20~23行代码用于提交form并跳转到pageScope2.jsp页面。pageScope2.jsp的源代码如下：

```
------------------------ pageScope2.jsp---------------------------
01  <%@ page pageEncoding="UTF-8"%>
02  <!-- 导入引用的Bean -->
03  <jsp:useBean id="circle2" class="com.eshore.pojo.Circle"/>
04  <!DOCTYPE HTML>
05  <html>
06    <head>
07      <title>测试scope为page</title>
08    </head>
09    <body>
10         pageScope2.jsp页面信息：<br/>
11            该Bean类对象为：<%=circle2.toString()%><br/>
12         <br/>圆的半径为：<jsp:getProperty property="radius" name="circle2"/>
13         <br/>圆的周长为：<jsp:getProperty property="circumference" name="circle2"/>
14         <br/>圆的面积为：<jsp:getProperty property="circleArea" name="circle2"/>
15    </body>
16  </html>
```

在上述代码中，第 3 行代码引用的 Bean 的 scope 与 pageScope.jsp 中的相同且 id 值也相同，第 12~14 行代码引用 Bean 的属性值。页面运行效果如图 9.5 所示。

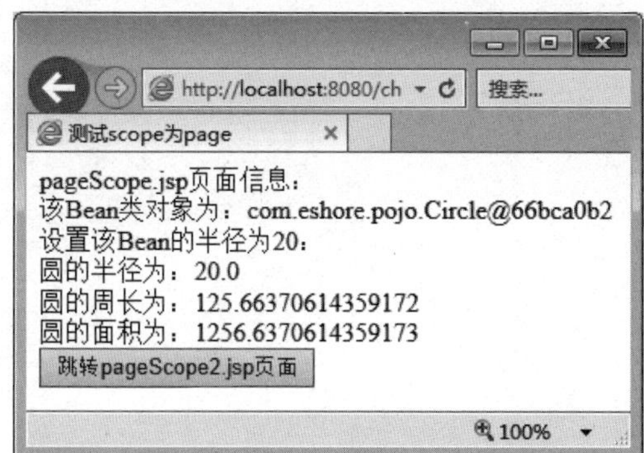

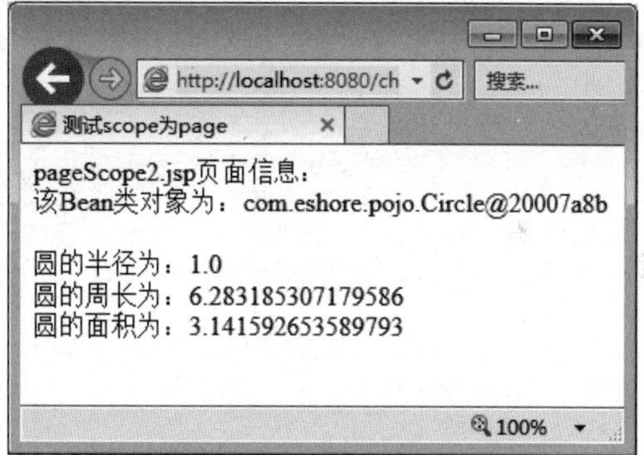

图9.5 pageScope.jsp和pageScope2.jsp运行效果图

从图9.5中可以看出，Bean类的作用域位于同一页面，若跳转到其他页面则失效。

【例 9.6】 演示 Bean 的 session 生命周期。

将例 9.5 中 pageScope.jsp 的源代码更改为如下形式：

```
------------------------ sessionScope.jsp---------------------------
<%@ page pageEncoding="UTF-8"%>
<!-- 导入引用的Bean -->
<jsp:useBean id="circle2" class="com.eshore.pojo.Circle" scope="session"/>
<!DOCTYPE HTML>
<html>
  <head>
    <title>测试scope为session</title>
  </head>
  <body>
       sessionScope.jsp页面信息：<br/>
         该Bean类对象为：<%=circle2.toString()%><br/>
         设置该Bean的半径为20：
       <%
```

```
              circle2.setRadius(20);
           %>
           <!-- 获取Bean的属性值 -->
          <br/>圆的半径为：<jsp:getProperty property="radius" name="circle2"/>
          <br/>圆的周长为：<jsp:getProperty property="circumference" name="circle2"/>
          <br/>圆的面积为：<jsp:getProperty property="circleArea" name="circle2"/>
          <!-- 跳转sessionScope2.jsp页面 -->
          <form action="sessionScope2.jsp" method="get">
               <input type="submit" value="跳转sessionScope2.jsp页面"/><br/>
          </form>
    </body>
  </html>
```

将pageScope2.jsp中的源代码更改为如下形式：

```
----------------------- sessionScope2.jsp--------------------------
  <%@ page pageEncoding="UTF-8"%>
  <!-- 导入引用的Bean -->
  <jsp:useBean id="circle2" class="com.eshore.pojo.Circle" scope="session"/>
  <!DOCTYPE HTML>
  <html>
    <head>
      <title>测试scope为session</title>
    </head>
    <body>
           sessionScope2.jsp页面信息：<br/>
             该Bean类对象为：<%=circle2.toString()%><br/>
          <br/>圆的半径为：<jsp:getProperty property="radius" name="circle2"/>
          <br/>圆的周长为：<jsp:getProperty property="circumference" name="circle2"/>
          <br/>圆的面积为：<jsp:getProperty property="circleArea" name="circle2"/>
    </body>
  </html>
```

在上述两段代码中，主要更改的是 Bean 的范围以及跳转的页面。运行效果如图 9.6 所示。

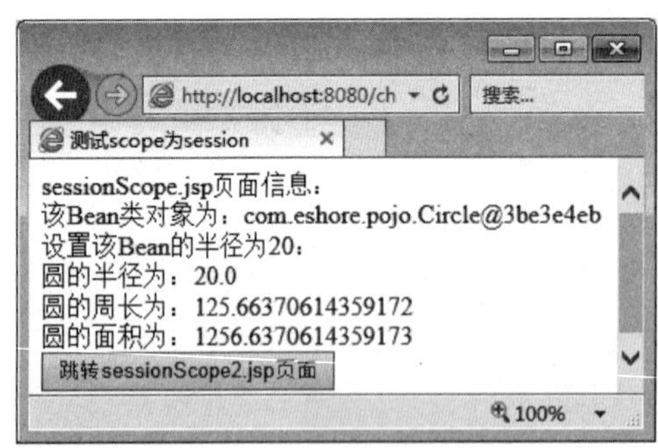

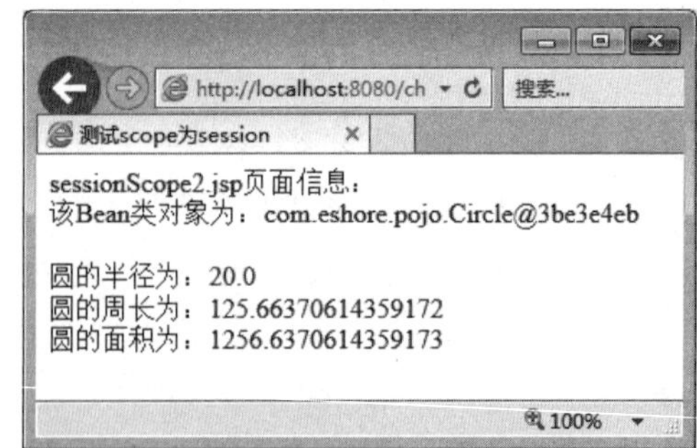

图9.6 sessionScope.jsp和sessionScope2.jsp运行效果图

从图9.6中可以看出，引用的Bean是同一个对象，所以取得的属性值也是一样的。

【例 9.7】 演示 Bean 的 application 生命周期。

在 applicationScope.jsp 页面中设置 scope 的范围为 application，先获取系统默认的半径，然后更改半径值，刷新页面，就能发现半径更改了。applicationScope.jsp 的源代码如下：

```
-----------------------applicationScope.jsp-------------------------
<%@ page pageEncoding="UTF-8"%>
<!-- 导入引用的Bean -->
<jsp:useBean id="circle2" class="com.eshore.pojo.Circle" scope="application"/>
<!DOCTYPE HTML>
<html>
  <head>
    <title>测试scope为application</title>
  </head>
  <body>
        applicationScope.jsp页面信息：<br/>
          application访问Bean类对象为：<%=circle2.toString()%><br/>
          获取Bean的半径：<jsp:getProperty property="radius" name="circle2"/><br/>
          设置该Bean的半径为30：
        <%
           circle2.setRadius(30);
        %>
        <!-- 获取Bean的属性值 -->
       <br/>圆的半径为：<jsp:getProperty property="radius" name="circle2"/>
       <br/>圆的周长为：<jsp:getProperty property="circumference" name="circle2"/>
       <br/>圆的面积为：<jsp:getProperty property="circleArea" name="circle2"/>
  </body>
</html>
```

上述代码的运行结果如图 9.7 所示。

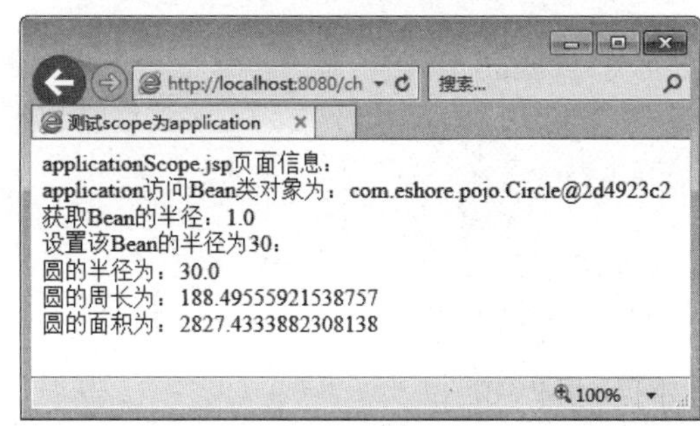

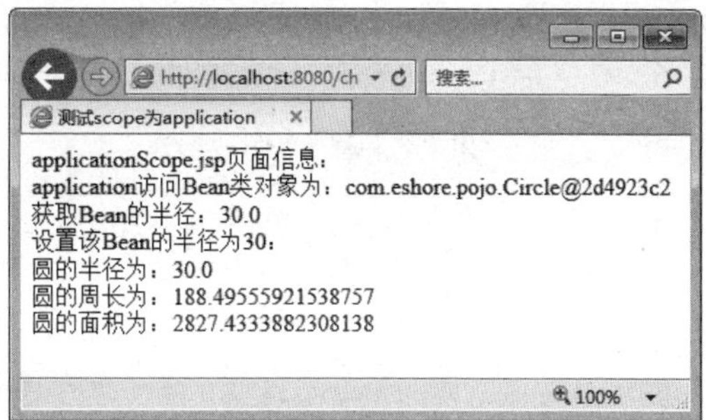

图9.7　applicationScope.jsp页面运行效果

从上述例子可以发现，在同一范围内的Bean对象，引用的是同一对象。

9.5　用户登录验证

在前面的章节中曾经介绍过用户登录验证，本节将介绍如何利用Java Bean的方式进行登录验证。用户在表单中填入用户名与密码，如果用户存在，就跳转到欢迎界面，否则显示错误信息。

在本程序中Java Bean是极其重要的一部分，不仅要完成数据的校验，还要进行错误信息的显示。本例为了简单，所有的错误信息均存放在Map容器中。

【例 9.8】 利用 Java Bean 进行用户登录验证。

User 类包含基础的用户名、密码属性，并用 Map 来保存错误信息。源代码如下：

```
-----------------------User.java--------------------------
package com.eshore.pojo;

import java.io.Serializable;
import java.util.HashMap;
import java.util.Map;

public class User implements Serializable{

    private String username="";                     //用户名
    private String passwd="";                       //密码
    Map<String,String> userMap = null;              //存放用户
    Map<String,String> errorsMap = null;            //存放错误信息

    public User() {                                 //无参数的构造方法
        super();
        this.username = "";
        this.passwd="";
        userMap = new HashMap<String,String>();
        errorsMap = new HashMap<String,String>();
        //添加用户，模拟从数据库中查询出的数据
        userMap.put("zhangsan", "123zs");
        userMap.put("lisi", "1234zs");
        userMap.put("wangwu", "1234ww");
        userMap.put("zhaoqi", "1234zq");
        userMap.put("zhengliu", "1234zl");
        // TODO Auto-generated constructor stub
    }
    //用户名和密码等数据验证
    public boolean isValidate(){
        boolean flag = true;
        //用户名验证
        if(!this.userMap.containsKey(this.username)){
            flag = false;
            errorsMap.put("username", "该用户不存在！");
            this.username = "";
        }
        //根据用户名进行密码验证
        String password = this.userMap.get(this.username);
        if(password==null||!password.equals(this.passwd)){
            flag = false;
            this.passwd = "";
            errorsMap.put("passwd", "密码错误，请输入正确密码！");
            this.username = "";
        }
        return flag;
    }
    //获取错误信息
    public String getErrors(String key){
        String errorV = this.errorsMap.get(key);
        return errorV==null?"":errorV;
    }
```

```
52        //以下是属性的get和set方法，但必须是public
53        public String getUsername() {
54            return username;
55        }
56
57        public void setUsername(String username) {
58            this.username = username;
59        }
60
61        public String getPasswd() {
62            return passwd;
63        }
64        public void setPasswd(String passwd) {
65            this.passwd = passwd;
66        }
67    }
```

在上述代码中，第 21~25 行代码利用 Map 保存用户数据；第 29~46 行代码验证页面输入的数据，如果验证失败，就将错误信息保存在错误 Map 中；第 48~51 行代码根据错误 Key 获取错误信息。

在登录页面 login.jsp 中引用 User 类，并用表单提交的方式设定 User 属性值，源代码如下：

```
-----------------------login.jsp--------------------------
<%@ page pageEncoding="UTF-8"%>
<jsp:useBean id="user" class="com.eshore.pojo.User" scope="session"/>
<!DOCTYPE HTML>
<html>
  <head>
    <title>用户登录</title>
  </head>

  <body>
        <p>用户登录</p>
        <!-- 用form表单提交用户名与密码 -->
        <form action="check.jsp" method="post">
            <table border="1" width="250px;">
                <tr>
                    <td width="75px;">用户名：</td>
                    <td ><input name="username" value="<jsp:getProperty
                        name="user" property="username"/>"/>
                    <!-- 用户错误信息 --><span style="color:red">
 <%=user.getErrors("username")%></span> <br/></td>
                </tr>
                <tr>
                    <td width="75px;">密  码：</td>
                    <td ><input type="password" name="passwd"
               value="<jsp:getProperty name="user" property="passwd"/> "/>
                    <!-- 密码错误信息 --><span style="color:red">
  <%=user.getErrors("passwd") %></span><br/> </td>
                </tr>
                <tr>
                    <td colspan="2">
                        <input type="submit" value="提交"/>  
                        <input type="reset" value="重置"/>
                    </td>
```

```
33              </tr>
34            </table>
35          </form>
36      </body>
37    </html>
```

login.jsp 页面的主要功能是显示表单和错误信息，例如代码中的第 23 行和第 30 行所示。第 02 行代码用于定义 User 的有效范围为 session。

在检验页面 check.jsp 中，同样定义一个范围为 session 的 User，对调用类的验证方法进行判断，源代码如下：

```
-----------------------check.jsp--------------------------
01  <%@ page pageEncoding="UTF-8"%>
02  <jsp:useBean id="user" class="com.eshore.pojo.User" scope="session"/>
03  <!DOCTYPE HTML>
04  <html>
05    <head>
06      <title>验证用户</title>
07    </head>
08    <body>
09      <!-- 设置user属性，判断是否合法
10          合法则跳转成功，否则跳转到登录页面-->
11      <jsp:setProperty property="*" name="user"/>
12      <%
13          if(user.isValidate()){
14      %>
15      <jsp:forward page="success.jsp"/>
16      <%
17          }else{
18      %>
19      <jsp:forward page="login.jsp"/>
20      <% } %>
21    </body>
22  </html>
```

在上述代码中，第 11 行代码设置 User 的属性值；第 12~20 行代码判断用户的合法性，如果验证通过，就跳转到 success.jsp 页面，如果失败，就跳转到 login.jsp 页面提示错误信息。

success.jsp 页面用于显示欢迎信息，源代码如下：

```
-----------------------success.jsp--------------------------
<%@ page pageEncoding="UTF-8"%>
<% request.setCharacterEncoding("UTF-8"); %>
<jsp:useBean id="user" class="com.eshore.pojo.User" scope="session"/>
<!DOCTYPE HTML>
<html>
  <head>
    <title>登录成功</title>
  </head>
  <body>
      <div style="text-align: center;">
          <h4>欢迎您：
              <SPAN style="color: red">
               <jsp:getProperty property="username" name="user"/>
              </SPAN>用户！
```

```
15            </h4>
16         </div>
17      </body>
18   </html>
```

在上述代码中，第 14 行用于显示登录成功的用户。本例子的运行结果如图 9.8 和图 9.9 所示。

图9.8　登录页面和错误提示页面

图9.9　用户登录验证页面

注意：在实际开发中，应该灵活运用 Java Bean 进行开发，根据具体业务的需要，尽量将代码解耦合。

9.6　DAO 设计模式

在前面章节中已介绍了Bean属性的概念以及Java Bean的用法，若想提高开发效率，实现模块化开发，又该如何做呢？DAO设计模式应运而生，虽然现在有很多成熟的开发框架，例如Spring MVC、Struts等框架，但它仍然是一个值得大家学习的设计模式。本节将详细介绍该设计模式，使得读者对DAO设计模式有一个整体的框架概念，能编写出适合自己的简易框架。

9.6.1　DAO设计模式简介

信息系统的开发架构如图9.10所示。

图9.10　信息系统开发架构图

各层级的介绍如下：

- 客户层：实际上就是客户端浏览器。
- 显示层：利用JSP和Servlet进行页面显示。
- 业务层：对数据层的原子性DAO操作进行整合。
- 数据层：对数据库进行原子操作，例如增加、删除、修改等。
- 数据库：顾名思义就是保存数据库的信息。

DAO是Data Access Object的简称，主要是对数据进行操作，对应上面的层级就是数据层。在数据操作过程中，主要是以面向接口编程为主。一般将DAO划分为以下几个部分。

- VO（Value Object）：一个用于存放网页的数据，比如网页要显示一条用户的信息，则这个类就是用户类，主要由属性以及属性的setter和getter方法组成，VO类中的成员变量与表中的字段是相对应的。
- DatabaseConnection：用于打开和关闭数据库操作的类。
- DAO接口：用于声明数据库的操作，定义对数据库的原子性操作，例如增加、修改、删除等。
- DAOImpl：实现DAO接口的类，但是不负责数据库的打开和关闭。
- DAOProxy：也是实现DAO接口，主要完成数据库的打开和关闭。
- DAOFactory：工厂类，通过getInstance()取得DAO的实例化对象。

9.6.2 DAO命名规则

在开发过程中，命名规范非常重要。命名必须具有可读性、可维护性。DAO 命名规则如下：

- DAO命名为XxxDao，有的开发人员喜欢在它前面加一个I表示是接口类，例如UserDao或者IUserDao。
- DAOImpl命名为XxxDaoImpl，表示是接口实现类，例如UserDaoImpl。
- DAOProxy命名为XxxDaoProxy或者XxxService，例如UserDaoProxy或者UserService。
- DAOFactory命名为XxxFactory，例如UserDaoFactory。
- VO的命名与表名一致，VO中的属性与表字段一致。

9.6.3 DAO开发

DAO 开发是指对数据的增、删、改、查操作。在 MySQL 中新建一个产品表 product，建表脚本如下：

```
DROP TABLE IF EXISTS 'product';
CREATE TABLE 'product' (
  'product_id' varchar(20) NOT NULL,
  'product_name' varchar(50) DEFAULT NULL,
  'price' decimal(6,2) DEFAULT NULL,
  'info' varchar(100) DEFAULT NULL,
  PRIMARY KEY ('product_id')
) ENGINE=InnoDB DEFAULT CHARSET=utf8;
```

【例 9.9】 VO 类利用例 9.1 中的 Product 类来演示 DAO 开发过程。

根据上述内容，分别新建 ProductDao、ProductDaoImpl、ProductService、DaoFactory、DBConnection类。目录结构如图9.11所示。

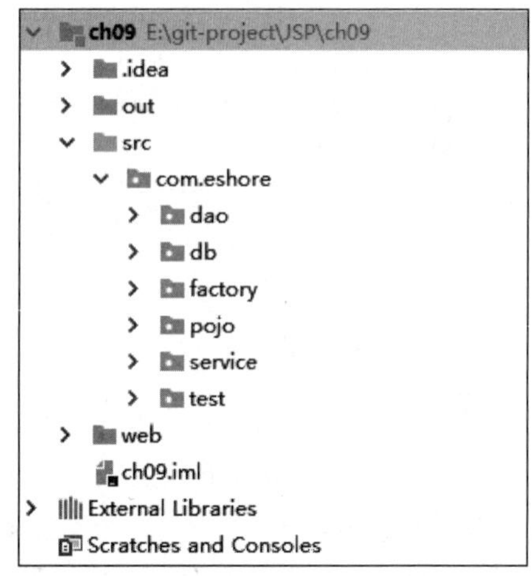

图9.11 DAO目录结构

首先，新建 DBConnection 类，源代码如下：

```
------------------------DBConnection.java--------------------------
01  package com.eshore.db;
02
03  import java.sql.Connection;
04  import java.sql.DriverManager;
05
06  public class DBConnection {
07      private static final String Driver = "com.mysql.cj.jdbc.Driver";
08      private static final String URL = "jdbc:mysql://localhost:3306/testweb";
09      private static final String USER = "root";
10      private static final String PASSWORD = "root";
11      private Connection conn = null;
12
13      public DBConnection() throws Exception {          // 进行数据库连接
14          try {
15              Class.forName(Driver);                    // 用反射加载数据库驱动
16              this.conn = DriverManager.getConnection(URL, USER, PASSWORD);
17          } catch (Exception e) {
18              throw e;                                  // 抛出异常
19          }
20      }
21      public Connection getConnection() {
22          return this.conn;                             // 取得数据库的连接
23      }
24      public void close() throws Exception {           // 关闭数据库
25          if (this.conn != null) {
26              try {
27                  this.conn.close();
28              } catch (Exception e) {
29                  throw e;
30              }
31          }
32      }
33  }
```

在上述代码中，主要是执行对数据库的连接配置。第 21 行代码用于获取当前的数据库连接。

注意：有关数据库的操作将在第 13 章进行详细讲解。

接着，新建 DAO 接口类 ProductDao。接口类在 DAO 设计模式中的地位是极其重要的。在定义接口类之前，要分析业务的需求，分析清楚系统需要哪些功能、方法。在本例中只是完成新增、查询等简单功能。ProductDao 接口类的源代码如下：

```
------------------------ProductDao.java--------------------------
package com.eshore.dao;
import java.util.List;
import com.eshore.pojo.Product;
public interface ProductDao {
    /**
    * 数据库 新增数据
    *@param product 要增加的数据对象
    *@return 是否增加成功的标记
    *@throws Exception如果有异常，将直接抛出
    */
   public boolean addProduct(Product product)throws Exception ;
```

```
    /**
    * 查询全部的Product数据
    *@param product_name 产品名称
    *@return 返回全部的查询结果，每一个product对象表示表的一行记录
    *@throws Exception如果有异常，将直接抛出
    */
   public List<Product> findAll(String product_name)throws Exception;
    /**
    * 根据产品编号查询产品
    *@param  product_id 产品编号
    *@return  产品的vo对象
    *@throws Exception如果有异常，将直接抛出
    */
   public Product findByProductId(String product_id)throws Exception;
}
```

在上述 DAO 的接口类中，定义了 addProduct()、findAll()、findByProductId()共 3 个功能。addProduct()方法执行数据库的插入操作；findAll()方法主要完成数据的查询操作，利用 List 保存返回的数据；findByProductId()方法根据产品编号返回单个 Product 对象。

在DAO接口定义完成后需要定义其实现类，为了降低耦合度，其实现类有两种：一种只是数据操作实现类，另一种是业务操作实现类。

数据操作实现类 ProductDaoImpl 主要是负责具体的数据库操作，源代码如下：

```
----------------------ProductDaoImpl.java--------------------------
package com.eshore.dao;

import java.sql.Connection;
import java.sql.PreparedStatement;
import java.sql.ResultSet;
import java.util.ArrayList;
import java.util.List;

import com.eshore.pojo.Product;

public class ProductDaoImpl implements ProductDao {

    private Connection conn = null;                         // 数据库连接对象
    private PreparedStatement pstmt = null;                 // 数据库操作对象
    // 通过构造方法取得数据库连接
    public ProductDaoImpl(Connection conn) {
        this.conn = conn;
    }

    public boolean addProduct(Product product) throws Exception {
        boolean flag = false;                                   // 定义标识
        String sql = "insert into product(product_id,product_name,price,info)
        values(?,?,?,?)";
        this.pstmt = this.conn.prepareStatement(sql);  // 实例化PrepareStatement对象
        this.pstmt.setString(1,product.getProduct_id());    // 设置产品ID
        this.pstmt.setString(2,product.getProduct_name()); // 设置产品名称
        this.pstmt.setDouble(3, product.getPrice());        // 设置产品价格
        this.pstmt.setString(4,product.getInfo());          // 设置产品信息
```

```

        if (this.pstmt.executeUpdate() > 0) {  // 更新记录的行数大于0
            flag = true;                        // 修改标识
        }
        this.pstmt.close();                     //关闭PreparedStatement操作
        return flag;
    }

    public List<Product> findAll(String product_name) throws Exception {
        List<Product> list = new ArrayList<Product>(); //定义集合，接收返回的数据
        String sql = "select product_id,product_name,price,info from product ";
        if(product_name!=null&&!"".equals(product_name)){
            sql = "select product_id,product_name,price,info
            from product where product_name like? ";
            this.pstmt.setString(1, "%" + product_name + "%"); // 设置查询产品名称
        }else{
            this.pstmt = this.conn.prepareStatement(sql);      // 实例化PreparedStatement
        }
        ResultSet rs = this.pstmt.executeQuery(); // 执行查询操作
        Product product = null;
        while (rs.next()) {
            product = new Product();                 // 实例化新的product对象
            product.setProduct_id(rs.getString(1));
            product.setProduct_name(rs.getString(2));
            product.setPrice(rs.getDouble(3));
            product.setInfo(rs.getString(4));
            list.add(product);                       // 向集合中增加product对象
        }
        this.pstmt.close();
        return list;                                 // 返回全部结果
    }

    public Product findByProductId(String product_id) throws Exception {
        Product product = null;
        String sql = "select product_id,product_name,price,info from
        product where product_id=?";
        this.pstmt = this.conn.prepareStatement(sql);
        this.pstmt.setString(1, product_id);                 // 设置产品编号
        ResultSet rs = this.pstmt.executeQuery();
        if (rs.next()) {
            product = new Product();
            product.setProduct_id(rs.getString(1));
            product.setProduct_name(rs.getString(2));
            product.setPrice(rs.getDouble(3));
            product.setInfo(rs.getString(4));
        }
        this.pstmt.close();
        return product;          // 如果查询不到结果，则返回null，默认值为null
    }

}
```

在上述代码中，第 13~14 行代码定义了数据库操作的接口对象 Connection 和 PreparedStatement；第 21 行代码定义一个成功标识，如果添加数据成功，就返回 true，否则返回 false；第 23 行代码实例化 PrepareStatement 对象，然后依次插入数据（如第 24~28 行代码所示）；第 30、31 行代码判断是否成功插入数据，如果返回值大于 0，就表示插入成功，否则表示插入失败。

第37~59行代码是查询数据方法：首先定义List集合对象，如第38行代码；在查询时定义产品名称为模糊查询条件，如第39行代码；第41~42行代码执行数据库查询操作，并将查询出的结果进行循环遍历，保存在List集合中。

第61~77行代码根据产品ID值查询Product对象。其中，第63~65行设置查询条件语句并实例化PreparedStatement对象；第67行调用executeQuery()执行查询，如果此产品编号存在，就实例化Product对象并设定属性值；若不存在此产品，则返回null。

在上述代码中的数据操作实现类中没有针对数据库的打开和连接操作，只是由构造方法取得连接的数据库，而对数据库的打开和关闭由业务操作实现类完成。业务操作类 ProductService 的源代码如下：

```
-------------------------ProductService.java---------------------------
package com.eshore.service;

import java.util.List;

import com.eshore.dao.ProductDao;
import com.eshore.dao.ProductDaoImpl;
import com.eshore.db.DBConnection;
import com.eshore.pojo.Product;

public class ProductService implements ProductDao{
    private DBConnection dbconn = null;           // 定义数据库连接类
    private ProductDao dao = null;                // 声明DAO对象
    // 在构造方法中实例化数据库连接，同时实例化DAO对象
    public ProductService() throws Exception {
        this.dbc = new DBConnection();
        this.dao = new ProductDaoImpl(this. dbconn.getConnection());
        // 实例化ProductDao的实现类
    }

    public boolean addProduct(Product product) throws Exception {
        boolean flag = false;                           // 标识
        try {
            if (this.dao.findByProductId(product.getProduct_id()) == null) {
                // 如果要插入的产品编号不存在
                flag = this.dao.addProduct(product);  // 新增一条产品信息
            }
        } catch (Exception e) {
            throw e;
        } finally {
            this. dbconn.close();
        }
        return flag;
    }
```

```
34
35      public List<Product> findAll(String keyWord) throws Exception {
36          List<Product> all = null;                    // 定义产品返回的集合
37          try {
38              all = this.dao.findAll(keyWord);         // 调用实现方法
39          } catch (Exception e) {
40              throw e;
41          } finally {
42              this. dbconn.close();
43          }
44          return all;
45      }
46
47      public Product findByProductId(String product_id) throws Exception {
48          Product product = null;
49          try {
50              product = this.dao.findByProductId(product_id);
51          } catch (Exception e) {
52              throw e;
53          } finally {
54              this. dbconn.close();
55          }
56          return product;
57      }
58
59  }
```

在上述代码中，第14~18行代码实例化数据库连接类以及ProductDao的实现类，第20~57行代码用于实现ProductDao接口中的方法。在每个方法操作完成之后必须记得关闭数据库。实现类编写完成后，可以编写DAO工厂类（用来获得业务操作类），在后续的客户端中就可以直接通过工厂类获得DAO接口的实例对象。DAO工厂类DAOFactory的源代码如下：

```
-----------------------DAOFactory.java--------------------------
01  package com.eshore.factory;
02  import com.eshore.dao.ProductDao;
03  import com.eshore.service.ProductService;
04  public class DAOFactory {
05      public static ProductDao getIEmpDAOInstance()throws Exception {
06        return new ProductService();                          //取得业务操作类
07    }
08  }
```

在上述代码中，第05、06行代码取得业务操作类。现在已经编写好了所需的类，再编写一个测试类来测试方法是否可用。测试添加产品类TestInsertProduct.java的源代码如下：

```
-----------------------TestInsertProduct.java--------------------------
package com.eshore.test;

import com.eshore.factory.DAOFactory;
import com.eshore.pojo.Product;

public class TestInsertProduct {
    public static void main(String[] args){
        Product product = null;
        try{
```

```
            for(int i=0;i<5;i++){
                product = new Product();
                product.setProduct_id("350115001010"+i);
                product.setProduct_name("水杯"+i);
                product.setPrice(100+i);
                product.setInfo("这是一个精美的杯子"+i);
                DAOFactory.getIEmpDAOInstance().addProduct(product);
            }
        }catch(Exception e){
            e.printStackTrace();

        }
    }
}
```

在上述代码中，循环添加了 5 个产品信息，通过调用数据操作实例来添加产品方法。运行后，在数据库表中添加了 5 条数据，如图 9.12 所示。从结果来看，程序是完全可以运行的，而且利用对象来操作数据库，插入的代码简单、易懂。

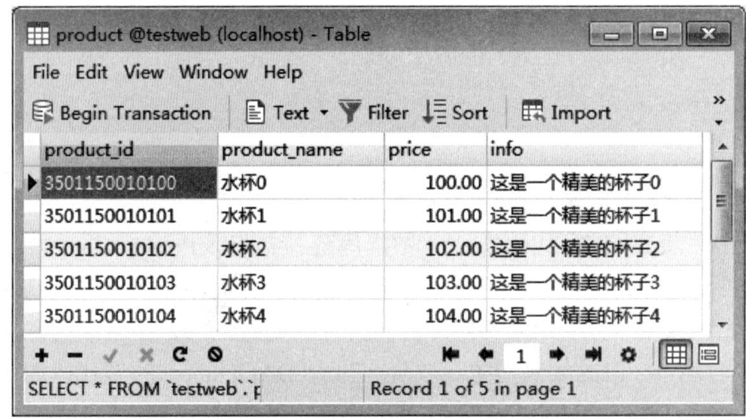

product_id	product_name	price	info
3501150010100	水杯0	100.00	这是一个精美的杯子0
3501150010101	水杯1	101.00	这是一个精美的杯子1
3501150010102	水杯2	102.00	这是一个精美的杯子2
3501150010103	水杯3	103.00	这是一个精美的杯子3
3501150010104	水杯4	104.00	这是一个精美的杯子4

图9.12　插入产品信息效果图

9.6.4　JSP调用DAO

当正确编写一个DAO后，就可以结合JSP一起实现前台的显示。下面演示一下如何在JSP中使用DAO。

【例 9.10】　在 JSP 中增加产品信息。

product_add.jsp 用于增加产品信息，源代码如下：

```
-----------------------product_add.jsp---------------------------
<%@ page pageEncoding="UTF-8"%>
<!DOCTYPE HTML>
<html>
  <head>
    <title>添加产品信息</title>
  </head>
  <body>
      <form action="product_insert.jsp" method="post">
          产品编号：<input name="product_id"/><br>
```

```
10          产品名称：<input name="product_name"/><br>
11          产品价格：<input name="price"/><br>
12          产品信息：<textarea rows="" cols="" name="info"></textarea><br/>
13          <input type="submit" value="添加">  
14          <input type="reset" value="重置">
15        </form>
16    </body>
17  </html>
```

在上述代码中，第08~15行代码利用表单form提供产品的提交信息，并跳转到product_insert.jsp页面。product_insert.jsp的源代码如下：

```
-----------------------product_ insert.jsp--------------------------
01  <%@ page import="java.util.*,com.eshore.pojo.Product" pageEncoding=
"UTF-8"%>
02  <%@ page import="com.eshore.factory.DAOFactory" %>
03  <%
04  request.setCharacterEncoding("utf-8");//解决中文乱码
05  %>
06
07  <!DOCTYPE HTML>
08  <html>
09    <head>
10      <title>执行添加产品</title>
11    </head>
12    <body>
13        <%
14          Product product = new Product();            //实例化Product对象
15          product.setProduct_id(request.getParameter("product_id"));
16          product.setProduct_name(request.getParameter("product_name"));
17          product.setPrice(Double.parseDouble(request.getParameter
("price")));
18          product.setInfo(request.getParameter("info"));
19          boolean flag = DAOFactory.getIEmpDAOInstance().
20                      addProduct(product); //执行添加操作
21          if(flag){
22         %>
23            <h4>添加产品信息成功</h4>
24          <%
25          }else{
26           %>
27             <h4>添加产品信息失败</h4>
28        <%} %>
29    </body>
30  </html>
```

在上述代码中，第 14 行定义了 Product 对象，然后将表单提交的参数依次设置到 Product 对象中；第 19、20 行代码调用 DAOFactory 的添加方法，第 21~28 行代码判断是否添加成功。代码的运行效果如图 9.13 所示。

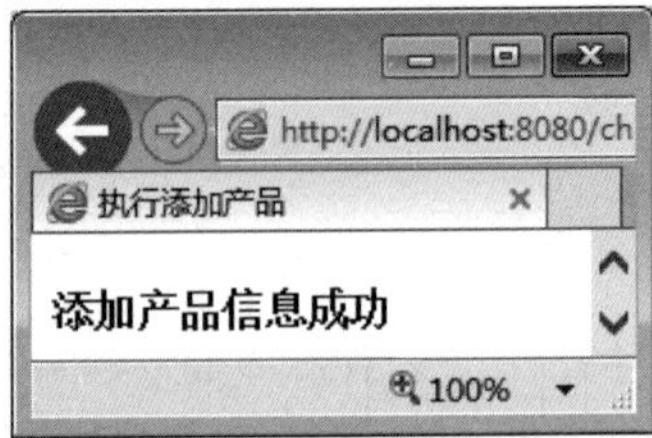

图9.13　JSP中增加产品信息效果图

【例 9.11】　在 JSP 中查询产品信息。

product_list.jsp 用于列出产品信息，源代码如下：

```
-----------------------product_list.jsp--------------------------
<%@ page import="java.util.*,com.eshore.pojo.Product" pageEncoding="UTF-8"%>
<%@ page import="com.eshore.factory.DAOFactory" %>
<%
request.setCharacterEncoding("utf-8");//解决中文乱码
%>
<!DOCTYPE HTML>
<html>
  <head>
    <title>查询产品列表</title>
  </head>

  <body>
      <%
        String product_name = request.getParameter("product_name");//
        List<Product> list = DAOFactory.getIEmpDAOInstance().findAll(product_name);
      %>
      <form action="product_list.jsp" method="post">
            请输入产品名称：<input name="product_name"/>
        <input type="submit" value="提交">
      </form>
      <table border="1">
          <tr>
              <td>产品编号</td>
              <td>产品名称</td>
              <td>产品价格</td>
              <td>产品信息</td>
          </tr>
          <%
              for(int i=0;i<list.size();i++){
                      Product p = list.get(i); //取出每一个产品
           %>
           <tr>
              <td><%=p.getProduct_id() %></td>
              <td><%=p.getProduct_name() %></td>
              <td><%=p.getPrice() %></td>
              <td><%=p.getInfo() %></td>
```

```
37             </tr>
38             <%} %>
39         </table>
40     </body>
41 </html>
```

在上述代码中，第 13~16 行用于获取产品列表，第 28~38 行代码用于循环遍历查询结果。如果输入产品名称，就可以实现模糊查询功能，如果不输入，就输出所有的产品列表。程序的运行效果如图 9.14 所示。

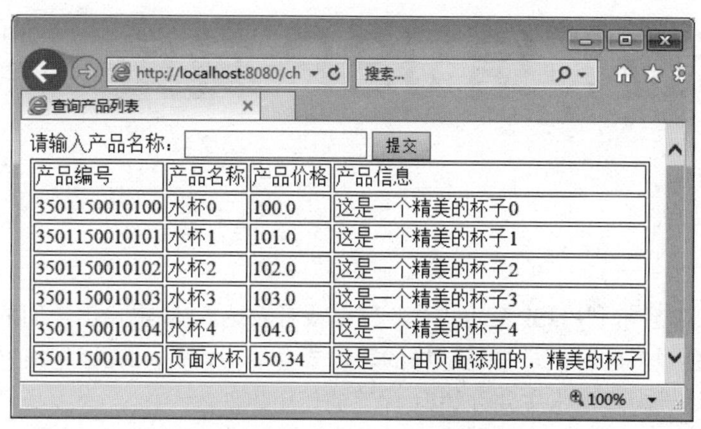

产品编号	产品名称	产品价格	产品信息
3501150010100	水杯0	100.0	这是一个精美的杯子0
3501150010101	水杯1	101.0	这是一个精美的杯子1
3501150010102	水杯2	102.0	这是一个精美的杯子2
3501150010103	水杯3	103.0	这是一个精美的杯子3
3501150010104	水杯4	104.0	这是一个精美的杯子4
3501150010105	页面水杯	150.34	这是一个由页面添加的，精美的杯子

图9.14　JSP中查询产品信息效果图

9.7 小　结

本章主要介绍Java Bean在JSP中的应用，首先介绍Java Bean的基本概念，接着介绍如何访问Bean属性，随后介绍Bean的作用域，最后介绍DAO设计模式。DAO设计模式在Web开发中经常被使用，对于构建分层的Web应用，它是十分重要的开发模式，希望读者能够熟练掌握。即使开发简单的JSP网站，若能够熟练掌握Bean，对于开发者而言也是十分有帮助的。

9.8 习　题

（1）Java Bean具有哪些特性？它能给JSP开发者带来什么便利？

（2）定义Java Bean需要注意哪些规范？

（3）如何访问Bean属性？

（4）Bean的作用域有哪些？它们的作用是什么？

（5）利用Java Bean技术实现网页的注册验证。

（6）DAO的设计模式是什么？应如何应用？

（7）应用DAO设计模式实现存储学生基本信息的功能，并在页面中实现查询、新增、修改等操作。

第 10 章

EL 标签：给 JSP 减负

在JSP页面中，经常利用JSP表达式<%=变量或者表达式%>来输出声明的变量以及页面传递的参数，当变量很多的时候，书写这样的表达式会显得累赘，EL标签很好地解决了这个问题，它简化了表达式。本章将主要介绍EL标签的使用方法。

本章主要涉及的知识点有：

- EL标签语法
- EL标签用法
- EL标签的操作符
- EL标签的隐含变量

10.1 EL 标签语法

EL 标签的语法形式如下：

```
${参数名}
```

例如：${param.name}用于获得参数 name 的值，等同于<%=request. getParameter('name')%>。

从形式和用法上看，这种EL表达式简化了JSP原有的表达式，也是目前开发中经常使用的方式。

【例 10.1】 EL 标签用法示例。

ex10_1.jsp 输出基本运算值，其源代码如下：

```
----------------- ex10_1.jsp ----------------
<%@ page pageEncoding="UTF-8"%>
<!DOCTYPE HTML>
<html>
 <head>
  <title>EL标签示例</title>
```

```
  </head>
  <body>
     <p style="text-align: center;">2+3=${2+3}</p> //输出2+3的值
  </body>
 </html>
```

ex10_1.jsp 的运行结果如图 10.1 所示。

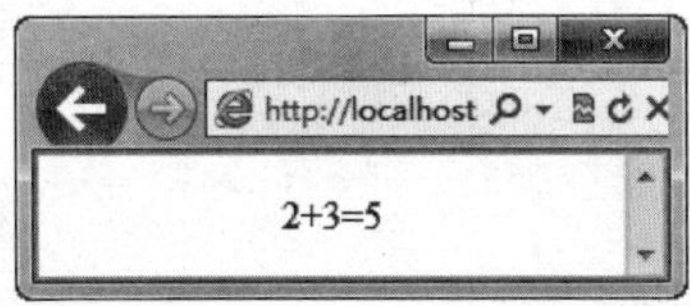

图10.1　ex10_1.jsp运行结果

从结果可以看出，页面得到了正常的结果，相当于<%=2+3%>这样的表达式输出。

10.2　EL 标签的功能

上一节介绍了EL标签的语法，让读者初步了解其使用规范，本节主要讲述其功能点。总体而言，EL标签提供了更为简洁、方便的形式来访问变量，不但可以简化JSP页面代码，而且使得开发者的逻辑更加清晰。

EL 标签具有以下功能：

- 可以访问JSP中不同域的对象。
- 可以访问Java Bean中的属性。
- 可以访问集合元素。
- 支持简单的运算符操作。

下面将分别介绍这 4 个功能。

1. 访问 JSP 中不同域的对象

在Web中有4个作用域，分别是page、request、session和application。EL标签可以对这4个作用域的参数进行访问。

【例 10.2】　EL 标签访问示例。

ex10_2.jsp 演示 EL 标签的访问方法，其源代码如下：

```
----------------- ex10_2.jsp -----------------
<%@ page pageEncoding="UTF-8"%>
<!DOCTYPE HTML>
<html>
 <head>
  <title>EL标签访问示例</title>
 </head>
 <body>
   <%
     pageContext.setAttribute("name","Smith");
```

```
10       request.setAttribute("age",20);
11       session.setAttribute("address","china");
12       application.setAttribute("sex","male");
13      %>
14    <h3 style="text-align: center;">访问演示</h3>
15    <table border="1" width="100%">
16     <tr>
17      <td align="center">姓名</td>
18      <td align="center">年龄</td>
19      <td align="center">性别</td>
20      <td align="center">地址</td>
21     </tr>
22     <!--范围.参数名称 -->
23     <tr>
24      <td align="center">${pageScope.name}</td>
25      <td align="center">${requestScope.age}</td>
26      <td align="center">${sessionScope.address}</td>
27      <td align="center">${applicationScope.sex}</td>
28     </tr>
29     <!-- 直接写参数名称 -->
30     <tr>
31      <td align="center">${name}</td>
32      <td align="center">${age}</td>
33      <td align="center">${address}</td>
34      <td align="center">${sex}</td>
35     </tr>
36     </table>
37    </body>
38   </html>
```

在上述代码中，有两种方法取得参数值：第 23~28 行代码是用 EL 隐含变量获取参数值；第 30~35 行直接写明参数名，EL 标签会依序从 page、request、session、application 范围查找，若找到 name 参数名，就直接回传参数值，不再继续查找，若没有参数，则返回 null。运行结果如图 10.2 所示。

图10.2　ex10_2.jsp页面运行结果

2. 访问 Java Bean 中的属性

在Java Bean中，经常将Java Bean用作内部变量（数据成员）。在JSP表达式中访问这些Java Bean中的变量比较麻烦，但是应用EL标签却极其简单、方便。

【例 10.3】 利用 EL 标签访问 Java Bean 中的属性。

通过 ex10_3.jsp 页面演示如何使用 EL 标签访问 Java Bean 中的属性和方法，源代码如下：

```
----------------- ex10_3.jsp ----------------
<%@ page pageEncoding="UTF-8"%>
<%@page import="com.eshore.pojo.Users" %>
<!DOCTYPE HTML>
<html>
 <head>
  <title>EL标签访问示例</title>
 </head>
```

```
08    <body>
09      <%
10        Users user = new Users();
11        user.setAddress("中国");
12        user.setAge(20);
13        user.setName("王五");
14        request.setAttribute("user",user);
15       %>
16     用户信息：${user }; <br/>
17     用户年龄：${user.age } ,用户姓名:${user.name }
18    </body>
19   </html>
```

在上述代码中，第 14 行向页面传递参数 user 对象，第 16 行取得 user 对象，第 17 行确定 user 对象中的 age 和 name 属性值。页面运行结果如图 10.3 所示。

图10.3　ex10_3.jsp页面运行结果

User 对象的 POJO 如下：

```
public class User {
   private String name;              //姓名
   private int age;                  //年龄
   private String address;           //地址
   public String getName() {
      return name;
   }
   public void setName(String name) {
      this.name = name;
   }
   public int getAge() {
      return age;
   }
   public void setAge(int age) {
      this.age = age;
   }
   public String getAddress() {
      return address;
   }
   public void setAddress(String address) {
      this.address = address;
   }
   @Override
   public String toString() {
      return "用户："+name+" 年龄 "+age+"，来自"+address;
   }
}
```

3. 可以访问集合元素

利用EL标签可以很方便地访问集合中的元素。

4. 支持简单的运算符操作

在EL标签中，可以使用一些简单的运算符进行运算操作。

10.3 EL 标签的操作符

上一节介绍了 EL 标签的两大功能，也介绍了一些示例，使读者了解了其基本的访问方法。本节主要介绍 EL 标签的操作符，它可以帮助实现各种所需的功能。操作符大致可以分为算术运算符、逻辑运算符、关系运算符以及其他运算符。

- 算术运算符：算术运算符有5个，即“+”、“-”、“*”、“/”或“div”、“%”或“mod”。
- 逻辑运算符：逻辑运算符有3个，即“&&”或“and”、“||”或“or”、“!”或“not”。
- 关系运算符：关系运算符有6个，即“==”或“eq”、“!=”或“ne”、“<”或“lt”、“>”或“gt”、“<=”或“le”、“>=”或“ge”。
- 其他运算符：empty运算符、条件运算符、“()”运算符、“[]”运算符。

下面通过例子说明运算符的使用方法。

【例 10.4】 EL 标签运算符示例。

通过 ex10_4.jsp 演示说明运算符的使用方法，源代码如下：

```
----------------- ex10_4.jsp ----------------
<%@ page pageEncoding="UTF-8"%>
<%@page import="com.eshore.pojo.Users" %>
<!DOCTYPE HTML>
<html>
 <head>
  <title>EL标签运算符示例</title>
 </head>
 <body>
 <%
     Users user = new Users();
     user.setAddress("中国");
     request.setAttribute("user",user);
     request.setAttribute("str",null);
     String[] arr = new String[]{"第1个","第2个"};
     request.setAttribute("arr",arr);
    %>
 <table border="1">
   <tr>
     <td>算术运算符：</td>
     <td>逻辑运算符：</td>
     <td>关系运算符：</td>
     <td>其他运算符：</td>
```

```
    </tr>
    <tr>
      <td>
   加：3+3 = ${2+3}<br/>
   减：4-3 = ${4-3}<br/>
   乘：2*3 = ${2*3}<br/>
   除：3/6 = ${3/6}<br/>
   求模：10%3 = ${10%3}</td>
      <td>
   逻辑与：${2<15 && 15<20}<br/>
   逻辑与： ${2<15 and 15 <20}<br/>
   逻辑或： ${2<15||15<20}<br/>
   逻辑或：${2<15 or 15<20}<br/>
   逻辑否： ${!(2<15)}<br/>
   逻辑否： ${not(2<15)}<br/></td>
      <td>
   符号左右两端是否相等：2==15 : ${2 == 15 }或${2 eq 15 }<br/>
   符号左右两端是否不相等：2!=15：${2 != 15 }或${2 ne 15 }<br/>
   符号左边是否小于右边： 2<15：${2 < 15 }或${2 lt 15 }<br/>
   符号左边是否大于右边 ：2>15：${2 > 15 }或${2 gt 15 }<br/>
   符号左边是否小于或者等于右边： 2<=15：${2 <= 15 }或${2 le 15 }<br/>
   符号左边是否大于或者等于右边： 2>=15：${2 >= 15 }或${2 ge 15 }<br/></td>
      <td>
   str是否为空：${empty str }<br/>
   user对象是否为空：${empty user }<br/>
   2小于15输出yes否则输出no：${2<15?'yes':'no' }<br/>
     输出user对象的address属性：${user.address}<br/>
     输出arr数组的第一个值：${arr[0]}<br/></td>
    </tr>
  </table>
 </body>
</html>
```

注意："."与"[]"运算符的区别在于"[]"访问数组或者集合；"."访问 Java Bean 属性或者 Map Entry。变量必须利用"[]"运算符，"."则不行。

页面ex10_4.jsp的运行结果如图10.4所示。

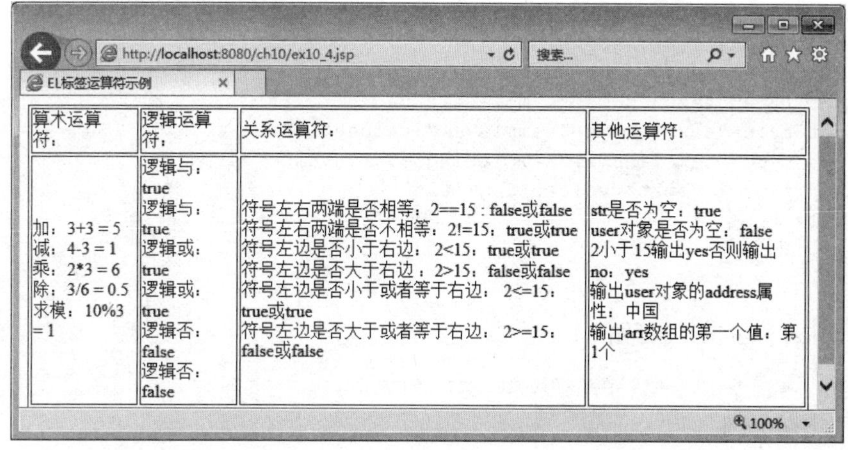

图10.4　ex10_4.jsp页面的运行结果

10.4 EL 标签的隐含变量

上一节介绍了EL标签操作符的使用方法，让读者了解到使用EL标签可以给JSP开发带来巨大的便利，而且使逻辑更加清晰。本节将介绍EL标签的隐含变量，这些隐含变量与JSP中的隐含对象很相似，下面逐一介绍。

10.4.1 隐含变量pageScope、requestScope、sessionScope、applicationScope

隐含变量pageScope、requestScope、sessionScope、applicationScope分别对应JSP隐含变量page、request、session、application，利用JSP中对应的作用域发送请求的参数变量，可以用相应的EL标签变量获取参数值，如例10.2所示。

10.4.2 隐含变量param、paramValues

隐含变量 param、paramValues 包含请求参数集合变量，param 是取得某一个参数，paramValues 是取得参数集合中的变量值，它们分别相当于 JSP 中的 request.getParameter(String name)和 request.getParameterValues(String name)。

【例 10.5】 参数 param、paramValues 的应用示例。

通过 ex10_5.jsp 演示隐含变量 param、paramValues 的使用方法，源代码如下：

```
----------------- ex10_5.jsp ----------------
<%@ page pageEncoding="UTF-8"%>
<!DOCTYPE HTML>
<html>
  <head>
    <title>EL标签隐含对象param、paramValues使用示例</title>
  </head>
  <body>
    <form action="test.jsp">
    <input type="text" name="sampleVal" value="1">
    <input type="text" name="sampleVal" value="17">
    <input type="text" name="sampleVal" value="16">
    <input type="text" name="sampleSingleVal" value="single">
    <input type="submit" value="Submit">
    </form>
  </body>
</html>
```

跳转页面test.jsp的源代码如下：

```
-----------------------test.jsp--------------------------
<%@ page pageEncoding="UTF-8"%>
<!DOCTYPE HTML>
<html>
  <head>
```

```
    <title>跳转页面</title>
  </head>
  <body>
  请求参数值：${param.sampleSingleVal} ,
  ${paramValues.sampleVal[1]}</br>
  </body>
</html>
```

运行结果如图 10.5 所示。

图10.5 ex10_5.jsp运行结果

10.4.3 其他变量

1. header、headerValues 变量

header、headerValues这两个变量用于获取请求HTTP表头信息，header表示取得HTTP表头信息，headerValues表示取得表头数组信息，它们分别对应于request.getHeader()和request.getHeaders()。

2. cookie 变量

cookie变量用于取得所有请求的cookie参数，参数中的每个对象对应jakarta.servlet.http.Cookie。例如，要获取cookie中名称为username的值，可以直接使用${cookie.username}。

3. initParam 变量

initParam用于取得应用程序的初始化参数，相当于application.getInitParameter()方法，但通常很少使用。

例如，一般页面获取初始化的方法String url= application.getInitParameter("url")可以用${initParam.url}来代替。

4. pageContext 变量

pageContext 变量用于取得其他相关用户的请求或页面的详细信息，其等同于 JSP 中的 PageContext 对象。下面列出几个常用的方法：

```
${pageContext.request.queryString}          取得请求的参数名
${pageContext.request.requestURL}           取得请求的URL
${pageContext.request.contextPath}          取得服务应用的名称
${pageContext.request.method}               取得HTTP 的提交方法
${pageContext.request.protocol}             取得使用的协议（HTTP/1.1、HTTP/1.0）
${pageContext.request.remoteUser}           获取登录用户名称
${pageContext.request.remoteAddr }          获取登录用户IP 地址
${pageContext.session.new}                  判断session 是否为新的会员
${pageContext.session.id}                   取得session 的ID号
${pageContext.servletContext.serverInfo}    取得主机端的服务信息
```

【例 10.6】 其他变量的应用示例。

通过 ex10_6.jsp 演示隐含变量的使用方法，源代码如下：

```
----------------- ex10_6.jsp ----------------
<%@ page pageEncoding="UTF-8"%>
<!DOCTYPE HTML>
```

```
<html>
 <head>
  <title>EL标签隐含对象使用示例</title>
 </head>
 <body>
 会话ID：${pageContext.session.id}</br>
 Accept-Encoding报头：${header["Accept-Encoding"] }</br>
 connection报头:${header["connection"] }
 </body>
</html>
```

ex10_6.jsp 页面的运行结果如图 10.6 所示。

图10.6 ex10_6.jsp页面运行结果

从以上讲解的EL隐含变量可以看出，它与JSP中隐含对象的大部分处理方法相对应，所以基本可以应用EL隐含变量来取代JSP中隐含对象的大多数方法，既方便又简单。

10.5 禁用 EL 标签

上一节介绍了EL标签中的隐含变量，使读者了解了隐含变量的基本用法。虽然EL标签给编码带来了便利，但有时并不想使用EL标签，例如让某页面禁用EL标签或者在页面中禁用表达式。本节将介绍如何禁用EL标签。

10.5.1 在整个Web应用中禁用

EL标签是在JSP 2.0之后才有的新特性，因此如果指定JSP版本为较低版本（Servlet 2.3或者更早），那么JSP页面将不再支持EL标签。这个功能可以通过修改web.xml文件来实现。

修改后的 web.xml 配置为：

```
<?xml version="1.0" encoding="UTF-8"?>
<!DOCTYPE web-app
  PUBLIC "-//Sun Microsystems, Inc.//DTD Web Application 2.3//EN"
  "http://java.sun.com/dtd/web-app_2_3.dtd">
<web-app>
 <welcome-file-list>
  <welcome-file>index.jsp</welcome-file>
 </welcome-file-list>
</web-app>
```

与能支持 EL 标签的 web.xml 相比，修改后的 web.xml 不用 DTD 对 XML 文档进行约束，而是利用 XSD 进行约束，且 DTD 的版本是 2.3。

能支持 EL 标签的 web.xml 如下：

```
<?xml version="1.0" encoding="UTF-8"?>
<web-app version="5.0"
    xmlns="https://jakarta.ee/xml/ns/jakartaee"
    xmlns:xsi="http://www.w3.org/2001/XMLSchema-instance"
    xsi:schemaLocation="https://jakarta.ee/xml/ns/jakartaee
    https://jakarta.ee/xml/ns/jakartaee/web-app_5_0.xsd">
 <welcome-file-list>
  <welcome-file>index.jsp</welcome-file>
 </welcome-file-list>
</web-app>
```

10.5.2 在单个页面中禁用

如果想在某个页面中禁用 EL 标签，那么可以直接通过 page 指令的 isELIgnored 属性实现。命令如下：

```
<%@page isELIgnored="true".%>
```

10.5.3 在页面中禁用个别表达式

通常情况下，在一个应用中要么全部允许使用EL标签，要么都不允许，因此很少出现部分页面使用而另一部分不用的情形。但EL标签允许在一个页面中部分支持标签、部分不支持，若要实现此功能，则只需在不需要解析的EL表达式的“$”符号前加入一个反斜杠即可，如“\$”。

10.6 小　　结

本章主要介绍了EL标签的语法、功能、操作符的运算和隐含变量。在JSP开发中，EL标签可以使业务逻辑与代码分工更明确，便于代码的维护，它在开发中占据着比较重要的位置。

10.7 习　　题

（1）使用EL标签编写四则混合运算，并分析运算符的优先级。

（2）编写一个JSP页面，应用EL标签输出用户信息。

（3）编写一个JSP页面，应用EL标签隐含对象、输出表头等信息。

第 11 章 JSTL 标签库

上一章介绍了EL标签的用法和示例，EL标签能简化JSP开发的代码量。本章介绍另外一种标签——JSTL标签，它不但可以简化JSP代码量，而且能使得JSP开发者的维护工作更加轻松。JSTL标签常与EL标签一起使用。

本章主要涉及的知识点有：

- JSTL标签技术
- 5类标签库的使用：core标签库、fmt标签库、fn标签库、XML标签库和SQL标签库
- 各种标签之间的区别

11.1 JSTL 标签概述

JSTL标签是一组与HTML标签相似但又比HTML标签强大的功能标签，编程人员可以通过它编写出动态的JSP页面。它包括5类标签库：core标签库、fmt标签库、fn标签库、XML标签库和SQL标签库。这5类标签库基本覆盖了Web开发中所涉及的标签技术。本节将通过一个实例来介绍JSTL标签的基本用法，使读者对JSTL标签的使用有初步的了解。

11.1.1 JSTL的来历

JSTL（JSP Standard Tag Library）是一个开源的JSP标准标签库，是由Apache的jakarta小组开发并维护的，目前还在不断完善中。

使用 JSTL 标签具有以下优点：

- 接口统一，便于各服务器之间的移植。
- 简化了Web程序的开发，本需要由大量的Java代码完成的功能可以用少量的JSTL代码代替。
- JSTL标签代码的可读性强，易于理解。
- 利用JSTL标签编写的Web程序维护相对简单。

JSTL标签中的标签库主要有以下5类：

- core标签库（核心标签库），包括通用标签（输出标签）、流控制标签和循环控制标签等。
- fmt标签库，包括格式化、国际化标签等。
- fn标签库，函数标签库。
- XML标签库，关于XML操作的标签库。
- SQL标签库，操作数据库的标签库。

11.1.2　一个标签实例带你入门

若想在 Tomcat10 及以上版本中使用 JSTL，需要引入 jstl 和 jstl-api 两个 JAR 包，地址如下：

- https://repo.maven.apache.org/maven2/org/glassfish/web/jakarta.servlet.jsp.jstl/
- https://repo.maven.apache.org/maven2/jakarta/servlet/jsp/jstl/jakarta.servlet.jsp.jstl-api/

从上面的地址中下载最新的 JAR 包并复制到 Web 应用的 lib 目录下，JSTL 就可以在当前的 Web 应用中使用；若想令 Tomcat 服务器中所有的 Web 应用都能使用 JSTL，则将这两个 JAR 包复制到 Tomcat 安装目录的 lib 下即可。

下面通过一个简单例子来初步了解JSTL该如何应用。

【例 11.1】　通过简单的 JSTL 示例输出内容。

theFirstJSTL.jsp 是用户显示页面，源代码如下：

```
----------------- theFirstJSTL.jsp----------------
<%@ page pageEncoding="UTF-8"%>
<%@taglib prefix="c" uri="http://java.sun.com/jsp/jstl/core" %><!--引入标签-->
<!DOCTYPE HTML>
<html>
  <head>
    <title>一个简单的JSTL示例</title>
  </head>
  <body>
     <div style="text-align: center;">
     <c:out value="一个简单的JSTL示例"></c:out> <!-- 输出内容 -->
     <br/>
     <c:out value="《JSP+Servlet+Tomcat开发入门》"></c:out>
     </div>
  </body>
</html>
```

在上述代码中，利用语句<%@taglib prefix="c" uri="http://java.sun.com/jsp/jstl/core" %>声明本页面需要用到的标签库中的标签，prefix 是页面中标签的前缀，uri 是 JSTL 中 c.tld 文件声明的 URI 地址，这个 URI 地址可以修改。

程序中使用输出标签<c:out>，该标签的作用是将内容显示在页面中。程序的运行结果如图11.1所示。

图11.1　theFirstJSTL.jsp页面的运行结果

11.2 JSTL 的 core 标签库

上一节介绍了 JSTL 的来源及其使用示例，本节将介绍 JSTL 的核心标签库。核心标签库中的标签又可以分为表达式标签、流程控制标签、迭代标签和操作标签。

- 表达式标签：<c:out>、<c:set>、<c:remove>、<c:catch>。
- 流程控制标签：<c:if>、<c:choose>、<c:when>、<c:otherwise>。
- 迭代标签：<c:forEach>、<c:forTokens>。
- 操作标签：<c:import>、<c:url>、<c:redirect>。

下面详细介绍一些常用的标签。

11.2.1 <c:out>标签

<c:out>标签用于把表达式中的结果输出到页面中，其语法如下：

```
<c:out value="表达式" [escapeXML="true|false"]/>
```

在上述语法中，“[]”中的内容是可选项，“|”是“或”的意思，escapeXML 的默认值为 true，即转换特殊字符，例如将“<”转换为“<”，“>”转换为“>”等。

11.2.2 <c:if>标签

<c:if>标签用于条件判断，是流程控制标签，其语法如下：

```
<c:if test="判断条件" [var="varName"] [scope="{request|page|session|application}"]>
```

条件为真时执行的语句如下：

```
</c:if>
```

在上述标签语法中，test 参数是<c:if>标签必须设置的，var 参数是条件的执行结果，scope 是 var 的有效范围。

【例 11.2】 <c:if>、<c:out>示例。

if.jsp 是<c:if>的演示页面，验证输出数是否为偶数，源代码如下：

```
----------------- if.jsp-----------------
<%@ page pageEncoding="UTF-8"%>
<%@taglib prefix="c" uri="http://java.sun.com/jsp/jstl/core" %>
<!DOCTYPE HTML>
<html>
  <head>
    <title>&lt;c:if&gt;标签使用例子</title>
  </head>
  <body>
      <div style="text-align: center;">
          <c:if test="${2%2==0}" var="num">
```

```
                输出数为：偶数！
            </c:if>
            <br/>该数为偶数的检查结果为：
            <c:out value="${num}"/>
            <br/>
            <c:if test="${3%2!=0}" var="num">
                输出数为：奇数！
            </c:if>
            <br/>该数为奇数的检查结果为：
            <c:out value="${num}"/>
        </div>
    </body>
</html>
```

程序的运行结果如图 11.2 所示。

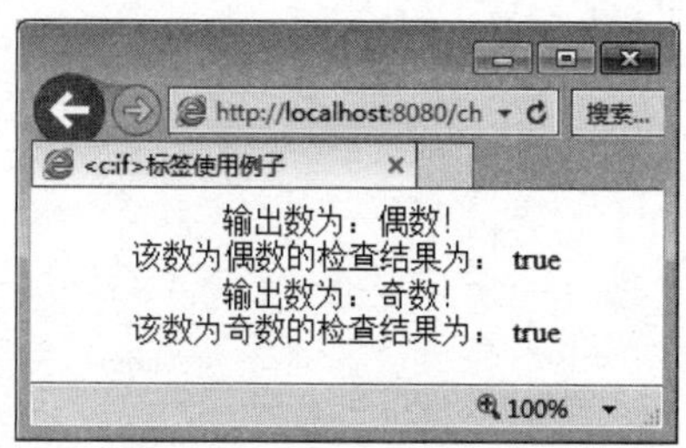

图11.2 if.jsp页面运行结果

通过程序可以看出，利用<c:if>标签可以非常简单地进行判断，但要注意别名的使用，别名相当于Java中的局部变量，别名的默认范围是page。

11.2.3 <c:choose>标签、<c:when>标签、<c:otherwise>标签

<c:choose>、<c:when>、<c:otherwise>标签是另外一组 JSTL 的流程控制标签，其语法如下：

```
<c:choose>
   <c:when  test="表达式">
      表达式为真时执行的语句
   </c:when>
   [<c:otherwise>
      表达式为假时执行的语句
   ]
</c:choose>
```

在上述语法中，<c:choose>是父标签，<c:when>和<c:otherwise>是子标签，<c:when>标签可以有 0 个或者多个，同样<c:otherwise>标签也可以有 0 个或者多个，但是<c:when>标签必须位于<c:otherwise>标签之前。当<c:when>标签中的判断语句为假时，才会执行<c:otherwise>中的内容。

注意：若<c:when>标签中有多个条件都为真，则只会执行条件最先为真的<c:when>中的内容。

【例 11.3】 <c:choose>、<c:when>、<c:otherwise>示例。

choose.jsp 是<c:choose>的演示页面，验证输出数是否为偶数，源代码如下：

```
------------------ choose.jsp-----------------
01  <%@ page pageEncoding="UTF-8"%>
02  <%@taglib prefix="c" uri="http://java.sun.com/jsp/jstl/core" %>
03  <!DOCTYPE HTML>
04  <html>
05    <head>
06      <title>&lt;c:choose&gt;标签使用例子</title>
07    </head>
08    <body>
09        <div style="text-align: center;">
10        输出数检查结果:
11        <c:set value="2" var="num"/><!-- set标签 设定输入值为2-->
12        <c:choose>
13        <c:when test="${num%2==0}">
14            偶数
15        </c:when>
16        <c:otherwise test="${num%2!=0}">
17            奇数
18        </c:otherwise >
19        </c:choose>
20        </div>
21    </body>
22  </html>
```

注意：第 12 行代码是<c:set>标签，用于为页面设定值。

程序的运行结果如图11.3所示。

程序中第一个<c:when>条件为真，所以输出第一个标签的内容。

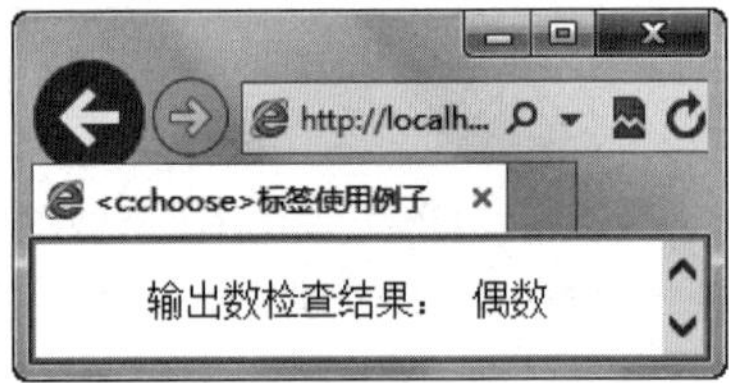

图11.3 choose.jsp页面的运行结果

11.2.4 <c:set>标签

<c:set>标签用于在某个范围中设定某个值（可以是对象或者参数），这个范围可以是 request、page、session、application，其语法如下：

```
<c:set value="表达式"  var="varname" [scope="request|page|session|application"]/>
```

在上述语法中，“[]”中的内容是可选项，“|”是“或”的意思，scope 的默认值是 page。

注意：<c:set>标签的语法也可以用“<c:set var="varname">表达式</c:set>”形式表示。

11.2.5 <c:forEach>标签

<c:forEach>标签是核心标签中的迭代标签，它的功能类似于 Java 中的 for 循环语句，其语法如下：

```
<c:forEach [var="varname"] [varStatus="varstatusName"] [begin="开始"] [end="结
束"] [step="step"]]
    Java程序或者HTML代码
</c:forEach>
```

或

```
<c:forEach item="collection" [var="varname" [varStatus="varstatusName"] [begin="开始"] [end="结束"] [step="step"]]
    Java程序或者HTML代码
</c:forEach>
```

在上述语法中，varname 用来存放当前迭代到的成员值，collection 用来迭代集合，varstatusName 用于存放当前迭代成员的状态信息，begin 是指迭代开始，end 是指迭代结束，step 是指迭代的步长。collection 集合可以是数组、Java 集合（List 容器和 Map 容器）。varstatusName 中存放的信息有 index（当前迭代的索引号）、count（当前迭代的次数）、first（是否第一次迭代）、last。

【例 11.4】 使用<c:forEach>遍历输出 List 集合。

secondForEach.jsp 是测试页面，用于输出 List 中的集合对象，源代码如下：

```
-----------------secondForEach.jsp---------------
01  <%@ page pageEncoding="UTF-8"%>
02  <%@taglib prefix="c" uri="http://java.sun.com/jsp/jstl/core" %>
03  <!DOCTYPE HTML>
04  <html>
05    <head>
06      <title>&lt;c:forEach&gt;标签使用例子</title>
07    </head>
08
09    <body>
10        <%
11             List<String> nameLists = new ArrayList<String>();
12             nameLists.add("Toms");
13             nameLists.add("Smith");
14             nameLists.add("John");
15             nameLists.add("Anna");
16             nameLists.add("James");
17             nameLists.add("Roses");
18             nameLists.add("Bruce");
19             request.setAttribute("nameLists",nameLists);
20         %>
21       <div style="text-align: center;">
22       输出集合中的内容：<hr/>
23       <c:forEach items="${nameLists}" var="name" varStatus= "currentStatus">
24       当前元素为：<c:out value="${name}"/>  
25       当前元素索引号为：<c:out value="${currentStatus.index}"/>   
26       当前迭代数为：<c:out value="${currentStatus.count}"/>  
27       <c:if test="${currentStatus.first}">第一次循环操作</c:if>
28       <c:if test="${currentStatus.last}">最后一次循环操作</c:if>
29       <hr/>
30       </c:forEach>
31       </div>
32    </body>
33  </html>
```

第 10~20 行代码在 JSP 页面中组合 List 集合，第 19 行代码向页面传递参数 nameLists 集合，第 23 行代码用 EL 标签获得参数 nameLists 的值并进行遍历。运行结果如图 11.4 所示。

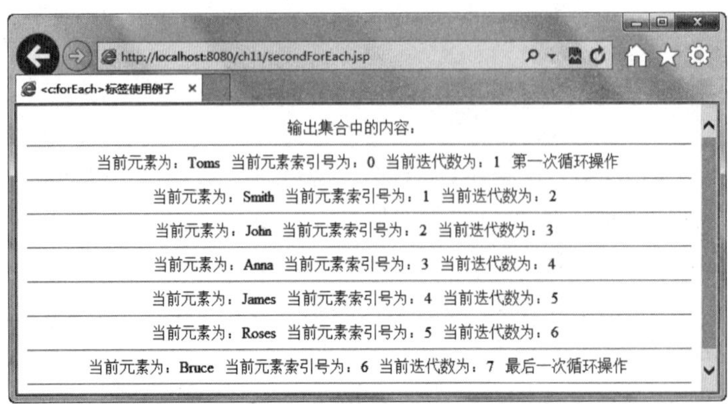

图11.4 secondForEach页面的运行结果

11.2.6 <c:forTokens>标签

<c:forTokens>标签用于对字符串进行分割，类似于Java中的split方法。语法形式如下：

```
    <c:forTokens items="字符串" delims="分隔符" [var="别名"] [varStatus="
varstatusName"] [begin="开始"]
        [end="结束"] [step="步长"]>
        Java代码，HTML代码等
    </c:forTokens>
```

在上述语法中，items中存放的是要进行处理的字符串，delims是进行拆分处理的分隔符，var用来指明迭代值的别名，varStatus是指当前迭代的状态信息，begin（可选项）是指迭代开始，end是指迭代结束，step是指迭代的步长。

11.2.7 <c:remove>标签

从<c:remove>标签的命名中可以看出，该标签的主要功能是删除，它可以删除某个范围中设定的值，其中范围可以是page、request、session、application，值可以是某对象或者系统中的某参数。语法形式如下：

```
    <c:remove var="varname" [scope="page|request|session|applicatioin"]/>
```

在上述语法中，varname是要删除的元素名；scope是删除的范围，默认为page。

11.2.8 <c:catch>标签

<c:catch>标签用于捕获嵌套在<c:catch>标签中的代码抛出的异常，并进行相应的处理，类似于Java中的try…catch方法。语法形式如下：

```
    <c:catch [var="varname"]>
        需要捕获异常的程序代码
    </c:catch>
```

在上述语法中，参数varname用于标识捕获的异常信息。

11.2.9 <c:import>标签与<c:param>标签

<c:import>标签的作用是把当前 JSP 页面之外的静态或者动态文件导入进来，甚至可以是其他网站的文件，其功能类似于 JSP 中的动作指令<jsp:include>。但是它们又有所不同，<jsp:include>只能导入当前 JSP 页面在同一 Web 应用下的文件，而<c:import>标签不是。其语法形式有两种，分别如下：

```
<c:import url="url" [context="context"] [var="varname"]
   [scope=page|request|session|application] [charEncoding="charencoding"]
   [<c:param/>标签语句]
</c:param>
```

或

```
<c:import url="url" [context="context"] [varReader="readerName"]
   [charEncoding="coding"]>
   [<c:param/>标签语句]
</c:import>
```

在上述语法中，利用第一种语法形式导入文件时，把文件的内容以字符串形式存入 varname 中；利用第二种语法形式导入文件时，文件的内容以 Reader 的方式向外提供读取。

url是输入的地址，既可以是网址、FTP服务的文件地址，也可以是Web应用中的文件。

<c:param>标签用于向导入的页面中传入参数，其语法形式如下：

```
<c:param name="paramName" value="paramValue"/>
```

paramName 是要传入的参数名称，paramValue 为传入的参数值。

11.2.10 <c:redirect>标签

<c:redirect>标签用于把客户端发送过来的请求重定向到另一个页面，其语法如下：

```
<c:redirect url="url" [context="context"]>
   [<c:param/>标签语句]
</ c:redirect>
```

在上述语法中，url是重定向的目标网页；<c:param>标签负责带入参数，为可选项；参数context的作用是当要重定向目标网址为其他Web应用的网页时指出其应用名。例如，当前Web应用为testJstl，若要重定向到Web应用targetJstl中的target.jsp文件，则语句如下：

```
<c:redirect url="/target.jsp" context=" targetJstl"/>
```

如果在传递参数时页面显示的中文为乱码，可以按照第 3 章中介绍的方式来解决，也可以修改 Tomcat 服务器中的 server.xml 配置。

在 Tomcat 的安装目录中的 conf 子目录中找到 server.xml 文件并打开，找到如下内容：

```
<Connector port="8080" protocol="HTTP/1.1"
           connectionTimeout="20000"
           redirectPort="8443" />
```

在上述内容中添加URIEncoding="utf8"，即修改为：

```
<Connector port="8080" protocol="HTTP/1.1"
           connectionTimeout="20000"
           redirectPort="8443" URIEncoding="utf8" />
```

重新启动 Tomcat，就可以发现中文显示正常了。

11.2.11 <c:url>标签

<c:url>标签用于生成一个 URL，其语法形式如下：

```
<c:url  value="url" [context="context"] var="varname"
   [scope="page|request|session|application"]>
   [<c:param/>标签语句]
</ c:url>
```

在上述语法中，url 是要生成的 URL；参数 context 作用于 URL，为其他应用的名字，与<c:import>标签中的 context 属性一样都是可选项；参数 varname 是生成 URL 字符串的变量名称；scope 是其作用范围，默认值为 page。

注意：如果生成的 URL 中的中文是乱码，就可以转换字符集。

11.3 JSTL 的 fmt 标签库

上一节主要介绍了 JSTL 的核心标签库，让读者初步了解了其标签库中各标签的用法。本节将介绍 JSTL 的格式化标签库。格式化标签库在 JSP 开发中也是经常被使用的，它可以通过很简单的方式将日期、数字进行转换。fmt 标签大致有以下几种：

- 国际化标签：<fmt:requestEncoding>和<fmt:setLocale>。
- 消息标签：<fmt:bundle>、<fmt:message>、<fmt:setBundle>和<fmt:param>。
- 数字和日期格式化标签：<fmt:formatNumber>、<fmt:formatDate>、<fmt:parseDate>、<fmt:parseNumber>、<fmt:setTimeZone>和<fmt:timeZone>。

11.3.1 国际化标签

1. <fmt:requestEncoding>标签

<fmt:requestEncoding>标签用于设置请求中数据的字符集，它的作用与 JSP 中设定字符集的语句 request.setCharacterEncoding("charsetName")相同。语法形式如下：

```
<fmt:requestEncoding value="charsetName"/>
```

在上述语句中，参数charsetName是要设置的字符集名称，其语句如下：

```
<fmt:requestEncoding value="utf-8"/>
```

设置完字符集后，就不必再用 request 为每个参数进行字符集的编码转换了。

2. <fmt:setLocale>标签

<fmt:setLocale>标签用于设置用户的语言、国家或地区，其语法如下：

```
<fmt:setLocale value="localcode" [scope=page|request|session|application]
[variant=" variant"] />
```

在上述语法中，参数 localcode 代表语言代码，例如 zh、en，也可以在后面加上国家或者地区的两位数代码，中间用“_”连接，例如：zh_TW（中国台湾地区）、zh_HK（中国香港地区）。参数 variant 用于设置浏览器的类型，例如 win 代表 Windows。Scope 用于设置有效范围，默认值为 page。

11.3.2　消息标签

<fmt:bundle>、<fmt:message>、<fmt:setBundle>、<fmt:param>这4个标签为fmt标签库中的消息标签，其数据来源都是.properties文件。

1. <fmt:bundle>标签

<fmt:bundle>标签的作用是绑定数据源（.properties）文件，语法如下：

```
<fmt:bundle basename="resourceName" prefix="pre">
  代码块
</fmt:bundle>
```

在上述语法中，参数 basename 是要绑定的数据源.properties 文件的文件名，参数 prefix 是要获取的.properties 文件的前缀。若设置 prefix 属性，则嵌套的<fmt:message>标签中的 key 属性就可以省略 prefix 属性设置的前缀部分，其功能主要是针对具有相同前缀的多个关键字的情况。

【例 11.5】　<fmt:bundle>示例。

bundle.jsp 是用于测试绑定源文件的页面，其作用是读取源文件信息，源代码如下：

```
----------------- bundle.jsp-----------------
<%@ page pageEncoding="UTF-8"%>
<%@taglib prefix="c" uri="http://java.sun.com/jsp/jstl/core" %>
<%@taglib prefix="fmt" uri="http://java.sun.com/jsp/jstl/fmt" %>
<fmt:requestEncoding value="utf-8"/>
<!DOCTYPE HTML>
<html>
  <head>
    <title>&lt;fmt:bundle&gt;标签使用例子</title>
  </head>

  <body>
          <c:out value="读取资源文件(myresource.properties)" />
          <br>
          <fmt:bundle basename="myresource"  prefix="my.">
              <fmt:message key="author" var="author" />
             <fmt:message key="teacher" var="teacher" />
          </fmt:bundle>
         作者:<c:out value="${author}" />
         老师:<c:out value="${ teacher }" />
```

```
            <br>
        </body>
    </html>
```

注意：参数 basename 中的文件名不能带扩展名。

在上述代码中，利用标签<fmt:bundle>绑定数据源中的myresource.properties文件，设定前缀是“my.”，利用标签<fmt:message>取出值，运行结果如图11.5所示。

图11.5 bundle.jsp页面的运行结果

如果.properties 文件中存在中文，那么在读取的时候会显示乱码，必须对.properties 文件进行编码转换。可以用 JDK 中提供的转换工具进行转换，转换命令如下：

```
native2ascii -encoding utf8 源.properties文件名  目标.properties文件名
```

注意：执行这个命令时，需要确定是否设置了 JDK 环境变量。

myresource.properties 文件的源文件信息如下：

```
my.author=linl
my.teacher=\u8C2D\u5DE5
```

2. <fmt:message>标签

<fmt:message>标签的作用是从指定的资源文件中进行调用，语法如下：

```
<fmt:message key="messageName" [var="varname"] [bundle="resourceName"]
   [scope="page|request|session|applicatioin"]>
   [<fmt:param>标签]
</fmt:bundle>
```

在上述语法中，参数key是要从.properties文件中取出的键名称，参数bundle是要绑定的数据源.properties文件的文件名，参数varname用来保存取出的键值，scope用来设置标签的有效范围。

3. <fmt:param>标签

<fmt:param>标签需要与<fmt:message>联合使用，用于设定<fmt:message>标签指定键的动态值，语法如下：

```
<fmt:param value="keyvalue"/>
```

若<fmt:message>标签对应的键有多个参数，可以用多个<fmt:param>标签来设置其动态值。若.properties文件中有如下键值对：

```
messageT=Welcome  {0},today is {1,date}
```

则{0}表示第 1 个参数，{1,date}表示第 2 个参数，参数的格式为日期类型。

注意：参数是从 0 开始编号的。

若数据已经绑定，则设定的标签语句为：

```
<fmt:message key="messageT">
   <fmt:param>linl</fmt:param>
   <fmt:param value="${dateT}"/>
</fmt:message>
```

在上述语句中，第 1 个<fmt:param>对应.properties 中的{0}，第 2 个<fmt:param>对应.properties 中的{1,date}，变量 dateT 是设定好的日期变量。

4. <fmt:setBundle>标签

<fmt:setBundle>标签与<fmt:bundle>标签都用于设置默认的数据源，语法如下：

```
<fmt:setBundle basename="resourceName" [var="绑定数据源别名"]
   [scope="page|session|request|application"] >
   代码块
</fmt:setBundle>
```

在上述语法中，参数 basename 是要绑定的数据源.properties 文件的文件名，参数 var 代表绑定的数据源，scope 用于设置 var 参数的有效范围，默认值是 page。

对<fmt:setBundle>标签的示例说明可以参考例 11.5，只需将标签<fmt:bundle>更改成<fmt:setBundle>即可，其运行结果是一样的。

11.3.3　数字和日期格式化标签

1. <fmt:formatNumber>标签

<fmt:formatNumber>标签用于显示不同地区的各种数据格式，语法如下：

```
<fmt: formatNumber  value="numberValue" [type="number|percent|currency"]
     [pattern="pattern"] [currencyCode="currenCyCode"]
[currencySymbol="currencySymbol"]
     [groupingUsed="true|false"] [maxIntegerDigits="maxIntegerDigits"]
     [minIntegerDigits="minIntegerDigits"] [maxFractionDigits="
maxFractionDigits"]
     [minFractionDigits=" minFractionDigits"] [var="varName"]
     [scope="page|session|request|application"] >
</fmt: formatNumber >
```

参数说明如表 11.1 所示。

表 11.1　formatNumber 参数说明

参　数	说　明
value	要格式化的数字
type	设定数字的单位，有 3 种：number、currency、percent

（续表）

参　数	说　明
pattern	设定显示的模式
currencyCode	设置 ISO-4217 编码
currencySymbol	设置货币符号
groupingUsed	设置是否在显示数字时隔开显示
maxIntegerDigits	设置最多的整数位，若设定的数值少于数字的实际位数时，则数字的左边位数会截去相应的位数。例如，123456，maxIntegerDigits 设定为 4，则结果为 3456
minIntegerDigits	设置最少的整数位，若设定的数值多于数字的实际位数时，则会在数字的左边补 0。例如，123，minIntegerDigits 设定为 4，则结果为 0123
maxFractionDigits	设置最多小数位数，若设定的数值小于数字的实际小数位数，则会从右边删掉多于位数。例如，123.56，maxFractionDigits 设定为 1，则结果为 123.5
minFractionDigits	设置最少小数位数，若设定的数值大于数字的实际小数位数，则会从右边补 0。例如，123.56，minFractionDigits 设定为 4，则结果为 123.5600
var	代表格式化后的数字，若设定了该参数，则需要使用<c:out>标签输出
scope	设定参数 var 的有效范围，默认为 page

2. <fmt:parseNumber>标签

<fmt:parseNumber>标签用于把字符串中的数字、货币、百分比转换成数字数据类型，语法如下：

```
    <fmt:parseNumber value="number" [integerOnly="true|false"]
parseLocale="parseLocal"]
       pattern="customPattern"  scope="page|request|session|application"
       type="number|currency|percent" var="varname"/>
```

参数说明如表 11.2 所示。

表 11.2　parseNumber 参数说明

参　数	说　明
value	要解析的数字
type	设定数字的单位，有 3 种：number、currency、percent
pattern	设定显示的模式
integerOnly	设置是否只输出整数部分
parseLocale	解析数字时所用的区域
var	代表格式化后的数字，若设定了该参数，则需要使用<c:out>标签输出
scope	设定参数 var 的有效范围，默认为 page

3. <fmt:formatDate>标签

<fmt:formatDate>标签用于格式化日期，使得可以利用不同方式输出日期和时间，语法如下：

```
<fmt:formatDate value="datevalue" [dateStyle="default|short|medium|long|full"]
   [pattern="customPattern"]  [scope="page|request|session|application"]
   [timeStyle=" default|short|medium|long|full "]  [timeZone="timeZone"]
   [type="time|date|both" ] [var="varname"]/>
```

参数说明如表 11.3 所示。

表 11.3　formatDate 参数说明

参　数	说　明
value	要显示的日期
type	设置输出的类别，如 time、date 和 both
dateStyle	设定日期输出的格式
pattern	自定义格式的模式
timeStyle	设定日期的输出风格
timeZone	设定日期的时区
var	代表格式化后的数字，若设定了该参数，则需要利用<c:out>标签输出
scope	设定参数 var 的有效范围，默认为 page

对 formatDate 输出模式的说明如表 11.4 所示。

表 11.4　formatDate 常用的日期和时间输出模式

模　式	示　例
yyyyMMdd	20140120
HH:mm	11:50
HH:mm:ss	11:50:20
yyyy.mm.dd G 'at' HH:mm:ss z	2014.01.20 公元 at 11:30:10 CST
yyMMddHHmmssZ	140120112030+0800

4. <fmt:parseDate>标签

<fmt:parseDate>标签用于把字符串类型的日期转换成日期数据类型，语法如下：

```
<fmt:parseDate [dateStyle="default|short|medium|long|full"]
   [parseLocale="parseLocale" ]  [pattern="customPattern"]
   [scope="page|request|session|application"]
   [timeStyle=" default|short|medium|long|full "]
   [timeZone="timeZone"]  [type="time|date|both" ]
   value="parseDateValue"  [var="varname"]/>
```

参数说明如表 11.5 所示。

表 11.5　parseDate 参数说明

参　数	说　明
value	要显示的日期
type	设置输出的类别，如 time、date 和 both
dateStyle	设定日期输出的格式
pattern	自定义格式的模式
timeStyle	设定日期的输出风格
timeZone	设定日期的时区
var	代表格式化后的数字，若设定了该参数，则需要利用<c:out>标签输出
scope	设定参数 var 的有效范围，默认为 page

5. <fmt:setTimeZone>标签和<fmt:timeZone>标签

<fmt:setTimeZone>标签用于设定默认的时区，语法如下：

```
<fmt:setTimeZone value="timeValue"
 [scope="page|request|session|application"] var="varname"/>
```

在上述语法中，参数 value 为要设置的时区，var 为存储新时区的变量名，scope 为变量 var 的有效范围。

<fmt:timeZone>标签用来指定时区，以供其他标签使用，语法如下：

```
<fmt:timeZone value="timeValue" >
   代码块
</fmt:timeZone>
```

11.4 JSTL 的 fn 标签库

上一节介绍了JSTL中的fmt标签库，从而让读者初步了解了JSTL中格式化标签库的使用方法。本节介绍JSTL的函数标签库。函数标签库在JSP开发中经常被使用，它可以使字符串的处理变得简单，就像在页面中利用Java方法处理一样。

若要使用 fn 标签库，则需要在 JSP 页面的头部加入如下语句：

```
<%@taglib prefix="fn" uri="http://java.sun.com/jsp/jstl/functions" %>
```

11.4.1 fn:contains()函数与fn:containsIgnoreCase()函数

fn:contains()函数用于判断一个字符串是否包含指定的字符串，语法如下：

```
boolean contains(sourceStr, testStr)
```

在上述语法中，sourceStr 是源字符串，testStr 是指定的字符串。

fn: containsIgnoreCase ()函数用于判断一个字符串是否包含指定的字符串，忽略大小写敏感，语法如下：

```
boolean containsIgnoreCase (sourceStr, testStr)
```

在上述语法中，sourceStr 是源字符串，testStr 是指定的字符串。

11.4.2 fn:startsWith()函数与fn:endsWith()函数

fn:startsWith()函数用于判断一个字符串是否以指定的前缀开始，语法如下：

```
boolean startsWith (sourceStr, startPrefix)
```

在上述语法中，sourceStr 是源字符串；startPrefix 是指定的开始前缀，与 Java 中的 startsWith 方法类似。

fn:endsWith()函数用于判断一个字符串是否以指定的后缀结尾，语法如下：

```
boolean endsWith (sourceStr, endPrefix)
```

在上述语法中，sourceStr 是源字符串；endPrefix 是指定的结束后缀，与 Java 中的 endsWith 方法类似。

11.4.3 fn:escapeXml()函数

fn: escapeXml()函数用于忽略 XML 标记中的字符，语法如下：

```
java.lang.String escapeXml(spcialString)
```

在上述语法中，spcialString 用于指定字符串。

11.4.4 fn:indexOf()函数与fn:length()函数

fn:indexOf()函数用于返回指定子字符串在此字符串中第一次出现处的索引，其用法与 Java 中的 indexOf()方法相同，语法如下：

```
int indexOf(sourceStr, specialStr)
```

在上述语法中，sourceStr 是源字符串，specialStr 是指定的字符串。

fn:length()函数用于返回指定集合或者字符串的长度，其用法与 Java 中的 length()方法相同，语法如下：

```
int length(sourceStr)
```

在上述语法中，sourceStr是要测试的字符串。

【例 11.6】 fn: indexOf()函数和 fn:length()函数示例。

indexOf.jsp 为测试页面，其功能是输出 indexOf 函数中指定的字符串索引和字符串长度，源代码如下：

```
----------------- indexOf.jsp----------------
<%@ page pageEncoding="UTF-8"%>
<%@taglib prefix="c" uri="http://java.sun.com/jsp/jstl/core" %>
<%@taglib prefix="fn" uri="http://java.sun.com/jsp/jstl/functions" %>
<!DOCTYPE HTML>
<html>
  <head>
    <title>fn:indexOf()函数和fn:length()函数使用例子</title>
  </head>
  <body>
    <h3>fn:indexOf()函数和fn:length()函数使用例子</h3>
    <c:set var="string1" value="This is the First my test String."/>
    <c:set var="string2" value="This <abc>is the Second my test String.</abc>"/>

    <p><h4>使用fn:indexOf()函数:</h4></p>
    <p>字符串1 : ${fn:indexOf(string1,"Fir")}</p>
    <p>字符串2 : ${fn:indexOf(string2,"my")}</p>

    <p><h4>使用fn:length()函数:</h4></p>
    <p>字符串1的长度 : ${fn:length(string1)}</p>
  </body>
</html>
```

在上述代码中，第 15、16 行代码用于输出指定字符串的索引，第 19 行代码用于输出字符串 1 的长度。运行结果如图 11.6 所示。

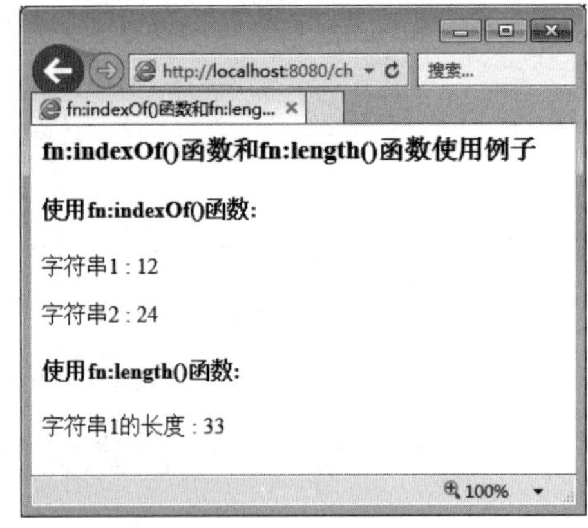

图11.6 例11.6的运行结果

11.4.5 fn:split()函数与fn:join()函数

fn:split()函数用于将字符串按指定的分隔符分割为一个子串数组，其用法与 Java 中的 split()方法相同，语法如下：

```
java.lang.String[] split(sourceStr, regex)
```

在上述语法中，sourceStr 是源字符串，regex 是指定的分隔符。

fn:join()函数用于将一个数组中的所有元素按指定的分隔符来连接成一个字符串，语法如下：

```
String join (array[], regex)
```

在上述语法中，array 是源数组，regex 是指定的分隔符。

11.5 JSTL 的 SQL 标签库

上一节介绍了JSTL的部分fn标签库，让读者初步了解JSTL中fn标签库的使用方法，本节将介绍JSTL的SQL标签库，即数据库标签库。有了SQL标签库，操作数据库就很方便了，虽然在大型的网站中不建议使用这种标签库，但是在小型网站中经常会用到。

若要使用 SQL 标签库，则需要在 JSP 页面的头部加入如下语句：

```
<%@taglib prefix="fn" uri="http://java.sun.com/jsp/jstl/sql" %>
```

SQL 标签库主要包括<sql:setDateSource>、<sql:query>、<sql:update>、<sql:dateParam>、<sql:param>等标签。

11.5.1 <sql:setDateSource>标签

<sql:setDateSource>标签用于设定操作的数据源，语法如下：

```
    <sql:setDataSource dataSource="dataSource" [scope="page|session|request|
application"]  [var="varname"]/>
```

或

```
    <sql:setDataSource driver="drivername" [password="password"]  url="jdbcURL"
user="username"
    var="varname" [scope="page|session|request|application"] />
```

参数说明如表 11.6 所示。

表 11.6　setDateSource 参数说明

参　数	说　明
driver	注册的 JDBC 驱动
url	数据库连接的 JDBC URL
user	连接数据库时使用的用户名
password	连接数据库时使用的密码
dataSource	已经存在的数据源
Var	代表数据源的变量
Scope	设定参数 var 的有效范围，默认为 page

设置<sql:setDateSource>标签的参数并不复杂，若要连接到本地MySQL数据库，则连接语句如下：

```
    <sql:setDateSource var="test" driver="com.mysql.cj.Jdbc.Driver"
url="jdbc:mysql://localhost:3306/test" user="lin" password="123456" />
```

在上述语句中，lin 是数据库用户名，password 是用户密码，driver 是数据库驱动。

11.5.2　<sql:query>标签

<sql:query>标签用于查询数据库中的数据，语法如下：

```
<sql:query var="varname"  [dataSource="dataSource"]  [maxRows="maxRows" ]
    [scope="page|session|request|application"]  sql="sqlQuery"
    [startRow="startRow"]/>
```

参数说明如表 11.7 所示。

表 11.7　query 参数说明

参　数	说　明
maxRows	设置最多可存放的记录条数
sql	查询的 SQL 语句
startRow	设置结果集从查询结果的第几条记录开始
dataSource	连接的数据源
var	代表 SQL 查询的结果
scope	设定参数 var 的有效范围，默认为 page

查询出的结果存放在 var 变量中，若要输出结果就要通过 var 的属性来输出，属性说明如表 11.8 所示。

表 11.8　var 属性说明

参　数	说　明
rows	以字段名称读取 query 结果记录集
rowsByIndex	以数字作为索引的查询结果
columnNames	设置结果集从查询结果的第几条记录开始

（续表）

参　数	说　明
rowCount	结果集中记录的条数
limitedByMaxRows	查询结果记录数是否因为 maxRows 而受到限制，若超过 maxRows，则返回 true，否则返回 false

11.5.3 <sql:update>标签

<sql:update>标签用于更新数据库中的数据，语法如下：

```
<sql:update var="varname"  [dataSource="dataSource"]
[scope="page|session|request|application"]  sql="sqlUpdate"/>
```

参数说明如表 11.9 所示。

表 11.9 <sql:update>标签参数说明

参　数	说　明
sql	更新的 SQL 语句，可以是 insert、update、delete 语句
dataSource	连接的数据源
var	用来存储所影响行数的变量
scope	设定参数 var 的有效范围，默认为 page

11.5.4 <sql:dateParam>标签与<sql:param>标签

<sql:dateParam>标签与<sql:query>标签、<sql:update>标签结合使用，用来提供日期和时间的动态值，语法如下：

```
<sql:dateParam value="value" type="DATE|time|timestamp" />
```

dataParam 参数说明如表 11.10 所示。

表 11.10 dataParam 参数说明

参　数	说　明
value	代表要设置的动态参数值
type	设置日期数据种类，有 date、time、timestamp 共 3 种类型

<sql:param>标签用来设置SQL语句中的动态值，语法如下：

```
<sql:param value="value" />
```

【例 11.7】 SQL 标签综合示例。

sqlZonghe.jsp 是 SQL 标签的综合示例页面，其源代码如下：

```
------------------sqlZonghe.jsp-----------------
<%@ page pageEncoding="UTF-8"%>
<%@ page import="java.util.Date" %>
<%@ taglib uri="http://java.sun.com/jsp/jstl/core" prefix="c"%>
<%@ taglib uri="http://java.sun.com/jsp/jstl/sql" prefix="sql"%>
<%@taglib prefix="fmt" uri="http://java.sun.com/jsp/jstl/fmt"%>
```

```
<html>
<head>
  <title>SQL标签综合示例</title>
</head>
<body>
<!-- 设置数据源-->
<sql:setDataSource var="test" driver="com.mysql.cj.jdbc.Driver"
    url="jdbc:mysql://localhost:3306/TEST"
    user="root"  password="root"/>
<!-- 将用户的年龄增加2岁 -->
<sql:update dataSource="${test}"   var="updatecount">
  update users set user_age=user_age+?
  <c:set value="2" var="count"/>
  <sql:param value="${count}"/>
</sql:update>
<!-- 给ID为1的用户设置日期 -->
<%
Timestamp nowdate = Timestamp.valueOf(LocalDateTime.now());
int userId = 1;
%>
<sql:update dataSource="${test}"   var="updatecount2">
  UPDATE users SET createtime = ? WHERE Id = ?
  <sql:dateParam value="${nowdate}" type="timestamp" />
  <sql:param value="${userId}" />
</sql:update>
<!-- 查询数据-->
<sql:query dataSource="${test}" var="result"
  sql="SELECT * from users;" >
</sql:query>
<!--显示数据  -->
<table border="1" width="100%">
<tr>
  <td colspan="7" align="center">
     共查询${result.rowCount}条用户记录
  </td>
</tr>
<tr>
  <th>用户ID</th>
  <th>用户姓名</th>
  <th>用户性别</th>
  <th>用户年龄</th>
  <th>联系电话</th>
  <th>出身地</th>
  <th>创建日期</th>
</tr>
<c:forEach var="user" items="${result.rows}">
<tr>
  <td><c:out value="${user.id}"/></td>
  <td><c:out value="${user.user_name}"/></td>
  <td><c:out value="${user.user_sex}"/></td>
  <td><c:out value="${user.user_age}"/></td>
  <td><c:out value="${user.user_phone}"/></td>
  <td><c:out value="${user.user_address}"/></td>
  <td><fmt:formatDate type="both" value="${user.createtime}" var=
"formatUsertime"/>
     <c:out value="${formatUsertime}"></c:out></td>
```

```
61  </tr>
62  </c:forEach>
63  </table>
64
65  </body>
66  </html>
```

在上述程序中，第 15~20 行代码给每个用户的年龄都增加 2 岁，第 26~30 行代码给指定用户设置创建日期。程序运行结果如图 11.7 所示。

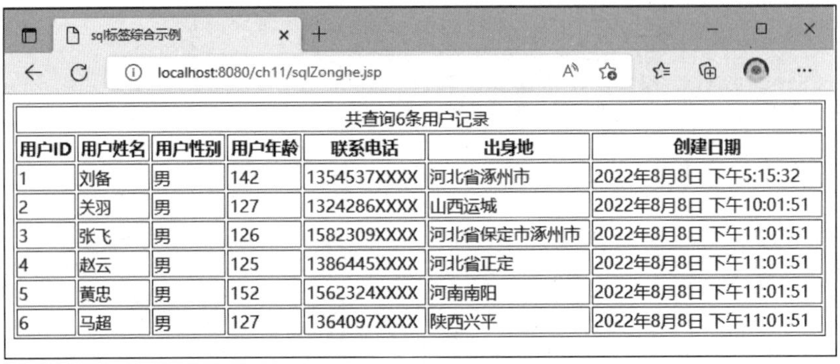

共查询6条用户记录						
用户ID	用户姓名	用户性别	用户年龄	联系电话	出身地	创建日期
1	刘备	男	142	1354537XXXX	河北省涿州市	2022年8月8日 下午5:15:32
2	关羽	男	127	1324286XXXX	山西运城	2022年8月8日 下午10:01:51
3	张飞	男	126	1582309XXXX	河北省保定市涿州市	2022年8月8日 下午11:01:51
4	赵云	男	125	1386445XXXX	河北省正定	2022年8月8日 下午11:01:51
5	黄忠	男	152	1562324XXXX	河南南阳	2022年8月8日 下午11:01:51
6	马超	男	127	1364097XXXX	陕西兴平	2022年8月8日 下午11:01:51

图11.7　sqlZonghe.jsp页面运行结果

11.5.5　<sql:transaction>标签

<sql:transaction>标签是事务标签，用来将<sql:query>标签、<sql:update>标签封装在单一事务中，以确保事务的一致性。<sql:transaction>标签的语法如下：

```
<sql:transaction dataSource="" isolation="">
    <sql:query>标签语句或者<sql:update>标签语句
</sql:transaction>
```

参数说明如表 11.11 所示。

表 11.11　transaction 参数说明

参　数	说　明
dataSource	代表要设置好的数据源
isolation	设置事务的隔离级别，有 4 个取值：READ_COMMITTED、READ_UNCOMMITTED、REPEATABLE_READ 或 SERIALIZABLE

11.6　JSTL 的 XML 标签库

上一节介绍了JSTL的部分SQL标签库，让读者了解了JSTL中SQL标签库的使用方法，以及如何在JSP中访问数据库。本节将介绍JSTL的XML标签库，有了XML标签，就可以轻松处理XML文件。XML标签库与核心标签库的功能相似，读者使用时可以参考核心标签库。

若要使用 XML 标签库，则需在 JSP 页面的头部加入如下语句：

```
<%@taglib prefix="x" uri="http://java.sun.com/jsp/jstl/xml" %>
```

XML 标签库主要包括<x:out>、<x:parse>、<x:set>、<x:choose>、<x:when>、<x:othersise>、<x:forEach>、<x:if>等标签。

11.6.1 <x:parse>标签

<x:parse>标签用来解析 XML 文件，语法如下：

```
<x:parse {doc="XMLDocument"| xml="XMLDocument"} [filter="filter" ]
    [systemId="systemId"] {var="varname" [scope="request|page|session|
application"]
    |varDom="" [scopeDom="request|page|session|application"] } />
```

参数说明如表 11.12 所示。

表 11.12 <x:parse>参数说明

参 数	说 明
var	代表已解析 XML 数据的变量
varDom	代表已解析 XML 数据的变量
xml	需要解析的 XML 文档的文本内容
doc	需要解析的 XML 文档的文本内容
filter	文档过滤器
systemId	XML 文档的 URI
scope	参数 var 的作用范围
scopeDom	参数 varDom 的作用范围

<x:parse>标签常与<c:import>标签结合使用，解析完的节点需要利用<x:out>标签输出。

【例 11.8】 利用<x:parse>标签获取新浪 RSS 新闻。

xparse.jsp 页面是获取新浪 RSS 新闻的测试页面，输出第一个 outline 中的 xmlUrl 属性值，源代码如下：

```
<%@ page pageEncoding="UTF-8"%>
<%@ taglib uri="http://java.sun.com/jsp/jstl/core" prefix="c"%>
<%@taglib prefix="x" uri="http://java.sun.com/jsp/jstl/xml"%>
<html>
<head>
   <title>&lt;x:parse&gt;标签示例</title>
</head>
<body>
    <h3>新浪RSS节点信息:</h3>
    <!-- 导入新浪RSS的XML信息 -->
     <c:import var="xinlangInfo" url="http://rss.sina.com.cn/
sina_all_opml.xml "/>
     <x:parse xml="${xinlangInfo}" var="output"/>
     <b>第一个outline的xmlUrl属性值</b>:
     <x:out select="$output/opml/body/outline/outline[1]/@xmlUrl" />
</body>
</html>
```

在上述代码中，第 11 行是导入新浪 RSS 的 XML 信息，第 12 行解析 XML 文件，第 14 行输出指定的内容。运行结果如图 11.8 所示。

图11.8　xparse.jsp页面的运行结果

11.6.2　<x:out>标签

<x:out>标签用于输出 XML 文件中指定的内容，语法如下：

```
<x:out select="expression" {escapeXml="true|false"}/>
```

在上述语句中，select 属性是 XPath 表达式；escapseXml 代表是否忽略 XML 特殊字符，默认值是 true，即转换特殊字符为实体代码。

11.6.3　<x:forEach>标签

<x:forEach>标签的功能与<c:forEach>相似，但<x:forEach>是针对 XML 文件内容的。<x:forEach>标签的语法如下：

```
    <x:forEach  select="expression" [var="varname"] [varStatus="varstatusName"]
[begin="开始"] [end="结束"]
        [step="step"]]
        Java程序或者HTML代码
    </x:forEach>
```

其参数的说明可以参见<c:forEach>标签（第 11.2.5 节）。

11.6.4　<x:if>标签

<x:if>标签与<c:if>标签相似，但<x:if>标签用于判断一个 XPath 表达式的值是真还是假。<x:if>标签的语法如下：

```
    <x:if  select="expression"  [var="varName"]
[scope="{request|page|session|application}"]>
        条件为真时执行的语句
    </x:if>
```

其中，参数 select 为判断的 XPath 表达式；参数 var 代表条件结果的变量；参数 scope 是 var 的有效范围，默认值是 page。

11.6.5 <x:choose>标签、<x:when>标签、<x:otherwise>标签

<x:choose>、<x:when>、<x:otherwise>标签与<c:choose>、<c:when>、<c:otherwise>标签相似，但前 3 个标签判断的是 XPath 表达式。语法如下：

```
<x:choose>
    <x:when  select="expression">
        表达式为真时执行的语句
    </x:when>
    [<x:otherwise>
        表达式为假时执行的语句
    ]
</x:choose>
```

其中，参数 select 为判断的 XPath 表达式。

11.6.6 <x:set>标签

<x:set>标签用于把 XML 文件中 XPath 表达式的值设置为一个变量，语法如下：

```
    <x:set select="expression"  var="varname" [scope="request|page|session|
application"]/>
```

在上述语句中，select 是 XPath 表达式；参数 var 代表 XPath 表达式值的变量；scope 是参数 var 的有效范围，即作用域。

注意：用<x:set>标签设定值之后，用<x:out>标签输出。

11.6.7 <x:transform>标签

<x:transform>标签可以将 XML 文档转换为 HTML 格式，语法如下：

```
    <x:transform {doc="XMLDocument" docSystemId="docSystemId" | xml="XMLDocument"
xmlSystemId=
    "xmlSystemId"}
    [result="result"] [scope="request|page|session|application"] [var="varname"]
xslt="xslt" [xsltSystemId=""]/>
```

参数说明如表 11.13 所示。

表 11.13 transform 参数说明

参　数	说　明
xml	需要转换的 XML 文档
doc	需要转换的 XML 文档
docSystemId	XML 文档的 URI
xmlSystemId	XML 文档的 URI
xslt	XSLT 样式表
xsltSystemId	源 XSLT 文档的 URI

（续表）

参　　数	说　　明
result	转换结果的对象
var	代表被转换的 XML 文档的变量
scope	参数 var 的作用范围

注意：XSL 是可扩展样式表语言（eXtensible Stylesheet Language），包括 3 个部分：XSLT（一种用于转换 XML 文档的语言）、XPath、XSL-FO（一种用于格式化 XML 文档的语言）。

11.7 小　　结

本章介绍了JSTL标签库的开发技术，并详细介绍了其中5类标签库的使用：core标签库、fmt标签库、fn标签库、SQL标签库和XML标签库。读者可以应用这些标签库开发出复杂的应用系统。

11.8 习　　题

（1）JSTL是什么的缩写？JSTL都有哪5类标签库？

（2）JSTL标签库给JSP开发带来什么好处？

（3）若想在JSP页面中使用JSTL核心标签库，则需要引入什么语句？

（4）使用JSTL的数据库标签库进行数据查询的步骤有哪些？

（5）现有如下报文信息：

```
username=Smith|password=aaaaa|age=30|number=15920578XXX
```

编写一个JSP程序，应用JSTL相关标签对上述报文进行解析，并显示解析结果。

（6）使用数据库标签对数据库表进行增、删、改、查等操作。

（7）编写一个XML文件，应用XML标签对它进行遍历解析。

第 12 章 自定义标签

在JSP页面中，最为理想的代码结构是页面中不含有Java代码，只含有HTML代码和部分标签代码，Java代码只存在于业务逻辑处理的后台中。在上一章中介绍的JSTL标签，使得JSP中的Java代码得到了简化，页面逻辑更加清晰，本章将介绍JSP的自定义标签，通过本章的学习，可以使JSP页面由标签组成，不留下Java代码。

本章主要涉及的知识点有：

- 如何自定义标签
- 标签库文件的描述
- 如何制定带参数的自定义标签
- 如何制定嵌套的自定义标签

12.1 编写自定义标签

所谓自定义标签就是由开发者自己定义的标签，该标签具有某些特殊功能。在JSP页面中使用自定义标签，可以实现在页面中无任何Java代码但却能实现业务的功能，它可以使JSP页面成为一个完完全全由各种标签组成的HTML文件。

从本节开始介绍如何自定义标签，以便读者对自定义标签有一个感性的认识。

12.1.1 版权标签

版权标签就是在JSP页面中显示版权信息。新建一个自定义版权标签的操作步骤如下。

1. 编写自定义标签实现类

CopyRightTag 类实现了生成版权信息并输出的功能：

```
------------------------CopyRightTag.java---------------------------
public class CopyRightTag extends TagSupport{
    @Override
    public int doStartTag(){
        String copyRight = "HazelCopyRight  &copy2022";
        try {
             pageContext.getOut().print(copyRight);
        } catch (IOException e) {
             e.printStackTrace();
        }
        return EVAL_PAGE;
    }
    public int doEndTag(){
        return EVAL_PAGE;
    }
}
```

注意：CopyRightTag 类若不是在编译器中编写，则需要进行编译，然后将编译好的 CLASS 文件放在项目的 WEB-INF\classes 下面。

2. 配置标签

在自定义标签实现类编译完成后，还需要配置标签，这与JSTL配置是一致的，只不过自定义标签需要自己编写TLD文件。自定义标签的TLD文件需要保存在项目的WEB-INF\目录下，为了管理方便，可以在目录下新建TLDS文件夹。

新建 copyright.tld 文件如下：

```
------------------------copyright. tld---------------------------
<?xml version="1.0" encoding="UTF-8" ?>
<taglib xmlns="https://jakarta.ee/xml/ns/jakartaee"
        xmlns:xsi="http://www.w3.org/2001/XMLSchema-instance"
        xsi:schemaLocation="https://jakarta.ee/xml/ns/jakartaee
https://jakarta.ee/xml/ns/jakartaee/web-jsptaglib_3_1.xsd"
        version="3.1">
    <tlib-version>1.0</tlib-version>
    <short-name>linl</short-name>
    <uri>/copyr-tags</uri>

    <tag>
        <!-- 自定义标签的功能描述 -->
        <description>自定义版权</description>
        <name>copyright</name>
        <!-- 自定义标签的实现类路径 -->
        <tag-class>com.eshore.CopyRightTag</tag-class>
        <!-- 正文内容类型  没有正文内容用empty表示 -->
        <body-content>empty</body-content>
    </tag>
</taglib>
```

在上述文件中有且只有一对 taglib 标签，而 taglib 标签下可以有多个 tag 标签，每个 tag 标签代表一个自定义标签。

上述自定义标签的含义是：定义一个copyright的自定义标签，它对应的实现类是com.eshore.CopyRightTag，该标签不需要标签内容，并且没有指定属性。

注意：标签文件的后缀为.tld，新建完成后要保存在 WEB-INF 文件夹下。

3. 使用自定义标签

最后一步就是如何使用标签，实际上使用自定义标签与使用JSTL标签是一样的。以下就是使用上述copyright标签的示例。

【例 12.1】 一个简单的 copyright 标签示例。

copyright.jsp 是引用标签示例，源代码如下：

```
------------------------copyright.jsp--------------------------
01  <%@ page import="java.util.*" pageEncoding="UTF-8"%>
02  <%@ taglib prefix="linl" uri="/copyr-tags" %>
03  <!DOCTYPE HTML>
04  <html>
05    <head>
06      <title>自定义版权标签示例</title>
07    </head>
08    <body>
09            <p>这里是正文的内容</p>
10            <linl:copyright/>
11    </body>
12  </html>
```

在上述代码中，第 2 行代码就是引用自定义标签 copyright，引用 copyright.tld 标签库时通过 prefix 属性指定了标签前缀为 linl，那么在页面中使用自定义标签时就可以直接使用标签<linl:copyright/>来引用标签库 copyright.tld 中的 copyright 标签。页面运行结果如图 12.1 所示。

图12.1 copyright.jsp页面运行结果

经过以上3个步骤，一个完整的自定义版权标签就设置完成并成功应用于页面中了。在后续的例子中也是按照这3个步骤进行的，但是不会详细描述标签的属性等内容。

12.1.2 tld标签库描述文件

tld标签库描述文件实质上是采用XML文件格式进行描述的，TLD文件中常用的元素有taglib、tag、attribute和variable。下面以copyright.tld为例逐个说明其含义。

1. 标签库元素<taglib>

<taglib>元素是用来设置整个标签库信息的，其属性说明如表12.1所示。

表 12.1　标签库中的元素属性说明

属　　性	说　　明
tlib-version	标签库版本号
jsp-version	JSP 版本号
short-name	当前标签库的前缀
uri	页面引用的自定义标签的 URI 地址
name	自定义标签名称
tag-class	自定义标签实现类路径
body-content	自定义标签正文内容（也称标签体）类型，若无则为 empty
description	自定义标签的功能描述
attribute	自定义标签功能的指定属性，可以有多个

2. 标签元素<tag>

<tag>元素用来定义标签的具体内容，属性说明如表 12.2 所示。

表 12.2　<tag>元素属性说明

属　　性	说　　明
name	自定义标签名称
tag-class	自定义标签实现类
body-content	自定义标签的正文内容（也称标签体）类型，有 3 个值：empty（表示无标签体）、JSP（表示标签体可以加入 JSP 程序代码）、tagdependent（表示标签体中的内容由标签自己处理）
description	自定义标签的功能描述
attribute	自定义标签功能的指定属性，可以有多个
variable	自定义标签的变量属性

3. 标签属性元素<attribute>

<attribute>元素用来定义标签<tag>中的属性，其属性说明如表 12.3 所示。

表 12.3　<attribute>元素属性说明

属　　性	说　　明
name	属性名称
description	属性描述
required	属性是否是必需的，默认值为 false
rtexprvalue	属性值是否支持 JSP 表达式
type	定义该属性的 Java 类型，默认值为 String
fragment	如果声明了该属性，那么属性值将被视为一个 JspFragment

注意： *在编写 attribute 属性时还要注意元素顺序。*

4. 标签变量元素<variable>

<variable >元素用来定义标签<tag>中的变量属性，其属性说明如表12.4所示。

表 12.4 <variable>元素属性说明

属　性	说　明
declare	变量声明
description	变量描述
name-from-attribute	指定的属性名称，其值为变量，在调用 JSP 页面时可以使用的名字
name-given	变量名（标签使用时的变量名）
scope	变量的作用范围，NESTED（开始和结束标签之间）、AT_BEGIN（从开始标签到页面结束）、AT_END（从结束标签之后到页面结束）
variable-class	变量的 Java 类型，默认值为 String

12.1.3 TagSupport类简介

在JSP 中，自定义标签库的实现类大多继承自TagSupport类来实现自身的方法。TagSupport类实现了IterationTag接口，有4个重要的方法，如表12.5所示。

表 12.5 TagSupport 类的 4 个重要方法

方　法	说　明
int doStartTag()	遇到自定义标签的开始标记时调用该方法，有 2 个可选值：SKIP_BODY（表示不用处理标签体，直接调用 doEndTag()方法）、EVAL_BODY_INCLUDE（正常执行标签体，但不对标签体做任何处理）
int doAfterBody()	重复执行标签体内容的方法，有 2 个可选值：SKIP_BODY（表示不用处理标签体，直接调用 doEndTag()方法）、EVAL_BODY_AGAIN（重复执行标签体内容）
int doEndTag()	遇到自定义标签的结束标记时调用该方法，有 2 个可选值：SKIP_PAGE（忽略标签后面的 JSP 内容，中止 JSP 页面执行）、EVAL_PAGE（处理标签后，继续处理 JSP 后面的内容）
void release()	释放获得的资源

TagSupport类中各方法的执行过程如下：

01 当在页面中遇到自定义标签的开始标记时，先建立一个标签处理对象，JSP容器回调setPageContext()方法，然后初始化自定义标签的属性值。

02 JSP容器运行doStartTag()方法，如果该方法返回SKIP_BODY，则表示JSP忽略此标签主体的内容；如果返回EVAL_BODY_INCLUDE，则表示JSP容器会执行标签主体的内容，接着运行doAfterBody()方法。

03 运行doAfterBody()方法，若返回EVAL_BODY_AGAIN，则表示JSP容器再次执行标签主体的内容；若返回SKIP_BODY，则JSP容器将运行doEndTag()方法。

04 运行doEndTag()方法，若返回SKIP_PAGE，则表示JSP容器会忽略自定义标签之后的JSP内容；若返回EVAL_PAGE，则运行自定义标签以后的JSP内容。

TagSupport 类的生命周期如图 12.2 所示。

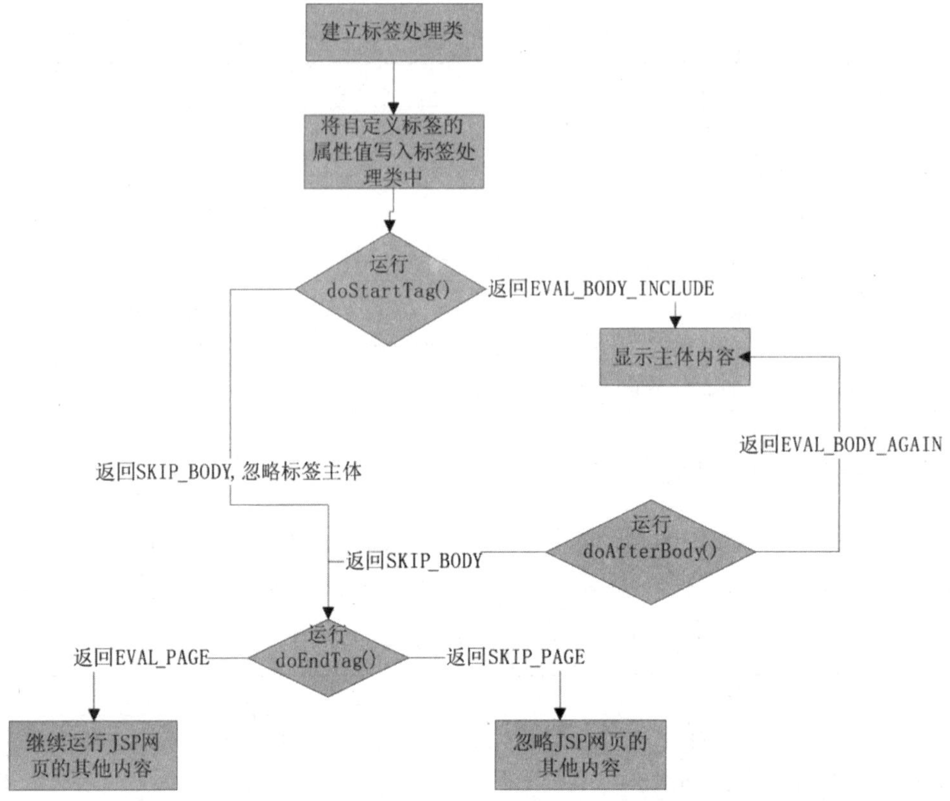

图12.2　TagSupport类的生命周期图

12.1.4　带参数的自定义标签

在自定义标签中，如果可以提供参数支持，那么自定义标签的应用范围就会大幅度提高。例如，对于上面显示版权标签的示例，如果能通过参数配置来指定版权所有人的名字，那么这个标签库就可以被任意使用，它将具有通用性和可配置性。

类似地，如果版权信息中的年份也可以进行指定，那么这个版权标签就不需要每年进行更新。假设满足上述需求的自定义标签名称为 copyrightV，其属性 user 用来指定版权所有人，startY 用来指定版权开始年份，那么在 JSP 页面中使用标签 copyrightV 的代码如下：

```
<%@ page import="java.util.*" pageEncoding="UTF-8"%>
<%@ taglib prefix="linl" uri="/copyrightv-tags" %>
<!DOCTYPE HTML>
<html>
  <head>
    <title>自定义版权标签示例</title>
  </head>

  <body>
          <p>这里是正文的内容</p>
          <linl:copyrghtV user="Hazel" startY="2013"/>
  </body>
</html>
```

1. 定义自定义标签的参数

如果需要自定义标签支持参数，那么必须在定义标签时添加参数，并在TLD文件中添加参数属性。

当前要实现的copyrightV标签有user和startY两个属性，因此在tag标签中必须添加两个attribute属性标签，而每个attribute属性标签都通过name标签指定属性的名字。

copyrightV.tld 文件的完整定义如下：

```
---------------------- copyrightV. tld--------------------------
01  <?xml version="1.0" encoding="UTF-8" ?>
02  <taglib xmlns="https://jakarta.ee/xml/ns/jakartaee"
03          xmlns:xsi="http://www.w3.org/2001/XMLSchema-instance"
04          xsi:schemaLocation="https://jakarta.ee/xml/ns/jakartaee
05          https://jakarta.ee/xml/ns/jakartaee/web-jsptaglib_3_1.xsd"
06          version="3.1">
07      <tlib-version>1.0</tlib-version>
08      <short-name>linl</short-name>
09     <uri>/copyrightv-tags</uri>
10
11      <tag>
12          <!-- 自定义标签的功能描述 -->
13          <description>自定义版权</description>
14          <name>copyrightV</name>
15          <!-- 自定义标签的实现类路径 -->
16          <tag-class>com.eshore.CopyRightTagV</tag-class>
17          <!-- 正文内容类型  没有正文内容用empty表示 -->
18          <body-content>empty</body-content>
19          <!-- 添加attribute属性 -->
20          <attribute>
21              <description>版权拥有者</description>
22            <name>user</name>
23            <required>true</required>
24            <rtexprvalue>true</rtexprvalue>
25            <type>java.lang.String</type>
26          </attribute>
27           <attribute>
28              <description>开始时间</description>
29            <name>startY</name>
30            <required>true</required>
31            <rtexprvalue>true</rtexprvalue>
32            <type>java.lang.String</type>
33          </attribute>
34      </tag>
35  </taglib>
```

在上述代码中，第 20~33 行是添加属性的代码，并配置其属性的各个元素，它们都是必需的元素，将 type 设置成 String 类型。至此，带参数的自定义标签的 TLD 文件配置完成。

2. 定义带参数的自定义标签实现类

若想实现带参数的自定义标签，还要在实现类中添加相应的属性代码。在处理类代码中，通过增加属性的 set 方法，系统就可以自动将标签中的属性值传递给标签类的实例。CopyRightTagV

类相对于 CopyRightTag 类增加了两个属性，因此需要增加两个私有成员变量，并分别为它们实现 set 方法。处理类 CopyRightTagV 的完整代码如下：

```
------------------------CopyRightTagV.java---------------------------
package com.eshore;
import java.io.IOException;
import java.text.SimpleDateFormat;
import java.util.Date;
import jakarta.servlet.jsp.tagext.BodyTagSupport;
import jakarta.servlet.jsp.tagext.BodyContent;
import jakarta.servlet.jsp.JspWriter;
public class CopyRightTagV extends BodyTagSupport{

    private String user;                        //用户名
    private String startY;                      //开始月份
    @Override
    public int doStartTag(){                    //开始标签
        SimpleDateFormat sdf = new SimpleDateFormat("yyyy");
        String endY = sdf.format(new Date());
        String copyRight = user+" CopyRight  "+startY+"-"+endY;
        try {
            pageContext.getOut().print(copyRight);
        } catch (IOException e) {
            e.printStackTrace();
        }
        return EVAL_PAGE;
    }
    public int doEndTag(){                      //结束标签

        return EVAL_PAGE;
    }
    public void setUser(String user) {
        this.user = user;
    }
    public void setStartY(String startY) {
        this.startY = startY;
    }
}
```

完成标签处理类的编写后，经过编译、项目部署，程序的运行结果如图 12.3 所示。

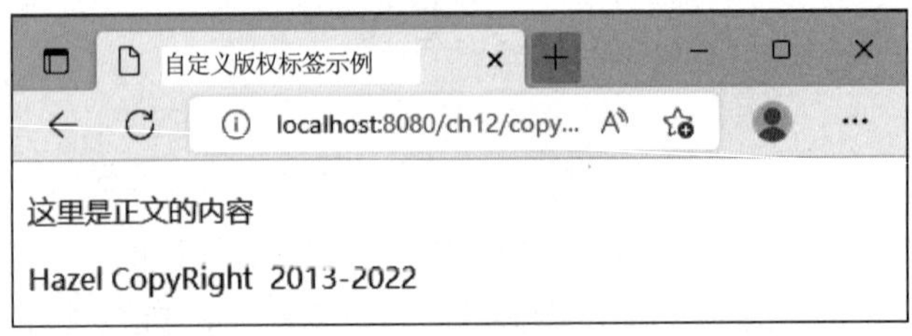

图12.3 页面运行结果

注意：处理类未实现变量的 get 方法，是因为自定义标签在实际项目中大多是将标签的属性值传递给标签实现类处理，get 方法很少被调用，所以一般不实现 get 方法。

12.1.5 带标签体的自定义标签

从tld标签库的描述文件中可以看出标签体除了名称和属性外，还可以有自定义标签，因为有了标签体，自定义标签的灵活性就更高了。

注意： 标签体的含义是指标签起始标记和标签结束标记之间的内容。若无标签体内容，就将起始标记和结束标记合二为一。

定义带标签体的自定义标签的操作步骤如下。

1. 定义包含标签体的 TLD 文件

若要实现自定义标签包含标签体，则必须修改标签库中的TLD定义。正如TLD描述文件中所述的那样，标签定义中的bodycontent属性用来说明当前自定义标签的标签体情况，在版权标签中它的值为empty，表明版权标签无标签体。如果希望标签体中可以包含页面代码，则可以将其值设置为JSP。

在标签库中添加版权标签 copyrightBodycontent，将 bodycontent 的值设定为 JSP，源代码如下：

```
-----------------------copyrightBodycontent. tld--------------------------
<?xml version="1.0" encoding="UTF-8" ?>
<taglib xmlns="https://jakarta.ee/xml/ns/jakartaee"
        xmlns:xsi="http://www.w3.org/2001/XMLSchema-instance"
        xsi:schemaLocation="https://jakarta.ee/xml/ns/jakartaee
        https://jakarta.ee/xml/ns/jakartaee/web-jsptaglib_3_1.xsd"
        version="3.1">
    <tlib-version>1.0</tlib-version>
    <short-name>linl</short-name>
   <uri>/copyrightBodycontent-tags</uri>

    <tag>
        <!-- 自定义标签的功能描述 -->
        <description>自定义版权</description>
        <name>copyright</name>
        <!-- 自定义标签的实现类路径 -->
        <tag-class>com.eshore.CopyRightBodyContentTag</tag-class>
        <!-- 正文内容类型  允许有JSP代码-->
        <body-content>JSP</body-content>
        <!-- 添加attribute属性 -->
        <attribute>
            <description>版权拥有者</description>
          <name>user</name>
          <required>true</required>
          <rtexprvalue>true</rtexprvalue>
          <type>java.lang.String</type>
        </attribute>
         <attribute>
             <description>开始时间</description>
          <name>startY</name>
          <required>true</required>
          <rtexprvalue>true</rtexprvalue>
          <type>java.lang.String</type>
        </attribute>
```

```
    </tag>
</taglib>
```

将 TLD 配置文件放在 WEB-INF/tlds 目录下，标签体的版权标签就定义完成了。

2. 通过定义自定义标签处理类来处理标签体

若想使得自定义标签能够处理标签体，则还需要修改标签处理类。从TagSupport类的生命周期可以看出，如果要处理标签体，就需要重写doAfterBody方法并使类继承自BodyTagSupport类。

添加了 doAfterBody 方法的版权标签处理类的源代码如下：

```
-----------------------CopyRightBodyContentTag.java---------------------------
package com.eshore;

import java.io.IOException;
import java.text.SimpleDateFormat;
import java.util.Date;

import jakarta.servlet.jsp.JspWriter;
import jakarta.servlet.jsp.tagext.BodyContent;
import jakarta.servlet.jsp.tagext.BodyTagSupport;

public class CopyRightBodyContentTag extends BodyTagSupport{

    private String user;                        //用户名
    private String startY;                      //开始月份
    @Override
    public int doStartTag(){                    //开始标签
        SimpleDateFormat sdf = new SimpleDateFormat("yyyy");
        String endY = sdf.format(new Date());
        String copyRight = user+" CopyRight  "+startY+"-"
+endY;
        try {
            pageContext.getOut().print(copyRight);
        } catch (IOException e) {
            e.printStackTrace();
        }
        return EVAL_PAGE;
    }
    public int doAfterBody(){                   //取得标签体

        BodyContent bc = getBodyContent();
        JspWriter out = getPreviousOut();
        try{
            //将标签体中的内容写入到JSP页面中
            out.write(bc.getString());
        }catch(IOException e){
            e.printStackTrace();
        }
        return SKIP_BODY;
    }
    public int doEndTag(){                      //结束标签
        return EVAL_PAGE;
    }
     public void setUser(String user) {
```

```
43            this.user = user;
44        }
45        public void setStartY(String startY) {
46            this.startY = startY;
47        }
48    }
```

在上述代码中，第 27~38 行用于添加 doAfterBody()方法，该方法用于获取标签体内容，并将标签体内容写到 JSP 页面中。至此，处理类的程序逻辑编写完成。

3. 使用带标签体的自定义标签

【例12.2】　使用带标签体的自定义标签。

copyrightBodycontent.jsp 是使用带标签体的版权标签，源代码如下：

```
------------------------ copyrightBodycontent.jsp--------------------------
01    <%@ page import="java.util.*" pageEncoding="UTF-8"%>
02    <%@ taglib prefix="linl" uri="/copyrightBodycontent-tags" %>
03    <!DOCTYPE HTML>
04    <html>
05      <head>
06        <title>自定义版权标签示例</title>
07      </head>
08
09      <body>
10              <p>这里是正文的内容</p>
11          <linl:copyright startY="2013" user="Hazel">
12              <a href="http://www.sina.com">新浪网</a>
13          </linl:copyright>
14      </body>
15    </html>
```

在上述代码中，第 12 行加入一个超链接，还可以加入其他 JSP 代码，它们都能被浏览器解析，并在页面上很好地展示出来。运行结果如图 12.4 所示。

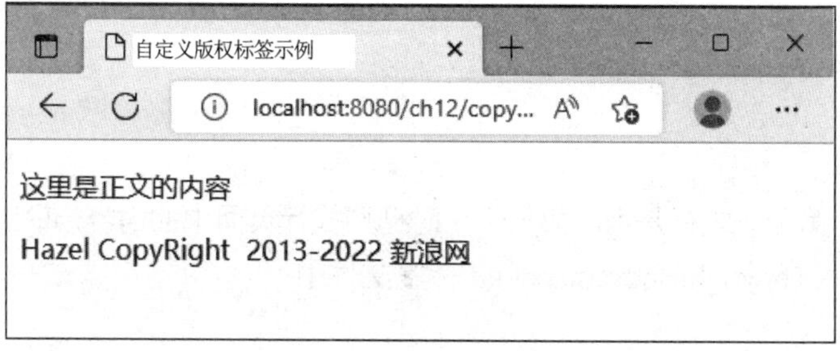

图12.4　copyrightBodycontent.jsp页面运行结果

12.1.6　多次执行的循环标签

自定义标签中的循环标签是指当标签执行 doAfterBody()方法的时候，其返回值是 EVAL_BODY_AGAIN（重复执行标签体内容）。

【例 12.3】 自定义标签的循环标签。

循环标签的实现类源代码如下：

```
---------------------- CopyRightTagLoop.java--------------------------
01  package com.eshore;
02  import java.io.IOException;
03  import java.text.SimpleDateFormat;
04  import java.util.Date;
05  import jakarta.servlet.jsp.JspWriter;
06  import jakarta.servlet.jsp.tagext.BodyContent;
07  import jakarta.servlet.jsp.tagext.BodyTagSupport;
08  public class CopyRightTagLoop extends BodyTagSupport{
09      private static final long serialVersionUID = 1L;
10      private int time;
11      @Override
12      public int doStartTag(){                 //开始标签
13          return EVAL_BODY_INCLUDE;
14      }
15      public int doAfterBody(){                //标签体内容
16          if(time>1){
17              time--;
18              return EVAL_BODY_AGAIN;
19          }else{
20              return SKIP_BODY;
21          }
22      }
23      public int doEndTag(){                   //结束标签
24          JspWriter out = pageContext.getOut();
25          try {
26               out.print("");
27          } catch (IOException e) {
28              e.printStackTrace();
29          }
30          return EVAL_PAGE;
31      }
32      public void setTime(int time) {
33          this.time = time;
34      }
35  }
```

在上述代码中，第 15~22 行是当次数大于 5 时循环执行页面中的内容，否则忽略标签体的内容。将下面代码加入 copyrightBodycontent.tld 标签文件中：

```
<tag>
     <!-- 自定义标签的功能描述 -->
     <description>自定义版权</description>
     <name>loop</name>
     <!-- 自定义标签的实现类路径 -->
     <tag-class>com.eshore.CopyRightTagLoop</tag-class>
     <!-- 正文内容类型  允许有JSP代码-->
     <body-content>JSP</body-content>
     <!-- 添加attribute属性 -->
     <attribute>
         <description>循环次数</description>
         <name>time</name>
```

```
13          <required>true</required>
14          <rtexprvalue>true</rtexprvalue>
15          <type>java.lang.Integer</type>
16      </attribute>
17  </tag>
```

在上述代码中，第 10~16 行用于增加 time 属性值参数，因为实现类中的 time 是 int 类型，所以 type 为 Integer 类。

copyrightloop.jsp 是输出循环标签页面，源代码如下：

```
-----------------------copyrightloop.jsp--------------------------
<%@ page import="java.util.*" pageEncoding="UTF-8"%>
<%@ taglib prefix="linl" uri="/copyrightBodycontent-tags" %>
<!DOCTYPE HTML>
<html>
  <head>
    <title>自定义版权标签示例</title>
  </head>

  <body>
        <p>这里是正文的内容</p>
        <linl:loop time="5" >
            <a href="http://www.sina.com">新浪网</a><br/>
        </linl:loop>
  </body>
</html>
```

上述代码表示循环 5 次输出标签体的内容，运行结果如图 12.5 所示。

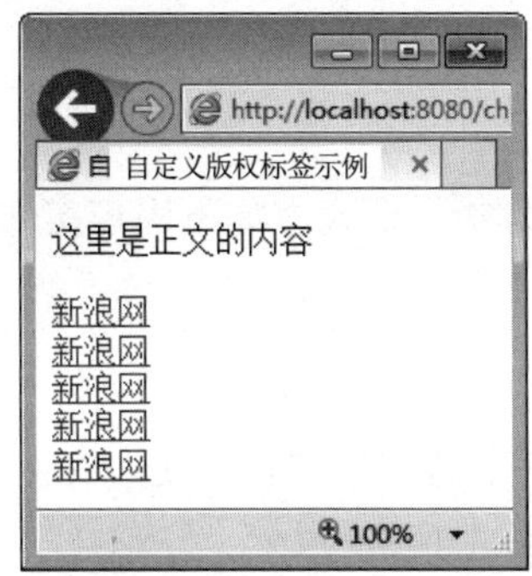

图12.5　copyrightloop.jsp页面运行结果

12.1.7　带动态属性的自定义标签

前面内容都是在标签中直接输入属性值，如果这个属性值可以动态输入，就更符合实际的开发需求了。以例 12.3 为例在页面 copyrightloop.jsp 中将标签<linl:loop>中的 time 值变成动态的输入，源代码如下：

```
<%@ page import="java.util.*" pageEncoding="UTF-8"%>
<%@ taglib prefix="linl" uri="/copyrightBodycontent-tags" %>
<!DOCTYPE HTML>
<html>
  <head>
```

```
06        <title>自定义版权标签示例</title>
07      </head>
08      <%
09          int num = (int)(Math.random()*6)+1;
10       %>
11      <body>
12            <p>这里是正文的内容</p>
13            <linl:loop time="<%=num %>" >
14                <a href="http://www.sina.com">新浪网</a><br/>
15            </linl:loop>
16      </body>
17    </html>
```

在上述代码中，第 08~10 行用于产生 1~6 的随机数，循环的次数就由随机数而定。

12.2 嵌套的自定义标签

嵌套的自定义标签是指自定义的标签相互嵌套，例如：

```
<linl:table var="item" items="${users}">
    <linl:showUserInfo user="${item}" />
</linl:table>
```

从上述的格式可以看出，它包括一个迭代标签<linl:table>和一个输出标签<linl:showUserInfo>，需要分别建立。下面讲解一下如何创建上述嵌套标签。

12.2.1 实例：表格标签

先建立实体类和对应的标签处理类。假设有如下用户实体类：

```
public class UserInfo {
    private String userName;                                          //用户名
    private int age;                                                  //年龄
    private String email;                                             //用户邮箱
    public UserInfo(String userName, int age, String email) {  //构造函数
        super();
        this.userName = userName;
        this.age = age;
        this.email = email;
    }
    public UserInfo() {
        super();

    }
    //省去属性的get和set方法
}
```

注意：上述代码中省略了属性的 get 和 set 方法。由于节省篇幅的原因，本书在不是非常必要的情况下将省略 get 和 set 方法，读者在编辑代码时可自行加上，或者参考本书给出的配套示例代码。

表格标签的处理类 UserInfoTag.java 的源代码如下：

```
------------------------UserInfoTag.java---------------------------
01  import jakarta.servlet.jsp.JspException;
02  import jakarta.servlet.jsp.JspWriter;
03  import jakarta.servlet.jsp.tagext.TagSupport;
04
05  public class UserInfoTag extends TagSupport{
06
07      private static final long serialVersionUID = 1L;
08      private UserInfo user;
09
10      @Override
11      public int doStartTag() throws JspException {            //开始标签
12
13          try {
14              if(user == null) {
15                  out.println("No UserInfo Found...");
16                  return SKIP_BODY;
17              }
18              String content = "<td>"+user.getUserName()+"</td>";
19              content+="<td>"+user.getAge()+"</td>";
20              content+="<td>"+user.getEmail()+"</td>";
21              this.pageContext.getOut().println("<tr>"+content+"</tr>");
22
23          } catch(Exception e) {
24              throw new JspException(e.getMessage());
25          }
26          return SKIP_BODY;
27      }
28
29      @Override
30      public int doEndTag() throws JspException {              //结束标签
31          return EVAL_PAGE;
32      }
33
34      @Override
35      public void release() {                                  //释放资源
36
37          super.release();
38          this.user = null;
39
40      }
41      //省去user属性的get和set方法
42  }
```

在上述代码中，第 14~21 行用于获得用户对象的值，并将值输出到页面中。

循环迭代标签 TableTag.java 的处理类如下：

```
------------------------TableTag.java---------------------------
public class TableTag extends TagSupport {

    private static final long serialVersionUID = 1L;
    private Collection items;
    private Iterator it;
    private String var;
```

```
07
08      @Override
09      public int doStartTag() throws JspException {          //开始标签
10
11          if(items == null || items.size() == 0) return SKIP_BODY;
12          it = items.iterator();
13          if(it.hasNext()) {
14              pageContext.setAttribute(var, it.next());
15          }
16          return EVAL_BODY_INCLUDE;
17      }
18      @Override
19      public int doAfterBody() throws JspException {          //标签体内容
20          if(it.hasNext()) {
21              pageContext.setAttribute(var, it.next());
22              return EVAL_BODY_AGAIN;
23          }
24          return SKIP_BODY;
25      }
26      @Override
27      public int doEndTag() throws JspException {          //结束标签
28          return EVAL_PAGE;
29      }
30      //省去属性的get和set方法
31  }
```

在上述代码中，第09~17行是开始标签，用于将集合容器中的内容输出到页面中；在第20~23行代码中，如果集合还有数据，那就继续遍历标签体的内容，否则返回页面。items属性值是遍历的集合对象，var是存放对象的别名。

12.2.2 嵌套标签的配置

嵌套标签的配置与一般的自定义标签配置基本一致，依次配置标签的名称、实现类、标签体、实现值等。

表格标签的配置如下：

```
<?xml version="1.0" encoding="UTF-8" ?>
<taglib xmlns="https://jakarta.ee/xml/ns/jakartaee"
        xmlns:xsi="http://www.w3.org/2001/XMLSchema-instance"
        xsi:schemaLocation="https://jakarta.ee/xml/ns/jakartaee
        https://jakarta.ee/xml/ns/jakartaee/web-jsptaglib_3_1.xsd"
        version="3.1">
    <tlib-version>1.0</tlib-version>
    <short-name>linl</short-name>
    <uri>/table-tags</uri>
    <!-- 显示表格标签内容 -->
    <tag>
        <name>showUserInfo</name>
        <tag-class>com.eshore.UserInfoTag</tag-class>
        <body-content>empty</body-content>
        <attribute>
            <name>user</name>
            <required>false</required>
```

```
18              <rtexprvalue>true</rtexprvalue>
19          </attribute>
20      </tag>
21      <!-- 遍历表格标签 -->
22      <tag>
23          <name>table</name>
24          <tag-class>com.eshore.TableTag</tag-class>
25          <body-content>JSP</body-content>
26          <!-- 属性 -->
27          <attribute>
28              <name>items</name>
29              <required>false</required>
30              <rtexprvalue>true</rtexprvalue>
31          </attribute>
32          <attribute>
33              <name>var</name>
34              <required>true</required>
35              <rtexprvalue>true</rtexprvalue>
36          </attribute>
37      </tag>
38  </taglib>
```

在上述代码中，第 26~36 行用于配置表格标签的 items 和 var 属性，可分别设置它们的属性值。

12.2.3 嵌套标签的运行效果

经过上面两小节的介绍，嵌套标签的主体步骤已经完成，下面只剩下如何使用它。实际上，使用嵌套标签也是十分简单的，与JSTL标签是相同的。

【例 12.4】 利用表格的方式输出用户信息。

页面 selfTableTag.jsp 利用表格输出用户，源代码如下：

```
-----------------------selfTableTag.jsp--------------------------
<%@ page import="java.util.*,com.eshore.*" pageEncoding="UTF-8"%>
<%@ taglib prefix="linl" uri="/table-tags" %>
<!DOCTYPE HTML>
<html>
  <head>
    <title>自定义表格标签示例</title>
  </head>
  <%
    //模拟从数据库中获取数据
    List<UserInfo> users = new ArrayList<UserInfo>();
    users.add(new UserInfo("张三", 20, "Zhangsan@163.com"));
    users.add(new UserInfo("李四", 26, "Lisi@sina.com"));
    users.add(new UserInfo("王五", 33, "Wangwu@qq.com"));
    pageContext.setAttribute("users", users);
   %>
  <body>
   <div style="text-align: center;">
      用户信息<br/>
   </div>
    <table width='400px' border='1' align='center'>
        <tr>
```

```
22                <td width='20%'>用户名</td>
23                <td width='20%'>年龄</td>
24                <td>邮箱</td>
25            </tr>
26            <!-- 使用标签输出用户信息 -->
27            <linl:table var="item" items="${users}">
28               <linl:showUserInfo user="${item}" />
29            </linl:table>
30        </table>
31     </body>
32   </html>
```

在上述代码中，第 02 行引入自定义标签，第 08~15 行是模拟从数据库中获取用户信息数据，第 26~29 行使用表格标签。页面的运行结果如图 12.6 所示。

图12.6　selfTableTag.jsp页面的运行结果

注意：本示例的缺点是表格的列数输出是固定的，实际上应该做成类似 HTML 标记中的 table 标签中的嵌套，效果有待改善。

12.3　SimpleTag 接口

SimpleTag接口极其简单，提供了doTag()方法去处理自定义标签中的逻辑过程、循环体以及标签体的过程，逻辑处理也简单，实现的接口也相对较少。

在SimpleTag接口中还提供了setJspBody()和getJspBody()方法，用于设置JSP的相关内容。JSP容器会依据setJspBody()方法产生一个JspFragment对象，它的基本特点是可以使处理JSP的容器推迟评估JSP标记属性，一般情况下JSP容器先设定JSP标记的属性，然后在处理JSP标签时使用这些属性；而JspFragment提供了动态属性，这些属性在JSP处理标记体时是可以被改变的。JSP将这样的属性定义为jakarta.servlet.jsp.tagext.JspFragment类型。当JSP标记设置成这种形式时，这种标记属性的处理方法类似于处理标记体。

【例 12.5】　利用 SimpleTag 接口改写版本标签。

首先，编写自定义标签处理类 SimpleTagCopyRight，源代码如下：

```
-----------------------SimpleTagCopyRight.java---------------------------
package com.eshore;
```

```
import java.io.IOException;
import java.text.SimpleDateFormat;
import java.util.Date;
import jakarta.servlet.jsp.JspException;
import jakarta.servlet.jsp.tagext.SimpleTagSupport;
public class SimpleTagCopyRight extends SimpleTagSupport{
    private String user;                                    //用户名
    private String startY;                                  //开始月份
    @Override
    public void doTag() throws JspException, IOException {      //开始标签
        SimpleDateFormat sdf = new SimpleDateFormat("yyyy");
        String endY = sdf.format(new Date());
        String copyRight = user+" CopyRight  "+startY+"-"
+endY;
        getJspContext().getOut().write(copyRight);
    }
    public void setUser(String user) {
        this.user = user;
    }
    public void setStartY(String startY) {
        this.startY = startY;
    }
}
```

注意：SimpleTag 的自定义标签需要继承 SimpleTagSupport 类。

在上述代码中，自定义标签只需重写一个doTag()方法，其余与继承TagSupport类相同。

其次，编写 TLD 文件 simpletagcopyright.tld，编写的方法与前面的例子一样，下面给出其标签文件内容。

```
<tag>
    <!-- 自定义标签的功能描述 -->
    <description>SimpleTag版权</description>
    <name>simpleTagcopyright</name>
    <!-- 自定义标签的实现类路径 -->
    <tag-class>com.eshore.SimpleTagCopyRight</tag-class>
    <!-- 正文内容类型  没有正文内容用empty表示 -->
    <body-content>empty</body-content>
    <!-- 添加attribute属性 -->
    <attribute>
        <description>版权拥有者</description>
        <name>user</name>
        <required>true</required>
        <rtexprvalue>true</rtexprvalue>
        <type>java.lang.String</type>
    </attribute>
    <attribute>
        <description>版权拥有者</description>
        <name>startY</name>
        <required>true</required>
        <rtexprvalue>true</rtexprvalue>
        <type>java.lang.String</type>
    </attribute>
</tag>
```

将上述内容添加到相应的标签文件中。

simpleTagCopyright.jsp 页面的源代码如下：

```
01  <%@ page import="java.util.*" pageEncoding="UTF-8"%>
02  <%@ taglib prefix="linl" uri="/simpleTagCopyright-tags" %>
03  <!DOCTYPE HTML>
04  <html>
05    <head>
06      <title>自定义版权标签示例</title>
07    </head>
08    <body>
09        <p>这里是正文的内容</p>
10        <linl:simpleTagcopyright startY="2013" user="Hazel"/>
11    </body>
12  </html>
```

在上述代码中，第 10 行使用版权标签，与原来的方法一样，其运行效果与图 12.3 相同。综上所述，在 JSP 2.X 中使用标签的方法更为简单，只需实现 doTag()方法即可，逻辑处理也更为简单。

12.4 小　　结

本章介绍了自定义标签的实现方法，通过本章的学习，读者能根据需求编写出合适的标签。在实际应用中，自定义标签应用最多的是自定义分页标签，读者可以根据分页原理自定义一个分页标签，这样在实现应用时可以大量减少开发工作。如果读者还需要进行更为高端的开发（例如开发自己的框架或者中间件等），那么自定义标签就更为重要了。

12.5 习　　题

（1）简述自定义标签处理类BodyTagSupport中的主要方法及其生命周期。

（2）在同一Web应用程序中，是否允许定义两个相同名称的标签？

（3）编写一个自定义标签来显示自己的签名。

（4）编写处理带标签体的自定义标签。

（5）编写自定义方法。

第 13 章 JDBC 详解

本章将介绍在Web中如何与数据库进行通信，包括对数据的CRUD操作。目前，主流的数据库都支持JDBC（Java Database Connectivity），使用JDBC连接某个数据库时，必须找到对应数据库的JDBC驱动包，这样就能连接到数据库，读者可以去MySQL官网下载其JDBC驱动包。

本章主要涉及的知识点有：

- JDBC简介
- MySQL的乱码解决方案
- JDBC的CRUD操作
- 结果集的处理

13.1 JDBC 简介

在Java中主要使用JDBC来访问数据库。JDBCAPI是Java语言访问数据库的一种规范，是Java数据库的编程接口，是一组标准的Java接口和类。使用这些接口和类，可以访问各种不同的数据库。

JDBC 的接口类位于 java.sql 包中，常用的类有 java.sql.Connection、java.sql.Statement、java.sql.ResultSet 等。一般的 JDBC 建立过程有以下 4 个步骤：

01 建立数据库的一个连接。

02 执行SQL语句。

03 处理数据库返回结果。

04 关闭数据库的连接。

下面以一个查询人员列表实例来说明操作 JDBC 的过程。

13.1.1 实例：列出人员信息

首先，建立一张人员信息表并插入数据。打开 MySQL 控制台或者 MySQL 工具，执行如下语句：

```
01  drop database if exists testweb;
02  create database testweb character set utf8;
03  use testweb;
04  drop table if exists person;
05  create table person(
06     id integer auto_increment comment 'id',
07     name varchar(20) comment '姓名',
08     age integer comment '年龄',
09     sex varchar(10) comment '性别',
10     birthday date comment '出生日期',
11     description text comment '备注',
12     create_time timestamp default current_timestamp(),
13     primary key(id)
14  );
15  insert into person(name,age,sex,birthday,description)
16    values('公孙胜','30','男','1984-02-18','绰号入云龙');
17  insert into person(name,age,sex,birthday,description)
18    values('李逵','26','男','1988-03-18','绰号铁牛');
19  insert into person(name,age,sex,birthday,description)
20    values('柴进','30','男','1984-01-18','绰号小旋风');
21  insert into person(name,age,sex,birthday,description)
22    values('秦明','24','男','1990-06-18','绰号霹雳火');
23  insert into person(name,age,sex,birthday,description)
24    values('林冲','30','男','1984-10-18','绰号豹子头');
```

在上述代码中，第 02 行用于以 UTF8 字符集建立 testweb 数据库，第 05~14 行用于创建 person 表，第 15~24 行向表中插入数据。

其次，建立Web项目，编写相应的Servlet类。例如编写PersonServlet.java类，源代码如下：

```
public void doPost(HttpServletRequest request, HttpServletResponse
response) throws ServletException, IOException {
    Connection con = null;
    Statement st = null;
    ResultSet rs = null;
    try {
        Class.forName("com.mysql.cj.jdbc.Driver");      //注册数据库
    } catch (ClassNotFoundException e) {
        e.printStackTrace();
        System.out.println("驱动程序加载错误");
    }
    try {
        con = DriverManager.
        getConnection("jdbc:mysql:     //localhost:3306/testweb",  "root",
"root");      //获取数据库连接
        st = con.createStatement();     //获取Statement
        rs = st.executeQuery("select * from person"); //执行查询，返回结果集
        response.setContentType("text/html;charset=utf-8");
        PrintWriter out = response.getWriter();
```

```
19          out .println("<!DOCTYPE HTML>");
20          //输出页面内容
21          out.println("<HTML>");
22          out.println("  <HEAD><TITLE>列出人员信息表</TITLE></HEAD>");
23          out.println("  <BODY>");
24          out.println("<div style='text-align: center;'><h4>人员信息列表</h4>");
25          out.println("  <table border=\"1\" width=\"100%\"
26                            cellpadding=\"2\" cellspacing=\"1\">");
27          out.println("<tr>");
28          out.println("<td>选择</td>");
29          out.println("<td>姓名</td>");
30          out.println("<td>年龄</td>");
31          out.println("<td>性别</td>");
32          out.println("<td>生日</td>");
33          out.println("<td>备注</td>");
34          out.println("</tr>");
35          while(rs.next()){
36             //遍历结果集ResultSet
37             int id = rs.getInt("id");                                     //获取ID
38             String name  = rs.getString("name");                          //获取姓名
39             int age = rs.getInt("age");                                   //获取年龄
40             String sex  = rs.getString("sex");                            //获取性别
41             Date birthday = rs.getDate("birthday");                       //获取生日
42             String description = rs.getString("description");            //获取备注
43             out.println("<tr>");
44             out.println("<td><input type=\"checkbox\"
45                                 name=\"id\" value=\""+id+"\"></td>");
46             out.println("<td >"+name+"</td>");
47             out.println("<td >"+age+"</td>");
48             out.println("<td >"+sex+"</td>");
49             out.println("<td >"+birthday+"</td>");
50             out.println("<td >"+description+"</td>");
51             out.println("</tr>");
52          }
53          out.println("</table></div>");
54          out.println("  </BODY>");
55          out.println("</HTML>");
56          out.flush();
57          out.close();
58       } catch (SQLException e) {
59          e.printStackTrace();
60       }finally{
61          try { //记住关闭连接
62             rs.close();
63             st.close();
64             con.close();
65          } catch (SQLException e) {
66             e.printStackTrace();
67          }
68       }
69   }
```

在上述代码中，第 07 行用于注册数据库连接，说明是 MySQL 数据库类型；第 13 行用于获得 MySQL 驱动；第 16 行用于获得数据集；第 35~52 行遍历数据结果集；第 61~67 行是关闭数据库连接。数据库连接一定要记得关闭，否则它会一直占用连接池。运行结果如图 13.1 所示。

图13.1　人员信息表效果图

注意：在上述源代码中，只给出 doPost 方法，其余的 doInit()、doDestroy()方法可以为默认的方法。

13.1.2　各种数据库的连接

不同数据库的类型虽然不同，但是使用JDBC连接的步骤却是一样的，只是在获取驱动的URL上有所不同。主流数据库的连接如表13.1所示。

表 13.1　主流数据库的连接

数据库类型	反射注册	连　接	备　注
MySQL	com.mysql.cj.jdbc.Driver	jdbc:mysql://IP:3306/database	IP 为连接的 IP 地址，默认端口是 3306，database 是连接的数据库
Oracle	oracle.jdbc.driver.OracleDriver	jdbc:oracle:thin:@IP:1521:database	IP 为连接的 IP 地址，默认端口是 1521，database 是连接的数据库
DB2	com.ibm.db2.jdbc.app.DB2Driver	jdbc:db2://IP:6789/database	IP 为连接的 IP 地址，默认端口是 6789，database 是连接的数据库
PostgreSQL	org.postgresql.Driver	jdbc:postgresql://IP:5432/database	IP 为连接的 IP 地址，默认端口是 5432，database 是连接的数据库
Sysbase	com.sybase.jdbc.SybDriver	jdbc:sybase:Tds:IP:5000/database	IP 为连接的 IP 地址，默认端口是 5000，database 是连接的数据库
SQLServer 2005	com.microsoft.sqlserver.jdbc.SQLServerDriver	jdbc:sqlserver://IP:1433;databaseName=database	IP 为连接的 IP 地址，默认端口是 1433，database 是连接的数据库

13.2 MySQL 的乱码解决方案

在MySQL控制台中，显示数据和插入数据时可能会出现乱码，在一般情况下都是字符集设定的问题，通过设定数据库的字符集可以解决这个问题。下面介绍几种修改乱码的方法。

MySQL从MySQL 4开始支持UTF-8等几十种编码方式，安装的时候默认是latin1编码方式，它不支持中文，因此需要修改编码方式。

UTF-8字符集能够编码目前世界上所有的语言，一般的网站也是采用UTF-8编码方式，它的缺点就是比较浪费空间，解析也比较复杂，但是为了能统一编码格式，我们都采用UTF-8进行编码。

注意：支持中文的编码方式还有 GBK、GB2312、GB18030、GB2312 是专门针对中文的编码方式，但是它支持的中文字符有限；GBK 比 GB2312 支持更多的中文和中文字符；GB18030 不仅可以支持中文字符，还可以支持少数民族语言。

13.2.1 在控制台中修改编码

在控制台中修改编码时，可以修改数据库编码、数据表编码、表字段编码。

修改数据库编码，输入如下命令：

```
ALTER DATABASE testweb CHARACTER SET utf8;
```

运行效果如图 13.2 所示。

修改数据表编码，输入如下命令：

```
ALTER TABLE  person  CHARACTER SET utf8;
```

运行效果如图 13.3 所示。

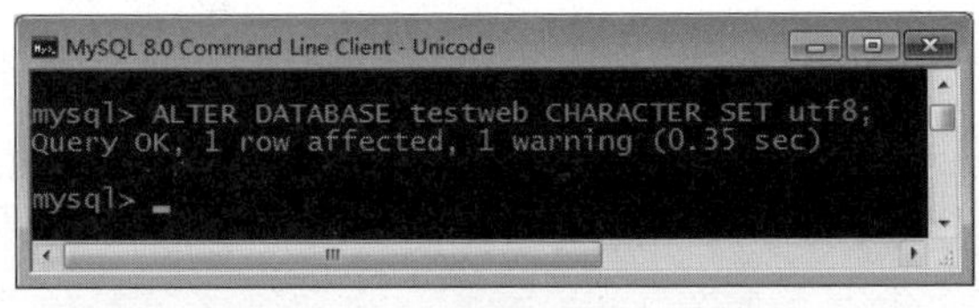

图 13.2　修改数据库编码效果图

MySQL 8.0 Command Line Client - Unicode
mysql> ALTER TABLE person CHARACTER SET utf8;
Query OK, 0 rows affected, 1 warning (0.18 sec)
Records: 0 Duplicates: 0 Warnings: 1
mysql>

图 13.3　修改数据表编码效果图

修改表字段编码，输入如下命令：

```
ALTER TABLE person CHANGE name name VARCHAR(100) CHARACTER SET utf8;
```

运行效果如图 13.4 所示。

修改完成后可以查看数据库字符集是否修改成功，表字段的字符集是否修改成功。

查看系统字符集的命令如下：

```
show variables like 'char%';
```

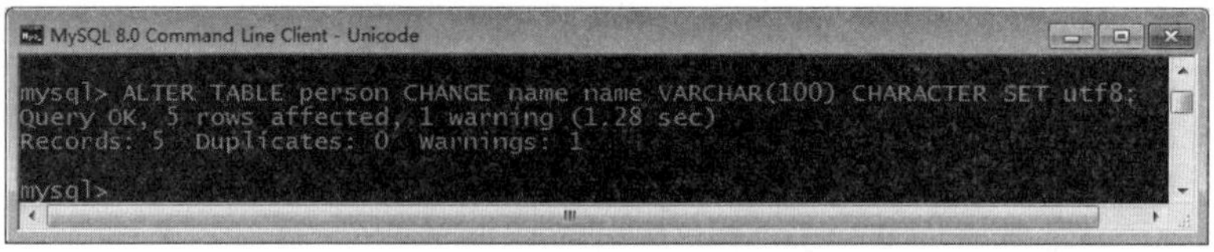

图13.4 修改数据表字段编码效果图

查看数据库字符集的命令如下：

```
show create database databasename;
```

查看数据表字符集的命令如下：

```
show full columns from tablename;
```

例如，执行“show full columns from person”命令，效果如图13.5所示。

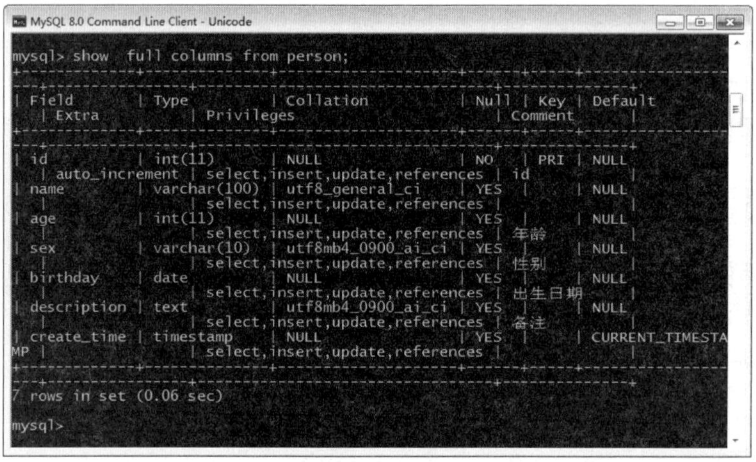

图13.5 查看数据表字符集效果图

13.2.2 在配置文件中修改编码

可以在控制台中修改编码，也可以在配置文件中永久性地修改数据库编码，利用UE工具或者记事本打开MySQL目录下的my.ini文件，Windows系统中该文件一般保存在C:\ProgramData目录下，找到如下语句（该类语句可能处于注释状态或设置为latin1）：

```
default-character-set=latin1
```

或

```
character-set-server= latin1
```

将两个编码方式都设置为utf8，处于注释状态的取消注释，然后重启MySQL服务。

13.2.3 利用图形界面工具修改编码

利用图形界面可以修改数据库中的各种参数，在图形界面中右击表名，在弹出的快捷菜单中选择Alter Table属性，在Collation中选择字符集，如图13.6所示。

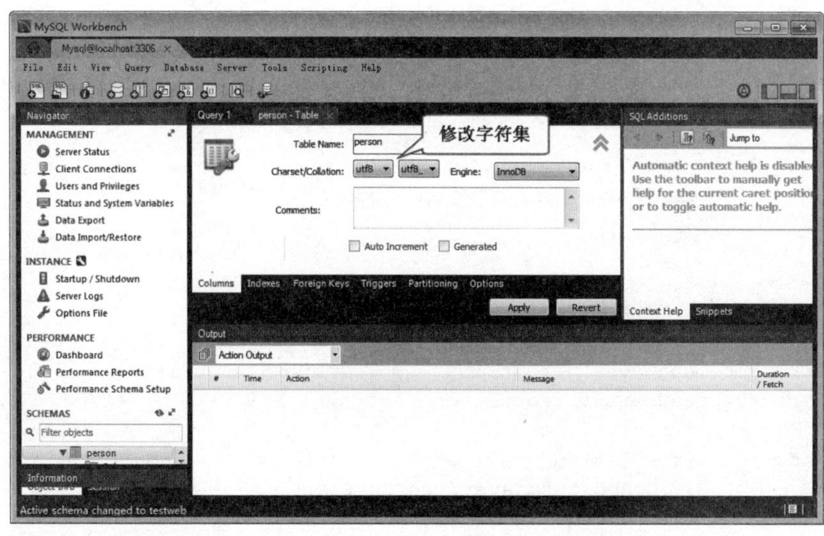

图13.6　利用图形界面工具修改字符集的效果图

13.2.4　在URL中指定编码方式

上面介绍了如何修改数据库的编码方式，在利用 JDBC 连接时，也可以指定编码方式，方法是在连接 URL 后面添加 unicode=true&characterEncoding=UTF-8，例如：

```
jdbc:mysql://localhost:3306/database?unicode=true&characterEncoding=UTF-8
```

一般情况下，经过上述步骤是不会有乱码出现的，如果在程序显示中还有乱码，就检查在请求中是否进行了编码转换。如果在控制台中还出现乱码或者插入不了中文数据，那么建议重装数据库，重装数据库时一定要删除干净 MySQL 文件夹。

13.3　JDBC 基本操作：CRUD

数据库的操作常被称为CRUD，主要是指对数据库进行创建（Create）、查询（Read）、更新（Update）、删除（Delete）等操作。本节将主要介绍如何利用JDBC进行数据库的CRUD操作。

13.3.1　查询数据库

在查询完数据库时，记得要关闭数据库，否则会占用连接，引发连接异常。

13.3.2　插入人员信息

在Java中插入人员信息，主要是要用到executeUpdate(String sql)方法。该方法用于执行对数据库的INSERT、UPDATE、DELETE操作，返回执行语句影响的行数，如果没有影响的行数，就返回0。

【例 13.1】 输出所有的 Locale 代码。

addPerson.jsp 是一个人员信息表单页面，在 action 中页面跳转到 OperateServlet.java 类中执行插入数据的后台操作，成功后返回列表信息。addPerson.jsp 页面的源代码如下：

```
------------------------addPerson.jsp--------------------------
<%@ page import="java.util.*" pageEncoding="UTF-8"%>
<!DOCTYPE HTML>
<html>
  <head>
    <title>增加人员列表</title>
    <!-- 调用日期控件的JS -->
    <script language="javascript" type="text/javascript"
    src="${pageContext.request.contextPath}/My97DatePicker/
     WdatePicker.js"></script>
    <script type="text/javascript"
     src="${pageContext.request.contextPath}/js/jquery-3.6.0.js"></script>
  </head>
  <body>
      <form action="${pageContext.request.contextPath}/servlet/
OperateServlet" method="post">
           <table>
                 <tr>
                     <td>姓名</td>
                     <td><input name="name"/></td>
                 </tr>
                 <tr>
                     <td>性别</td>
                     <td><select name="sex">
                            <option value="男">男</option>
                            <option value="女">女</option>
                       </select></td>
                 </tr>
                 <tr>
                     <td>年龄</td>
                     <td><input name="age"/></td>
                 </tr>
                 <tr>
                     <td>生日</td>
                     <!-- 调用日期控件 -->
                     <td><input id="d11" name="datePicker"
                        type="text" onClick="WdatePicker()"/></td>
                 </tr>
                 <tr>
                     <td>描述</td>
                     <td><textarea name="description"></textarea></td>
                 </tr>
                 <tr>
                     <td colspan="2">
                        <input type="submit" value="提交"/>
                     </td>
                 </tr>
            </table>
      </form>
  </body>
</html>
```

上述代码就是编写的基本表单页面，第 14 行填写跳转的页面或者 Servlet，在第 35 行中填写日期控件。

OperateServlet.java 类的 doPost 方法如下：

```
----------------------OperateServlet.java--------------------------
public void doPost(HttpServletRequest request, HttpServletResponse
response) throws ServletException, IOException {
    request.setCharacterEncoding("UTF-8");
    String name = request.getParameter("name");
    String age = request.getParameter("age");
    String sex = request.getParameter("sex");
    String birthday = request.getParameter("birthday");
    String description = request.getParameter("description");
    String sql = "insert into person(name,age,sex,birthday,description)
"+ "values('"+name+"','"+age+"','"+sex+"','"
+birthday+"', '"+description+"')";
    Connection con = null;
    Statement st = null;
    int  result = 0;
    try {
        Class.forName("com.mysql.cj.jdbc.Driver");    //注册数据库
        con = DriverManager.
        getConnection("jdbc:mysql://localhost:3306/testweb","root","root");
                                                    //获取数据库连接
        st = con.createStatement();                 //获取Statement
        result = st.executeUpdate(sql);        //执行查询，返回结果集
        response.setContentType("text/html;charset=utf-8");
        PrintWriter out = response.getWriter();
        out .println("<!DOCTYPE HTML>");
        //输出页面内容
        out.println("<HTML>");
        out.println("  <HEAD><TITLE>列出人员信息表</TITLE></HEAD>");
        out.println("  <BODY>");
        out.println("");
        out.println("<a href=\""+request.getContextPath()+
"/listPerson.jsp\">返回人员列表</a>");
        out.println("  </BODY>");
        out.println("</HTML>");
        out.flush();
        out.close();
    } catch (SQLException e) {
        e.printStackTrace();
    }catch (ClassNotFoundException e) {
        e.printStackTrace();
    }finally{
        try { //记住关闭连接
            st.close();
            con.close();
        } catch (SQLException e) {
            e.printStackTrace();
        }
    }
}
```

上述代码与 PersonServlet.java 类的差别不大，主要是修改了第 21 行代码，执行插入 executeUpdate()方法。

查询列表的 listPerson.jsp 页面的源代码如下：

```
-----------------------listPerson.jsp--------------------------
<%@ page contentType="text/html; charset=UTF-8"%>
<%@ page import="java.sql.*" %>
<!DOCTYPE HTML>
<HTML>
    <HEAD>
        <TITLE>人员信息列表</TITLE>
<script type="text/javascript"
src="${pageContext.request.contextPath}/js/jquery-3.6.0.js"> </script>
    </HEAD>
    <BODY>
        <div style="text-align: center;">
            <h4>人员信息列表</h4>
        </div>
        <%
          Connection con = null;
          Statement st = null;
          ResultSet rs = null;
          try {
                Class.forName("com.mysql.cj.jdbc.Driver");     //注册数据库
                con = DriverManager.
                getConnection("jdbc:mysql://localhost:3306/testweb", "root","root");
                                                           //获取数据库连接
                st = con.createStatement();                //获取Statement
                rs = st.executeQuery("select * from person");  //执行查询，返回结果集
         %>
         <a href="addPerson.jsp">新增人员信息</a>
         <br/>
         <br/>
        <table border="1" width="100%" cellpadding="2" cellspacing="1">
            <tr>
                <td>选择</td>
                <td>姓名</td>
                <td>年龄</td>
                <td>性别</td>
                <td>生日</td>
                <td>备注</td>
                <td>操作</td>
            </tr>
           <%
                while(rs.next()){
                    //遍历结果集ResultSet
                    int id = rs.getInt("id");          //获取ID
                    String name  = rs.getString("name");  //获取姓名
                    int age = rs.getInt("age");            //获取年龄
                    String sex  = rs.getString("sex");    //获取性别
                    Date birthday = rs.getDate("birthday");   //获取生日
                    String description = rs.getString("description"); //获取备注
```

```
48                          out.println("<tr>");
49                          out.println("<td><input type=\"checkbox\"
50  name=\"checkPerson\" value=\""+id+"\"></td>");
51                          out.println("<td >"+name+"</td>");
52                          out.println("<td >"+age+"</td>");
53                          out.println("<td >"+sex+"</td>");
54                          out.println("<td >"+birthday+"</td>");
55                          out.println("<td >"+description+"</td>");
56                          out.println("<td><a href=\"modify.jsp?id="+id+"\">修改
57                      </a>  "+
58                          "<a href=\"delete.jsp?id="+id+"\"
59                      onclick=\"return confirm('确定删除该记录？')\">删除</a> </td>");
60                         out.println("</tr>");
61                      }
62                  %>
63              </table>
64              <table>
65                  <tr>
66                     <td>
67                         全选<input  type="checkbox" onclick="selectPerson(this);"/>
68
69                     </td>
70                  </tr>
71              </table>
72              <%
73                  } catch (SQLException e) {
74
75                      e.printStackTrace();
76                  }finally{
77                      try {//记住关闭连接
78                          rs.close();
79                          st.close();
80                          con.close();
81                      } catch (SQLException e) {
82
83                          e.printStackTrace();
84                      }
85                  }
86              %>
87          </BODY>
88          <script type="text/javascript">
89             function selectPerson(obj){
90                 $('input[name="checkPerson"]').attr("checked",obj.checked);
91             }
92          </script>
93  </HTML>
```

上述代码实际上是将 PersonServlet.java 改写成 JSP 页面，并新增操作列，分别有修改、删除操作。第 89~91 行代码应用 jQuery 技术对数据的全选进行选择，在页面中输入访问地址 http://localhost:8080/ch13/addPerson.jsp，并填写人员基本信息后提交，执行成功后返回 OperateServlet 的输出页面，单击“返回人员列表”超链接，即可显示是否插入数据成功。页面效果如图 13.7 所示。

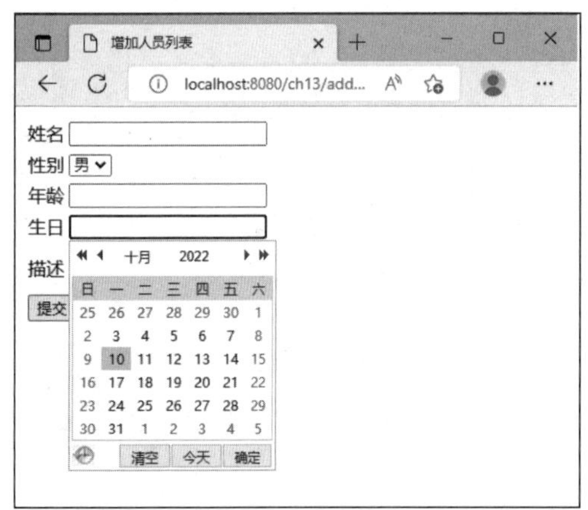

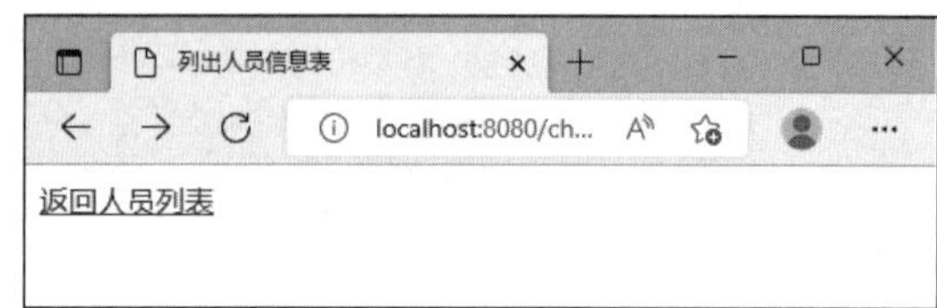

图13.7 表单页面与Servlet效果图

13.3.3 注册数据库驱动

注册数据库驱动有以下两种方式：

- 直接调用DriverManager注册

```
DriverManager.registerDriver(new com.mysql.cj.jdbc.Driver());
```

- 利用Java反射机制注册

```
Class.forName("com.mysql.cj.jdbc.Driver");
```

利用这两种方法注册数据库驱动都是可行的，在目前的开发过程中第 2 种方法的应用比较普遍。

13.3.4 获取自动插入的ID

在上述实例中，创建数据库表的主键采用的是自动增长的数值类型，它由数据库自己去计算与维护，开发者无须关心，但是在对数据进行修改、删除操作时，需要用到主键ID，因此我们要获取这些ID值，以方便对数据进行操作。

Statement提供了getGeneratedKeys()方法来获取数据库主键，遍历其结果集就可以获得自动插入的ID。

13.3.5 删除人员信息

删除人员与插入人员信息的Statement执行方法相同，都是利用executeUpdate(sql)方法，只是SQL语句不一样，删除的语句大致为delete from tableName whcrc id-1，其中where条件可以自定义，如果不指定条件，就为全表删除。

【例 13.2】 删除指定的人员信息。

在查询人员信息的JSP页面中，有一个删除操作，单击“删除”超链接后，在弹出的页面中单击“确定”按钮，程序即可根据用户需求执行删除数据的操作。运行效果如图13.8所示。

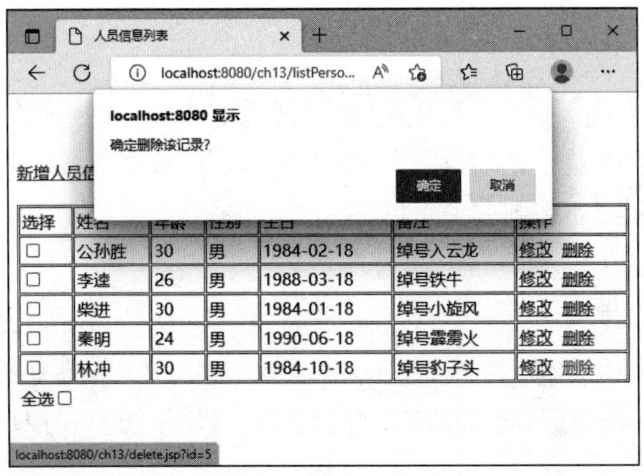

图13.8 删除操作效果图

delete.jsp 的页面源代码如下：

```
-----------------------delete.jsp--------------------------
<%@ page contentType="text/html; charset=UTF-8"%>
<%@ page import="java.sql.*" %>
<!DOCTYPE HTML>
<HTML>
    <HEAD>
        <TITLE>删除人员信息</TITLE>
    </HEAD>
    <BODY>
    <%
      Connection con = null;
      Statement st = null;
      try {
         Class.forName("com.mysql.cj.jdbc.Driver");  //注册数据库
         con = DriverManager.getConnection(      //获取数据库连接
                "jdbc:mysql://localhost:3306/testweb","root","root");
         st = con.createStatement();             //获取Statement
         request.setCharacterEncoding("UTF-8");
         String id = request.getParameter("id"); //获取页面参数id
         String sql = "delete from Person where id='"+id+"'"; //删除人员
         int  result = st.executeUpdate(sql);
      } catch (SQLException e) {

          e.printStackTrace();
      }finally{
         //记住关闭连接
         try {
            st.close();
            con.close();
         } catch (SQLException e) {

            e.printStackTrace();
         }
      }
    %>
    <a href="listPerson.jsp">返回人员信息列表</a>
    <br/>
```

```
        </BODY>
</HTML>
```

从上述代码中可以发现，删除的代码比插入的简单，只有 SQL 语句发生了变化，其余代码类似。在页面中执行删除操作，然后返回人员信息列表，就会发现系统少了一条记录。当然，也可以编写删除多条记录的方法来删除多条记录。

13.3.6 修改人员信息

修改操作是CRUD操作中最复杂的操作，一个完整的修改过程需要多个子过程来协助完成：首先，需要从数据库中查询出数据并呈现给用户进行查看、修改；然后，用户提交修改的内容以保存到数据库中。

在本例中，修改人员信息列表的流程如图13.9所示。首先，单击页面listPerson.jsp的修改操作，提交到modify.jsp页面，从数据库中查询出数据并显示到update.jsp上，执行保存方法。

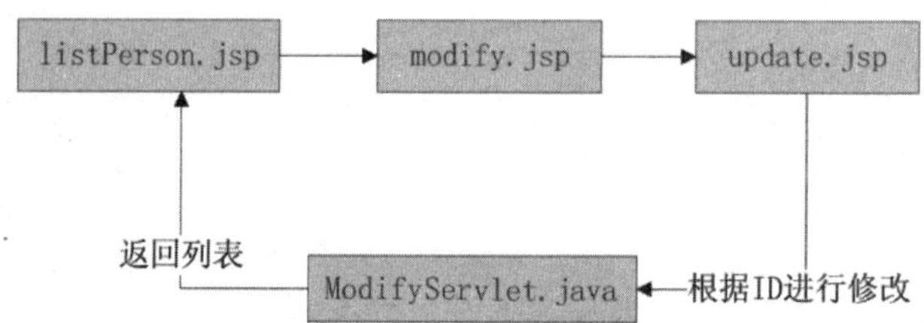

图13.9 修改人员信息流程图

【例 13.3】 修改指定的人员信息。

listPerson.jsp页面用于显示人员信息列表，其源代码在前面的例子中已经给出，这里不再赘述。

modify.jsp 是选择某人员，并从数据库中查询有无该人员的页面，源代码如下：

```
----------------------- modify.jsp---------------------------
<%@ page contentType="text/html; charset=UTF-8"%>
<%@ page import="java.sql.*" %>
<!DOCTYPE HTML>
<HTML>
        <HEAD>
                <TITLE>修改人员信息</TITLE>
        </HEAD>
        <BODY>
        <%
            Connection con = null;
            Statement st = null;
            ResultSet rs = null;
            try {
                Class.forName("com.mysql.cj.jdbc.Driver"); //注册数据库
                con = DriverManager.getConnection(           //获取数据库连接
                "jdbc:mysql://localhost:3306/testweb","root","root");
                st = con.createStatement();            //获取Statement
                request.setCharacterEncoding("UTF-8");
                                                       //获得修改人员的主键ID
                String id = request.getParameter("id");
                //查询该人员的SQL语句
```

```
22                  String sql = "select *  from Person where id='"+id+"'";
23                  rs = st.executeQuery(sql);
24                  if(rs.next()){
25                      //向页面中传递人员信息参数
26                      request.setAttribute("id",rs.getInt("id"));
27                      request.setAttribute("name",rs.getString("name"));
28                      request.setAttribute("sex",rs.getString("sex"));
29                      request.setAttribute("age",rs.getInt("age"));
30                      request.setAttribute("birthday", rs.getString
("birthday"));
31                      request.setAttribute("description",rs.getString
("description"));
32                      RequestDispatcher rd = request.getRequestDispatcher
("update.jsp");
33                      rd.forward(request,response);
34                  }else{
35                       out.println("没有找到id为"+id+"的人员记录");
36                  }
37              } catch (SQLException e) {
38
39                  e.printStackTrace();
40              }finally{
41                  try {//记住关闭数据库连接
42                      rs.close();
43                       st.close();
44                       con.close();
45                  } catch (SQLException e) {
46
47                       e.printStackTrace();
48                  }
49              }
50          %>
51              <br/>
52          </BODY>
53      </HTML>
```

在上述代码中，第 14~17 行代码与前面的例子相同，都是用于连接数据库；第 22 行是组合 SQL 语句；第 23 行执行查询结果语句，如果数据库中存在该数据，就向修改页面中传递人员信息，如果不存在，就打印出该人员不存在的信息。同样要记得关闭数据库连接。

update.jsp（修改页面）与 addPerson.jsp 页面差不多，主要的区别在于修改页面需要将传递的参数显示在页面中，以便客户修改，源代码如下：

```
----------------------- update.jsp--------------------------
<%@ page import="java.util.*" pageEncoding="UTF-8"%>
<!DOCTYPE HTML>
<html>
  <head>
    <title>修改人员</title>
    <!-- 调用日期控件的JS -->
   <script language="javascript" type="text/javascript"
     src="${pageContext.request.contextPath}/My97DatePicker/
WdatePicker.js"></script>
  </head>
  <%
```

```
    //获取传递的参数信息
    Integer id = (Integer)request.getAttribute("id");
    String name = (String)request.getAttribute("name");
    Integer age = (Integer)request.getAttribute("age");
    String sex = (String)request.getAttribute("sex");
    String birthday = (String)request.getAttribute("birthday");
    String description = (String)request.getAttribute("description");
  %>
  <body>
     <form action="${pageContext.request.contextPath}/servlet/
ModifyServlet?id=<%=id%>" method="post">
         <table>
              <tr>
                 <td>姓名</td>
                 <td><input name="name" value="<%=name %>" />
                 </td>
              </tr>
              <tr>
                 <td>性别</td>
                 <td><select name="sex">
                         <%
                          if("男".equals(sex)){
                         %>
                         <option value="男" selected>男</option>
                         <%
                          }else{
                         %>
                          <option value="男" >男</option>
                          <%
                          }
                          %>
                         <%
                          if("女".equals(sex)){
                         %>
                         <option value="女" selected>女</option>
                         <%
                          }else{
                          %>
                          <option value="女" >女</option>
                          <%
                          }
                          %>
                    </select></td>
              </tr>
              <tr>
                 <td>年龄</td>
                 <td><input name="age" value="<%=age%>"/></td>
              </tr>
              <tr>
                 <td>生日</td>
                 <!-- 调用日期控件 -->
                 <td><input id="d11" name="birthday" type="text"
                 onClick="WdatePicker()" value="<%=birthday %>"/></td>
              </tr>
              <tr>
                 <td>描述</td>
```

```
67                  <td><textarea name="description" ><%=description %>
</textarea>
68                  </td>
69              </tr>
70              <tr>
71                  <td colspan="2">
72                      <input  type="submit" value="提交"/>
73                  </td>
74              </tr>
75          </table>
76      </form>
77    </body>
78  </html>
```

在上述代码中，第 10~18 行代码嵌入 Java 代码，以获取人员的参数信息；第 22~75 行在 HTML 中嵌入 JSP 表达式，将人员信息赋值给 HTML 中相应的参数，最后利用 submit 提交 form 到指定的 ModifyServlet 中。

ModifyServlet.java 是将提交的人员信息更新到数据库操作的 Servlet，其 doPost 方法的源代码如下：

```
---------------------- ModifyServlet.java------------------------
public void doPost(HttpServletRequest request, HttpServletResponse
response) throws ServletException, IOException {
    request.setCharacterEncoding("UTF-8");
    String id = request.getParameter("id");
    String name = request.getParameter("name");
    String age = request.getParameter("age");
    String sex = request.getParameter("sex");
    String birthday = request.getParameter("birthday");
    String description = request.getParameter("description");
    String sql = "update Person
    set name='"+name+"',age='"+age+"',sex='"+sex+"',birthday='"
    +birthday+"',description='"+description+"' where id='"+id+"'";
    Connection con = null;
    Statement st = null;
    int  result = 0;
    try {

        Class.forName("com.mysql.cj.jdbc.Driver");    //注册数据库
        //DriverManager.registerDriver(new com.mysql.cj.jdbc. Driver());
        con =DriverManager.getConnection            //获取数据库连接
          "jdbc:mysql://localhost:3306/testweb","root","root");
                                                    //获取Statement
        st = con.createStatement();
        System.out.println(sql);
        result = st.executeUpdate(sql);        //执行查询，返回结果集
        response.setContentType("text/html;charset=utf-8");
        PrintWriter out = response.getWriter();
        out .println("<!DOCTYPE HTML>");
        //输出页面内容
        out.println("<HTML>");
        out.println("  <HEAD><TITLE>列出人员信息表</TITLE></HEAD>");
        out.println("  <BODY>");
        out.println("");
```

```
34              out.println("<a href=\""+request.getContextPath()+
35     "/listPerson.jsp\">返回人员列表</a>");
36              out.println("  </BODY>");
37              out.println("</HTML>");
38              out.flush();
39              out.close();
40          } catch (SQLException e) {
41
42              e.printStackTrace();
43          }catch (ClassNotFoundException e) {
44
45              e.printStackTrace();
46          }finally{
47              //记住关闭连接
48              try {
49                  st.close();
50                  con.close();
51              } catch (SQLException e) {
52
53                  e.printStackTrace();
54              }
55          }
56      }
```

在上述代码中，第 04~09 行是获取页面提交的参数值，第 10~12 行为更新的 SQL 语句，第 18~21 行注册数据库连接，第 28~39 行输出页面信息，第 48~54 行关闭数据库连接。执行成功后返回列表信息，查看是否更新成功，运行效果如图 13.10 所示。

单击“提交”按钮后，返回列表页面，可以发现该人员信息已经修改成功。

注意：修改操作是 CRUD 操作中最复杂的，因此需要注意流程的顺序以及信息的传递。

图13.10　更新页面效果

13.3.7　使用PreparedStatement

在上述代码中除了使用Statement来处理数据库的操作之外，还可以利用Statement的子类PreparedStatement来处理数据的CRUD。

PreparedStatement的优点在于允许使用不完整的SQL语句，可以用“？”来代替数据列值，执行前再将值设置进去。因为PreparedStatement使用“？”来代替SQL中所有可变的部分，只剩下不变的内容，所以JDBC将PreparedStatement中SQL的不变内容进行预先编译，这样下次执行相同的SQL语句时效率就会很高。PreparedStatement与Statement的主要区别如表13.2所示。

表 13.2　PreparedStatement 与 Statement 的主要区别

比较选项	PreparedStatement	Statement
预编译	可以	不可以
大文本数据	支持	不支持

（续表）

比较选项	PreparedStatement	Statement
二进制数据	支持	不支持
执行效率	相对较高	相对较低

【例 13.4】 使用 PreparedStatement 修改指定的人员信息。

针对例 13.3，在 ModifyServlet.java 中将 Statement 修改为 PreparedStatement，实现的程序如下：

```
-----------------------ModifyServletPrepared.java---------------------------
public void doPost(HttpServletRequest request, HttpServletResponse
response) throws ServletException, IOException {
    request.setCharacterEncoding("UTF-8");
    String id = request.getParameter("id");
    String name = request.getParameter("name");
    String age = request.getParameter("age");
    String sex = request.getParameter("sex");
    String birthday = request.getParameter("birthday");
    SimpleDateFormat sdf = new SimpleDateFormat("yyyy-MM-dd");
    String description = request.getParameter("description");
    String sql = "update Person set name=?,age=?,sex=?,birthday=?,
                 description=? where id=?";
    Connection con = null;
    PreparedStatement prest = null;
    int result = 0;
    try {
        Class.forName("com.mysql.cj.jdbc.Driver");    //注册数据库
        con = DriverManager.getConnection(            //获取数据库连接
                "jdbc:mysql://localhost:3306/testweb","root","root");
        prest = con.prepareStatement(sql);
        //获取PreparedStatement，并且预编译SQL语句
        prest.setString(1, name);
        prest.setInt(2, Integer.parseInt(age));   //设定第2个参数
        prest.setString(3, sex);                  //设定第3个参数
        Date date = new Date(sdf.parse(birthday).getTime());
        prest.setDate(4, date);                   //设定第4个参数
        prest.setString(5, description);          //设定第5个参数
        prest.setInt(6, Integer.parseInt(id));    //设定第6个参数
        result = prest.executeUpdate(sql);        //执行查询，返回结果集
        response.setContentType("text/html;charset=utf-8");
        PrintWriter out = response.getWriter();
        out .println("<!DOCTYPE HTML>");
        //输出页面内容
        out.println("<HTML>");
        out.println("  <HEAD><TITLE>列出人员信息表</TITLE></HEAD>");
        out.println("  <BODY>");
        out.println("");
        out.println("<a href=\""+request.getContextPath()+
"/listPerson.jsp\"> 返回人员列表</a>");
        out.println("  </BODY>");
        out.println("</HTML>");
        out.flush();
        out.close();
    } catch (SQLException e) {

        e.printStackTrace();
```

```
47        }catch (ClassNotFoundException e) {
48
49            e.printStackTrace();
50        }catch (ParseException e) {
51
52            e.printStackTrace();
53        }finally{
54            try {//记住关闭连接
55                prest.close();
56                con.close();
57            } catch (SQLException e) {
58
59                e.printStackTrace();
60            }
61        }
62    }
```

从上面的代码可以看出，主体的代码没有大的变化，第 20 行代码产生 PreparedStatement 对象，第 22~28 行为 SQL 语句中的参数赋值，执行效果跟 Statement 是一样的。

13.3.8 利用Statement与PreparedStatement批处理SQL

所谓的批处理SQL，就是批量执行SQL语句，PreparedStatement与Statement都可以进行批处理SQL，并且都是通过addBatch()方法添加SQL语句，通过executeBatch()方法批量执行SQL语句并返回int类型的数组，以表示SQL语句的执行情况。

注意：能够进行批量处理的 SQL 语句必须是 INSERT、UPDATE、DELETE 这样的语句，因为它们返回的是 int 类型。

下面演示PreparedStatement批处理SQL的实例。

【例13.5】 使用PreparedStatement批量插入人员信息。

```
String sql = "insert into Person(name,age,sex,birthday,description)
values(?,?,?,?,?)";
Connection con = null;
PreparedStatement prest = null;
try {
    Class.forName("com.mysql.cj.jdbc.Driver");     //注册数据库
    con = DriverManager.getConnection(             //获取数据库连接
            "jdbc:mysql://localhost:3306/testweb","root","root");
    prest = con.prepareStatement(sql);
    //获取PreparedStatement，并且预编译SQL语句
    for(int i=0;i<=10;i++){
        prest.setString(1, "李四"+i);            //设定第1个参数
        prest.setInt(2, 30);                     //设定第2个参数
        prest.setString(3, "男");                //设定第3个参数
        prest.setDate(4,                         //设定第4个参数
                new Date(System.currentTimeMillis()));
        prest.setString(5, "PreparedStatement 批量插入");   //设定第5个参数
        prest.addBatch();                       //添加SQL语句
    }
    int[] result = prest.executeBatch();     //执行批量插入，返回结果集
```

```
20 } catch (SQLException e) {
21
22     e.printStackTrace();
23 }catch (ClassNotFoundException e) {
24
25     e.printStackTrace();
26 }finally{
27     //记住关闭连接
28     try {
29         prest.close();
30         con.close();
31     } catch (SQLException e) {
32
33         e.printStackTrace();
34     }
35 }
```

从上述代码中的第 11~19 行可以看出，与执行单条 SQL 语句相比，在执行多条语句时，只要将 SQL 语句添加到 addBatch 方法中，再执行 executeBatch 方法，批处理 SQL 即可完成。如果有异常错误，就进行事务回滚，执行不成功。

13.4 结果集的处理

在上一节中，详细介绍了JDBC的CRUD基本操作，使得读者能清楚了解Java操作数据的基本流程。在查询结果数据中，数据都返回到ResultSet结果集中，遍历结果集就能得到数据。本节将介绍与结果集处理相关的内容。

13.4.1 查询多个结果集

查询多个结果集是指查询多张表，返回同一个 ResultSet 对象，在执行第一次查询时，Statement 会返回一个结果集对象，第二次查询时，它会返回一个全新的结果集对象。中间不要关闭 Statement，因为它会自动关闭上一次查询的结果集，以便下次返回全新表的数据。例如：

```
ResultSet rs = stmt.executeQuery("select * from 表一");
//遍历ResultSet对象
while(rs.next()){
}
rs = stmt.executeQuery("select * from 表二");
//遍历ResultSet对象
while(rs.next()){
}
```

13.4.2 可以滚动的结果集

在创建结果集对象时，可以设定滚动的结果集，形式如下：

```
conn.createStatement(ResultSet.TYPE_SCROLL_INSENSITIVE,
ResultSet.CONCUR_UPDATABLE);
```

ResultSet.TYPE_SCROLL_INSENSITIVE 用于指定结果集是可以滚动的，ResultSet.CONCUR_UPDATABLE用于指定结果集是可以直接修改的。

13.4.3 带条件的查询

在一般的Web网站中，带条件的查询是非常普遍的，因此从实用方面考虑，复杂的查询条件是网站所必备的。它的实现原理是将页面中的条件转换为SQL语句中的查询条件。

【例13.6】 带条件的查询。

```
-----------------------searchPerson.jsp---------------------------
<%@ page contentType="text/html; charset=UTF-8"%>
<%@ page import="java.sql.*" %>
<!DOCTYPE HTML>
<HTML>
    <HEAD>
        <TITLE>查询人员信息列表</TITLE>
        <script type="text/javascript"
 src="${pageContext.request.contextPath}/js/jquery-3.6.0.js"></script>
        <!-- 调用日期控件的JS -->
<script language="javascript" type="text/javascript"
src="${pageContext.request.contextPath}/My97DatePicker/WdatePicker.js"></script>
    </HEAD>
    <BODY>
        <user>
            <name>张三</name>
            <english-name>zhangsan</english-name>
            <age>20</age>
            <sex>男</sex>
            <address>广东省广州市</address>
            <description>他是一个工程师</description>
        </user>
        <user>
            <property name="name" value="张三"/>
            <property name="english-name" value="zhangsan"/>
            <property name="age" value="20"/>
            <property name="sex" value="男"/>
            <property name="address" value="广东省广州市"/>
            <property name="description" value="他是一个工程师"/>
        </user>

            <h4 style="text-align: center;">人员信息列表</h4>

        <%
          //获取页面查询条件
          request.setCharacterEncoding("UTF-8");
          String name = request.getParameter("name");
          String sex = request.getParameter("sex");
          String age = request.getParameter("age");
          String description = request.getParameter("description");
          String startTime = request.getParameter("startTime");
          String endTime = request.getParameter("endTime");
          Connection con = null;
```

```
            PreparedStatement st = null;
            ResultSet rs = null;
            //组合SQL的where条件
            String sql = "select * from person where 1=1 ";
            if(name!=null&&!"".equals(name)){
               sql+="and name like '%"+name+"%'";
            }
            if(sex!=null&&!"".equals(sex)){
               sql+="and sex ='"+sex+"'";
            }
            if(age!=null&&!"".equals(age)){
               sql+="and age ='%"+age+"'";
            }
            if(description!=null&&!"".equals(description)){
               sql+="and description like '%"+description+"%'";
            }
            if(startTime!=null&&!"".equals(startTime)){
               sql+="and birthday >= '"+startTime+"'";
            }
            if(endTime!=null&&!"".equals(endTime)){
               sql+="and birthday <= '%"+endTime+"'";
            }
            try {
                 Class.forName("com.mysql.cj.jdbc.Driver");     //注册数据库
                 con = DriverManager.getConnection(    //获取数据库连接
                 "jdbc:mysql://localhost:3306/testweb","root","root");
                 st = con.prepareStatement(sql);       //获取Statement
                 rs = st.executeQuery(sql);            //执行查询，返回结果集
          %>
          <form action="searchPerson.jsp" method="post">
             <table>
                 <tr>
                    <td>姓名：</td>
                    <td><input name="name"/></td>
                    <td>性别：</td>
                    <td><select name="sex" style="width:100">
                           <option value="">无限制</option>
                           <option value="男">男</option>
                           <option value="女">女</option>
                      </select></td>
                 </tr>
                 <tr>
                    <td>年龄：</td>
                    <td><input name="age"/></td>
                    <td>备注：</td>
                    <td><input name="description"/></td>
                 </tr>
                 <td colspan="4">出生日期:
                <label> 从：</label><input class="Wdate"
name="startTime" id="startBeginTime"
onFocus="WdatePicker({dateFmt:'yyyyMMdd HH:mm:ss',minDate: '1900-01-01'})"/>
                      到
                 <input class="Wdate" name="endTime"
id="endBeginTime" onFocus="WdatePicker({dateFmt:'yyyyMMdd HH:mm:ss',
minDate:'1900-01-01'})"/> </td>
                 </tr>
```

```
                <tr>
                    <td>
                        <input type="submit" value="提交">
                        <input type="reset" value="重置">
                    </td>
                </tr>
            </table>
        </form>
        <br/>
        <br/>
        <table border="1" width="100%" cellpadding="2" cellspacing="1">
            <tr>
                <td>选择</td>
                <td>姓名</td>
                <td>年龄</td>
                <td>性别</td>
                <td>生日</td>
                <td>备注</td>
                <td>操作</td>
            </tr>
            <%
                //遍历结果集ResultSet
                while(rs.next()){
                    int id = rs.getInt("id");                        //获取ID
                    String name2  = rs.getString("name");     //获取姓名
                    int age2 = rs.getInt("age");                     //获取年龄
                    String sex2  = rs.getString("sex");       //获取性别
                    Date birthday = rs.getDate("birthday");  //获取出生日期
                    String description2 = rs.getString("description");  //获取备注
                    out.println("<tr>");
                    out.println("<td><input type=\"checkbox\"
                  name=\"checkPerson\" value=\""+id+"\"></td>");
                    out.println("<td >"+name2+"</td>");
                    out.println("<td >"+age2+"</td>");
                    out.println("<td >"+sex2+"</td>");
                    out.println("<td >"+birthday+"</td>");
                    out.println("<td >"+description2+"</td>");
                    out.println("<td><a href=\"modify.jsp?id="+id+"\">
                  修改</a>  "+
                    "<a href=\"delete.jsp?id="+id+"\"
                    onclick=\"return confirm('确定删除该记录？')\">删除</a></td>");
                   out.println("</tr>");
                }
            %>
        </table>
        <table>
            <tr>
              <td>
              全选 <input  type="checkbox"  onclick="selectPerson(this);"/>

              </td>
            </tr>
        </table>
        <%
```

```
153                 } catch (SQLException e) {
154
155                     e.printStackTrace();
156                 }finally{
157                     //记住关闭连接
158                     try {
159                         rs.close();
160                         st.close();
161                         con.close();
162                     } catch (SQLException e) {
163
164                         e.printStackTrace();
165                     }
166                 }
167             %>
168         </BODY>
169         <script type="text/javascript">
170             function selectPerson(obj){
171                 $('input[name="checkPerson"]').attr("checked",obj.checked);
172             }
173         </script>
174     </HTML>
```

相比 listPerson.jsp 而言，上述代码增加了查询条件的 form 以及从页面获取参数的代码。第 35~41 行代码用于获取提交的参数值；第 46~64 行代码是组合查询的 SQL 条件语句，其中日期条件可以直接使用 “>=” “<=”，因为这是建立在 MySQL 数据库上的查询，如果是 Oracle 数据库就要进行转换，所以当遇到某些特殊类型查询的时候要考虑数据库是否支持；第 120~127 行代码用于遍历查询结果集。运行结果如图 13.11 所示。

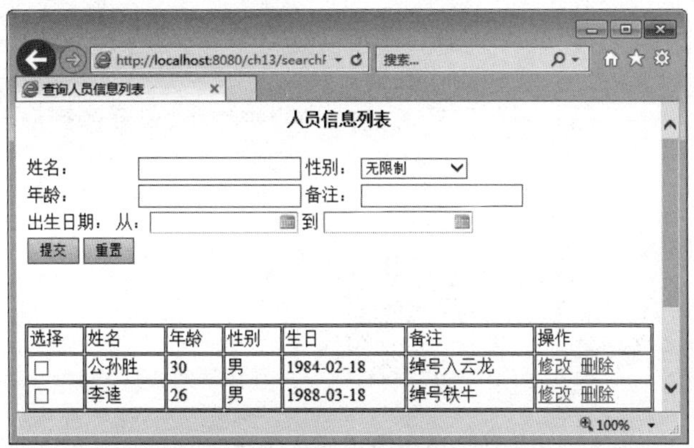

图13.11　searchPerson.jsp运行结果

注意：在进行条件查询时，应注意特殊类型的转换，例如日期函数以及 clob 类型、blob 类型的数据等。

13.4.4　ResultSetMetaData元数据

通过ResultSetMetaData可以获得元数据内容，通过它可以直接得知列名，可以动态显示、查询各列内容。下面我们就来了解一下元数据里面的内容。

【例13.7】 根据ResultSetMetaData元数据输出列值。

```
------------------------getMetaData.jsp----------------------------
<%@ page contentType="text/html; charset=UTF-8"%>
<%@ page import="java.sql.*" %>
<!DOCTYPE HTML>
<HTML>
    <HEAD>
        <TITLE>人员信息列表</TITLE>
    </HEAD>
    <BODY>
        <div style="text-align: center;">
            <h4>人员信息列表</h4>
        </div>
<%
    Connection con = null;
    Statement st = null;
    ResultSet rs = null;
    try {
        Class.forName("com.mysql.cj.jdbc.Driver");     //注册数据库
        con = DriverManager.getConnection(          //获取数据库连接
                "jdbc:mysql://localhost:3306/testweb","root","root");
        st = con.createStatement();                 //获取Statement
        rs = st.executeQuery("select * from person"); //执行查询，返回结果集
        ResultSetMetaData rsmd = rs.getMetaData();
        int columnCount= rsmd.getColumnCount();    //获取列数
        String[] columnNames = new String[columnCount];
        for(int i=0;i<columnCount;i++){             //获取列对应的列名
            columnNames[i] = rsmd.getColumnName(i+1);
        }
        out.println("<table border=\"1\" width=\"100%\"
        cellpadding=\"2\" cellspacing=\"1\"><tr>");
        for(int i=0;i<columnCount;i++){             //输出列名
            out.println("<td>"+columnNames[i]+"</td>");
        }
        out.println("</tr>");
        //遍历结果集ResultSet
        while(rs.next()){
            out.println("</tr>");
            //根据列名取得对应列的值
            for(int i=0;i<columnCount;i++){
                out.println("<td>"+rs.getString(columnNames[i])+"</td>");
            }
            out.println("</tr>");
        }
        out.println("</table>");
    } catch (SQLException e) {

        e.printStackTrace();
    }
    finally{      //记住关闭连接
        try {
            rs.close();
            st.close();
            con.close();
        } catch (SQLException e) {
```

```
54
55                 e.printStackTrace();
56             }
57         }
58  %>
```

在上述代码中，第 22 行用于获得元数据，第 23~26 行用于分析元数据的列数以及将列数对应的列名存放在数组中，第 28~47 行用于显示数据结果集。运行结果如图 13.12 所示。

人员信息列表

id	name	age	sex	birthday	description	create_time
1	公孙胜	30	男	1984-02-18	绰号入云龙	2022-10-10 16:53:45
2	李逵	26	男	1988-03-18	绰号铁牛	2022-10-10 16:53:45
3	柴进	30	男	1984-01-18	绰号小旋风	2022-10-10 16:53:45
4	秦明	24	男	1990-06-18	绰号霹雳火	2022-10-10 16:53:45
5	林冲	30	男	1984-10-19	绰号豹子头	2022-10-10 16:53:46

图13.12 getMetaData.jsp页面效果图

注意：在 ResultSetMetaData 元数据中不仅包含列名，还包含列类型、长度等其他信息。

13.4.5 直接显示中文列名

在查询 SQL 语句时，可以声明别名，例如：

```
select id as 主键,name as 姓名,age as 年龄,sex as 年龄 from person;
```

若使用元数据，则可以直接显示中文列名。如果取得的列名是乱码，则需要进行转码。例如：

```
String name = rs.getMetaData().getColumnName(2);          //取得列名
name= new String(name.getBytes("latin1"),"UTF-8");     //进行转码
```

13.5 小 结

本章介绍了页面与数据库之间的通信、MySQL数据库的乱码问题、数据库的CRUD操作，并通过示例讲解具体的操作过程，包括对数据库的查询，数据的插入、修改、删除等操作；还介绍了对结果集的操作，包括查询多个结果集、可以滚动的结果集、带条件的查询以及元数据的显示等。以上内容基本覆盖了Web开发中的基本问题和对数据的处理流程，读者可以根据本章的例子进一步了解Web系统的开发流程。

13.6 习　　题

（1）JDBC是什么？它的作用是什么？

（2）建立数据库连接的步骤是什么？

（3）数据库的更新操作包括哪几个？

（4）什么是元数据？它有什么作用？

（5）自定义学生成绩表，并编写JSP页面来查询学生成绩，查询方式可以是多样的。

第 14 章 XML 概述

在目前的开发系统中，总是会有很多XML文件，例如struts.xml、spring.xml、web.xml、server.xml以及自定义的XML文件，可以说XML文件无处不在。那么XML是什么呢？应该怎样去编写呢？如何应用呢？本章将解答这些问题。

本章主要涉及的知识点有：

- 什么是XML
- XML的基本用法
- XML的解析方法
- XML与Java类映射

14.1 初识 XML

XML（eXtensible Markup Language）是一种可扩展的标记语言，被设计用来传输和存储数据，是由万维网协会推出的一套数据交换标准。它可以用于定义Web网页上的文档元素，以及复杂数据的表述和传输。在W3C的官网（http://www.w3.org）上有更多的描述，读者也可以去W3C在线学校网站去学习XML的更多内容。

14.1.1 什么是XML

XML与HTML类似，设计的宗旨是传输数据，它没有规定的标签体，需要自定义标签，也是一种自我描述的语言，可以储存数据和共享数据。

XML与HTML的主要差异在于：HTML用来显示数据，XML用来传输和存储数据。

XML的最大特点是它的自我描述和任意扩展。当用它来描述数据时，用户可以根据需要，组织符合XML规范形式的任意内容，并且标签的名称也可以由用户指定。下面以例子来说明XML的定义格式：

```
<?xml version="1.0" encoding="UTF-8"?>
<user>
    <name>张三</name>
    <english_name>zhangsan</english_name>
    <age>20</age>
    <sex>男</sex>
    <address>广东省广州市</address>
    <description>他是一个工程师</description>
</user>
```

以上代码定义的是一个用户的基本信息，包括用户的姓名、英文名称、性别、年龄、住址、描述等。同样是上述内容，也可以利用另外的自定义形式进行描述，比如：

```
<?xml version="1.0" encoding="UTF-8"?>
<user>
    <property name="name" value="张三"/>
    <property name="english_name" value="zhangsan"/>
    <property name="age" value="20"/>
    <property name="sex" value="男"/>
    <property name="address" value="广东省广州市"/>
    <property name="description" value="他是一个工程师"/>
</user>
```

不论利用哪种结构格式，都能清楚地描述用户的基本信息，这就体现了 XML 的可扩展和自定义标签的特点。

14.1.2 XML的用途

XML的设计宗旨是用来传输和存储数据，不仅具有一般纯文本文件的用途，还具有自身的特点。下面介绍XML在开发系统的过程中的常见用途。

1．传输数据

通过XML可以在不同的系统之间传输数据。在开发过程中难免会遇到多个系统之间相互通信，且各系统的存储数据又多种多样的情况，对于开发者而言，这些工作量是巨大的，通过转换为XML格式来传输数据可以减少传输数据时的复杂性，并且还可以具备通用性。

例如，目前流行的SOA协议、Web Service服务、json、Ajax等，其实都是利用XML数据格式在不同的系统之间交互数据。

2．存储数据

存储数据是 XML 最基本的用途，因为它可以作为数据文件，所以当需要持久化保存数据时，可以利用 XML 数据格式进行存储，例如 web.xml、struts.xml、spring.xml 等。下面就是经常见到的 web.xml 文件内容：

```
<?xml version="1.0" encoding="UTF-8"?>
<web-app xmlns="https://jakarta.ee/xml/ns/jakartaee"
        xmlns:xsi="http://www.w3.org/2001/XMLSchema-instance"
          xsi:schemaLocation="https://jakarta.ee/xml/ns/jakartaee
            https://jakarta.ee/xml/ns/jakartaee/web-app_5_0.xsd"
       version="5.0">
  <welcome-file-list>
```

```
    <welcome-file>index.jsp</welcome-file>
  </welcome-file-list>
</web-app>
```

14.1.3 XML的技术架构

XML的技术架构如图14.1所示，它也展示了XML中用到的技术和术语。

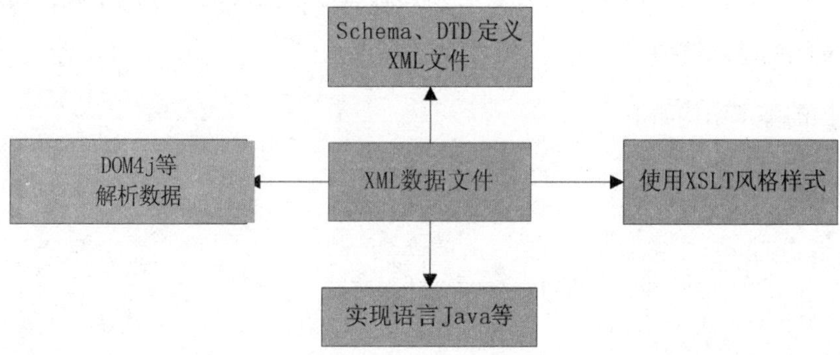

图14.1 XML的技术架构

- 数据定义Schema、DTD：XML数据文件也是要按照一定的协议进行定义的，有两种可遵循的定义规则，即DTD和Schema。DTD是早期的语言；Schema是后期发展的语言，也是现在用得最多的定义XML语言。
- 数据风格样式XSLT：XSLT（eXtensible Stylesheet Language Transformation）是可扩展样式转换，使用XSLT可以将XML中存放的内容按照指定的样式转换为HTML页面。
- 解析XML文件工具：目前比较盛行的工具是DOM、DOM4j、SAX，它们各有特点。
- 操作XML数据：目前将XML数据作为具体操作，其功能都是由额外的程序实现的，一般采用Java比较多，也可以使用JavaScript。

14.1.4 XML开发工具

XML其实就是一个文本文件，其开发工具很多，例如：普通的文本编辑器EditPlus、UEStudio、IntelliJ IDEA的XML编辑器，以及XMLSpy等。在IntelliJ IDEA中编辑XML的情况如图14.2所示。

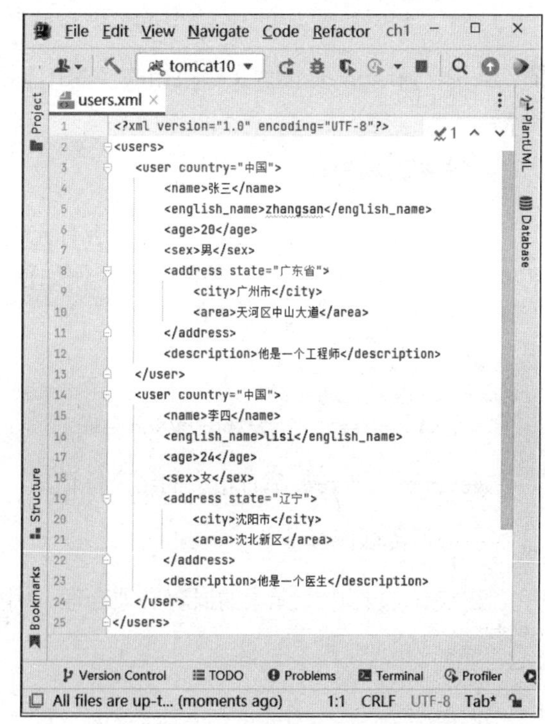

图14.2 IntelliJ IDEA中编辑XML示意图

14.2 XML基本语法

上一节介绍了XML的用途、技术架构和开发工具，从而使读者对XML有一些基本概念。本节将着重介绍XML的基本语法，开发者必须熟悉这些语法规范，才能正确使用XML。

1. XML文档的基本结构

XML文档的基本结构如下：

```
01  <?xml version="1.0" encoding="UTF-8"?>
02  <users>
03  <user>
04      <name>张三</name>
05      <english_name>zhangsan</english_name>
06      <age>20</age>
07      <sex>男</sex>
08      <address>广东省广州市</address>
09      <description>他是一个工程师</description>
10  </user>
11  </users>
```

如上XML文档首先必须要有XML文档的声明，如第01行的声明：

```
<?xml version="1.0" encoding="UTF-8"?>
```

其中，version是指该文档遵循的XML标准版本，encoding用于指明文档使用的字符编码格式。

2. 标记必须闭合

在XML文档中，除XML声明外，所有元素都必须有结束标识，如果XML元素没有文本节点时，就采用自闭合的方式关闭节点，例如使用<note/>这样的形式进行自闭合。

正常的标记闭合形式如下：

```
<age>20</age>
<sex>男</sex>
<address>广东省广州市</address>
```

age、sex、address都有相应的结束标记。

3. 必须合理嵌套

在XML文档中，元素的嵌套必须合理。例如，如下嵌套的例子就不合理：

```
<age>30<sex>
</age>女</sex>
```

这样会导致XML错误，且描述不清。

正确的描述如下：

```
<age>30</age>
<sex>女</sex>
```

4. XML 元素

XML元素是指成对标签出现的内容，且每个元素之间有层级关系。例如：<age>元素指的是<age>20</age>。

<user>元素指的是：

```
<user>
   <name>张三</name>
   <english_name>zhangsan</english_name>
   <age>20</age>
   <sex>男</sex>
   <address>广东省广州市</address>
   <description>他是一个工程师</description>
</user>
```

其中，<age>元素是<user>元素的子元素，<user>元素是<age>元素的父元素，两个<user>元素是并列关系。

元素的命名规则如下：

- 可以包含字母、数字和其他字符。
- 不能以xml开头，包括其大小写，例如XML、xMl等。
- 不能以数字或者标点符号开头，不能包含空格。
- XML文档除了XML以外，没有其他的保留字，任何的名字都可以使用，但是应该尽量使元素名字具有可读性。
- 尽量避免使用“-”和“.”，可能会引起混乱，但可以使用下画线。
- 在XML元素命名中不要使用“:”，因为XML命名空间需要用到这个特殊字符。

例如，下面这些命名是不合法的：

```
<2title>
<xmlTtle>
<titel name>
<.age>
```

正确的命名如下：

```
<title2>
<title_name>
```

5. XML 属性

XML 元素可以在开始标签中包含属性（Attribute），类似于 HTML。属性提供关于元素的额外（附加）信息，被定义在 XML 元素的标签中，且自身有对应的值。例如，<user>元素的属性名和属性值（字体加粗部分）如下：

```
<user  language="java">
   <name>张三</name>
   <english_name>zhangsan</english_name>
   <age>20</age>
   <sex>男</sex>
   <address>广东省广州市</address>
```

```
    <description>他是一个工程师</description>
</user>
```

属性的命名规则与元素的命名规则一样。

注意：属性值必须使用英文引号，英文单引号和英文双引号均可，如果属性值本身包含英文双引号，那么有必要使用英文单引号包围它或者使用实体（"）引用它。

6. 只有一个根元素

所有的 XML 文档有且只有一个根元素来定义整个文档。例如，在 web.xml 代码中，可以看到< web-app>就是它的根元素。下面的定义方式是错误的：

```
<?xml version="1.0" encoding="UTF-8"?>
<web-app xmlns="http://xmlns.jcp.org/xml/ns/javaee"
       xmlns:xsi="http://www.w3.org/2001/XMLSchema-instance"
         xsi:schemaLocation="http://xmlns.jcp.org/xml/ns/javaee
           http://xmlns.jcp.org/xml/ns/javaee/web-app_4_0.xsd"
       version="4.0">...
</web-app>
<web-app>
...
</web-app>
```

7. 大小写敏感

XML文档是大小写敏感的，包括标签名称、属性名和属性值等。例如，<age>与<Age>是不同的。

一般情况下，初学者常犯的错误就是因开始标记与结束标记的大小写不一致而导致的 XML 错误。例如：

```
<name>Jonh</Name>
```

<name>与</Name>不能相互匹配，从而导致<name>没有被正确关闭。

8. 空白被保留

空白被保留是指在 XML 文档中，空白部分并不会被解析器删除，而是被当作数据完整地保留。例如：

```
<description>好好学习    天天向上</ description>
```

“好好学习　　天天向上”中的空白会被当作数据保留。

9. 注释的写法

XML 的注释形式如下：

```
<!-- 注释单行 -->
<!--
    注释多行
-->
```

10. 转义字符的使用

在XML中有些特殊字符需要转义，例如“>”“<”“&”以及单引号、双引号等，其转义字符与HTML中的转义字符是一样的。

11. CDATA 的使用

CDATA 适用于需要原文保留的内容，尤其是在解析 XML 过程中产生歧义的部分。当某个节点的数据有大量需要转义的字符时，CDATA 就可以发挥其作用了，语法如下：

```
<![CDATA[
    内容
  ]]>
```

例如：

```
<![CDATA[
if(m>n){
  alert(m大于n);
}else if(m<n&&m!=0){
   alert(m小于n)
}
   ]]>
```

注意：CDATA 是不能嵌套的。

14.3　JDK 中的 XML API

JDK 中涉及 XML 的 API 有两个，分别是：

- The Java API For XML Processing：负责解析XML。
- Java Architecture for XML Binding：负责将XML映射为Java对象。

它们所涉及的类包有 javax.xml.*、org.w3c.dom.*、org.xml.sax.*和 javax.xml.bind.*。

经常用到的JDK XML API类如表14.1所示。

表 14.1　常用的 JDK XML API 类

类	说　明
javax.xml.parsers.DocumentBuilder	从 XML 文档获取 DOM 文档实例
javax.xml.parsers.DocumentBuilderFactory	从 XML 文档获取生成 DOM 对象树的解析器
javax.xml.parsers.SAXParser	获取基于 SAX 的解析器以解析 XML 文档实例
javax.xml.parsers. SAXParserFactory	获取基于 SAX 的解析器以解析 XML 文档
org.w3c.dom.Document	整个 XML 文档
org.w3c.dom.Element	XML 文档中的一个元素
org.w3c.dom.Node	Node 接口是整个文档对象模型的主要数据类型
org.xml.sax. XMLReader	是 XML 解析器的 SAX2 驱动程序必须实现的接口

14.4 最常见的 XML 解析模型

上一节介绍了XML的基本语法、命名规则和语法规范，使得读者能对XML的文档格式有了基本了解，本节将在此基础上讲解如何对XML进行解析。由于XML结构基本上是一种树型结构，因此处理XML的步骤都差不多，Java已经将它们封装成了现成的类库。目前流行的解析方法有DOM、SAX和DOM4j这3种，下面将逐一进行介绍。

14.4.1 DOM解析

DOM（Document Object Model，文档对象模型）是W3C组织推荐的处理XML的一种方式。它是一种基于对象的API，把XML内容加载到内存中，生成一个XML文档相对应的对象模型，根据对象模型，以树节点的方式对文档进行操作。下面以实例说明解析步骤。

【例 14.1】 DOM 解析 XML 文件。

假设 XML 文件如下：

```
<?xml version="1.0" encoding="UTF-8"?>
<users>
    <user country="中国">
        <name>张三</name>
        <english_name>zhangsan</english_name>
        <age>25</age>
        <sex>男</sex>
        <address state="广东省">
            <city>广州市</city>
            <area>天河区中山大道</area>
        </address>
        <description>他是一个工程师</description>
    </user>
    <user country="中国">
        <name>李四</name>
        <english_name>lisi</english_name>
        <age>30</age>
        <sex>女</sex>
        <address state="辽宁省">
            <city>沈阳市</city>
            <area>沈北新区</area>
        </address>
        <description>他是一个医生</description>
    </user>
</users>
```

编写解析类JAXBDomDemo的代码如下：

```
-----------------------JAXBDomDemo.java--------------------------
public class JAXBDomDemo {

    /**
```

```
     * 用DOM解析XML文件
     */
    public static void main(String[] args) {
        //创建待解析的XML文件，并指定目录
        File file = new File("F:\\users.xml");
        //用单例模式创建DocumentBuilderFactory对象
        DocumentBuilderFactory factory =
DocumentBuilderFactory.newInstance();
        //声明一个DocumentBuilder对象
        DocumentBuilder documentBuilder =null;
        try {
            //通过DocumentBuilderFactory构建DocumentBuilder对象
            documentBuilder = factory.newDocumentBuilder();
            //DocumentBuilder解析XML文件
            Document document = documentBuilder.parse(file);
            //获得XML文档中的根元素
            Element root = document.getDocumentElement();
            //输出根元素的名称
            System.out.println("根元素："+root.getNodeName());
            //获得根元素下的子节点
            NodeList childNodes = root.getChildNodes();
            //遍历根元素下的子节点
            for(int i=0;i<childNodes.getLength();i++) {
                //获得根元素下的子节点
                Node node = childNodes.item(i);
                if (node instanceof Element) {//判断是否为元素
                    System.out.println("节点的名称为" +
                                node.getNodeName());
                    //获得子节点的country属性值
                    String attributeV = node.getAttributes().
                            getNamedItem("country").getNodeValue();
                    System.out.println(node.getNodeName() +
                "节点的country属性值为" + attributeV);
                    //获得node子节点下的集合
                    NodeList nodeChilds = node.getChildNodes();
                    //遍历node子节点下的集合
                    for (int j = 0; j < nodeChilds.getLength(); j++) {
                        Node details = nodeChilds.item(j);
                        if (details instanceof Element) {//判断是否为元素
                            String name = details.getNodeName();
                            //判断如果是address元素，则获取其子节点
                            if ("address".equals(name)) {
                                NodeList addressNodes =
details.getChildNodes();
                        //遍历address元素的子节点
                        for (int k = 0; k < addressNodes.getLength(); k++) {
                        Node addressDetail = addressNodes.item(k);
                        if(addressDetail instanceof Element) {
                        System.out.println(node.getNodeName() + "节点的子
节点"
                            + name + "节点的子节点" +
                        addressDetail.getNodeName() + " 节点内容为："
                            + addressDetail.getTextContent());
                                    }
                                }
                        String addressAtt = details.getAttributes().
```

```
56                                  getNamedItem("state").getNodeValue();
57                              System.out.println(name + "节点的state属性值为"
58                                  + addressAtt);
59                                  }
60                                  System.out.println(node.getNodeName() +
61                              "节点的子节点" + details.getNodeName() +
62                              " 节点内容为：" + details.getTextContent());
63                              }
64                          }
65                      }
66                  }
67          } catch (ParserConfigurationException e) {
68              e.printStackTrace();
69          }catch (IOException e) {
70              e.printStackTrace();
71          }catch (SAXException e) {
72              e.printStackTrace();
73          }
74      }
75  }
```

在上述代码中，详细描述了解析步骤，其中第 09~15 行代码分别创建了解析工程类 DocumentBuilderFactory、DocumentBuilder 对象，第 17 行传入解析的 XML 文件，从第 19 行开始逐个遍历整个 Document 树，得到 XML 数据。代码运行结果如图 14.3 所示。

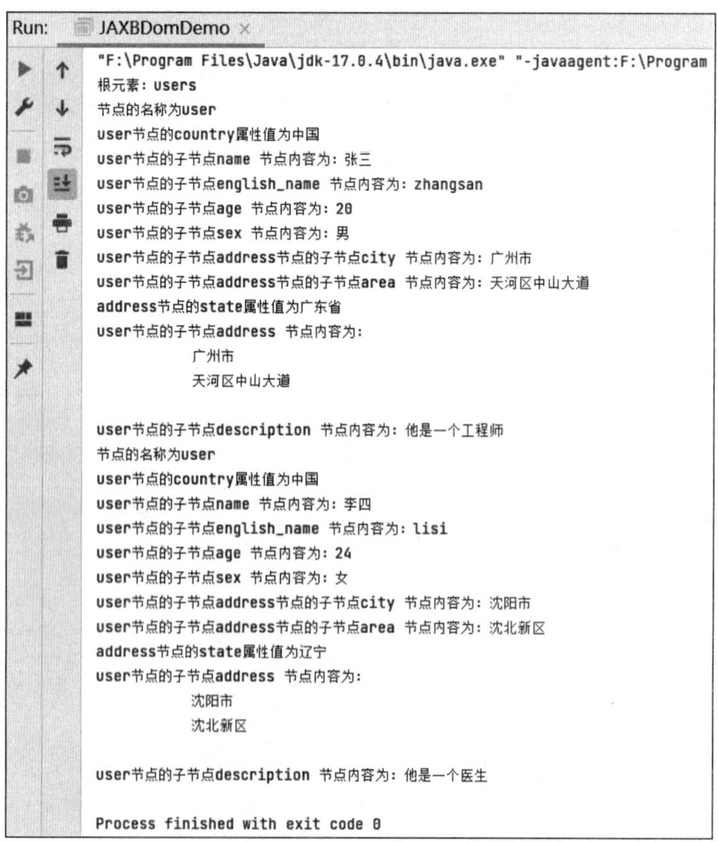

图14.3　DOM解析XML文件运行结果

通过上述代码，不难发现利用DOM解析XML时主要有以下几个步骤：

01 创建DocumentBuilderFactory对象。

```
//用单例模式创建DocumentBuilderFactory对象
DocumentBuilderFactory factory = DocumentBuilderFactory.newInstance();
```

02 通过DocumentBuilderFactory构建DocumentBuilder对象。

```
DocumentBuilder documentBuilder =factory.newDocumentBuilder();
```

03 DocumentBuilder解析XML文件变为Document对象。

```
Document document = documentBuilder.parse(file);
```

04 取得Document对象之后就可以利用Document中的方法获取XML数据。

14.4.2 SAX解析

SAX（Simple API for XML）是另外一种解析XML文件的方法，虽然不是官方标准，但它是XML社区上的事实标准，大部分XML解析器都支持它。SAX与DOM不同的是，它不是一次性地将XML加载到内存中，而是从XML文件的开始位置进行解析，根据已经定义好的事件处理器来决定当前解析的部分是否有必要存储。下面以例子说明SAX解析XML的过程。

【例 14.2】 SAX 解析 XML 文件。

还是以例 14.1 中的 XML 文件为源文件，编写解析类 JAXBSAXDemo 的代码如下：

```
------------------------JAXBSAXDemo.java---------------------------
import org.xml.sax.Attributes;
import org.xml.sax.SAXException;
import org.xml.sax.XMLReader;
import org.xml.sax.helpers.DefaultHandler;
import org.xml.sax.helpers.XMLReaderFactory;

public class JAXBSAXDemo extends DefaultHandler{

    private String preTag;

    //接收文档开始的通知
    @Override
    public void startDocument() throws SAXException {
        preTag = null;

    }
    //接收元素开始的通知
    @Override
    public void startElement(String uri, String localName, String qName,
            Attributes attributes) throws SAXException {
        if("user".equals(qName)) {
            System.out.println(qName+"节点的country属性值为:
"+attributes.getValue("country"));
        }
        if("address".equals(qName)){
```

```
26                System.out.println(qName+"节点的state属性值为:
"+attributes.getValue("state"));
27            }
28            preTag = qName;
29        }
30
31        //接收元素结束的通知
32        @Override
33        public void endElement(String uri, String localName, String qName)
34                throws SAXException {
35            preTag = null;
36        }
37
38        //接收元素中数据的通知，在执行完startElement和endElement方法之后执行
39        public void characters(char ch[], int start, int length)throws SAXException {
40            String value = new String(ch, start, length);
41            if("name".equals(preTag)) {
42                System.out.println("name节点的值为: "+value);
43            } else if("english_name".equals(preTag)) {
44                System.out.println("english_name节点的值为: "+value);
45            }else if("age".equals(preTag)){
46                System.out.println("age节点的值为: "+value);
47            }else if("sex".equals(preTag)){
48                System.out.println("sex节点的值为: "+value);
49            }else if("description".equals(preTag)){
50                System.out.println("description节点的值为: "+value);
51            }
52            if("city".equals(preTag)) {
53                System.out.println("city节点的值为: "+value);
54            } else if("area".equals(preTag)) {
55                System.out.println("area节点的值为: "+value);
56            }
57        }
58
59        public static void main(String[] args) throws Exception {
60            //由XMLReaderFactory类创建XMLReader实例
61            XMLReader xmlReader = XMLReaderFactory.createXMLReader();
62            //创建一个事件监听类
63            JAXBSAXDemo handler = new JAXBSAXDemo();
64            //XMLReader解析类设定事件处理类
65            xmlReader.setContentHandler(handler);
66            //XMLReader解析类解析XML文件
67            xmlReader.parse("F:\\users.xml");
68        }
69    }
```

上述代码介绍了如何利用 SAX 解析 XML 文件，其中第 61 行代码由 XMLReaderFactory 工厂类创建解析器 XMLReader，第 65 行创建事件处理类，第 67 行开始解析 XML 文件，从第 11~36 行分别重写 DefaultHandler 处理类中的文档开始通知、元素开始通知、元素结束通知等方法。上述代码的运行结果如图 14.4 所示。

通过上述代码可以看出，使用SAX解析XML时主要有以下几个步骤：

01 由XMLReaderFactory类创建XMLReader实例：

```
XMLReader xmlReader = XMLReaderFactory.createXMLReader();
```

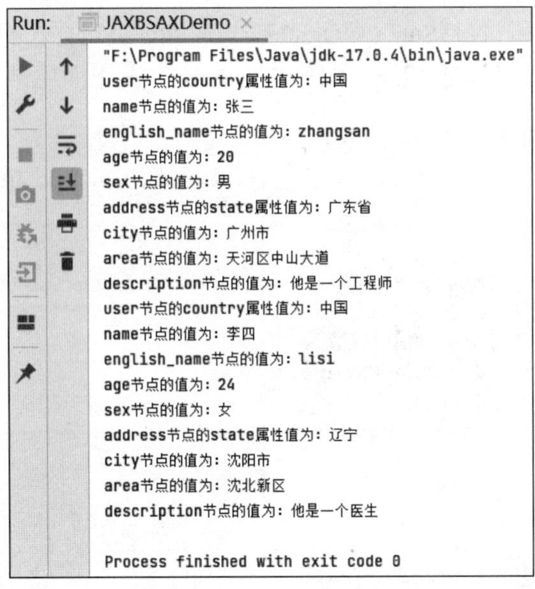

图14.4　SAX解析XML文件效果图

02 创建一个事件监听类：

```
JAXBSAXDemo handler = new JAXBSAXDemo();
```

03 为解析类设定事件处理类：

```
xmlReader.setContentHandler(handler);
```

04 解析XML文件：

```
xmlReader.parse("F:\\users.xml");
```

若想了解更多的 SAX 内容，可查询 org.xml.sax 的 API。

注意： 例 14.2 中应用的是 XMLReader 而不是 SAXParser，这是因为在 SAX 中实现解析的接口名称重命名为 XMLReader。在使用 SAX 解析 XML 资源文件时，默认使用 SAXParser 实现类，它继承自 AbstractSAXParser。同样地，工厂类也是使用 XMLReaderFactory 而不是 SASParserFactory 来创建解析类。

14.4.3　DOM4j解析

DOM在解析的时候把整个XML文件映射到Document的树型结构中，XML中的元素、属性、文本都能在Document中看清，但是它消耗内存、查询速度慢。SAX是基于事件的解析，解析器在读取XML时根据读取的数据产生相应的事件，由应用程序实现相应的事件处理，所以它的解析速度快，内存占用少，但是它需要应用程序自身处理解析器的状态，实现起来比较麻烦，而且它只支持对XML文件的读取，不支持写入。

DOM4j是一个简单、灵活的开源库，前身是JDOM。DOM4j使用接口和抽象类的基本方法，并使用了大量的Collections类，提供一些替代方法以允许更好的性能或更直接的编码方法。它不仅可以读取XML文件，还可以写入XML文件。目前越来越多的Java软件都在使用DOM4j来读写XML，例如Hibernate，包括Sun（被Oracle公司收购）公司自己的JAXM也使用了DOM4j。DOM4j的官方开源版本下载地址为https://dom4j.github.io/j。下面用例子来说明它的使用方法。

【例 14.3】 DOM4j 解析 XML 文件。

同样，XML 文件的内容还是例 14.1 中的内容，编写解析类 Dom4jDemo，代码如下：

```
-----------------------Dom4jDemo.java---------------------------
import java.io.File;
import java.util.List;

import org.dom4j.Document;
import org.dom4j.DocumentException;
import org.dom4j.Element;
import org.dom4j.io.SAXReader;

public class Dom4jDemo {

    /**
     * DOM4j解析XML文件
     */
    @SuppressWarnings("unchecked")
    public static void main(String[] args) {
        //创建待解析的XML文件，并指定目录
        File file = new File("F:\\users.xml");
        //指定XML解析器SAXReader
        SAXReader saxReader = new SAXReader();
        try {
            //SAXReader解析XML文件
            Document document = saxReader.read(file);
            //指定要解析的节点
            List<Element> list = document.selectNodes("/users/user" );
            for(Element element:list){
                //获得节点country属性值
                System.out.println("country----"+element.attributeValue
("country"));;
                //获得节点的子节点
                List<Element> childList = element.elements();
                //遍历节点的子节点
                for(Element childelement:childList){
                    //如果是address子节点，就遍历address的子元素
                    if("address".equals(childelement.getName())){
                        //获得节点state属性值
                        System.out.println("state----"+childelement.
attributeValue("state"));
                        //遍历address元素的子元素
                        List<Element> addressElements =
childelement.elements();
                        for(Element e:addressElements){
                            System.out.println(e.getName()+"----"+
e.getText());
                        }
                    }
                System.out.println(childelement.getName()+"----"+
childelement.getTextTrim());
                }
```

```
44                 }
45             } catch (DocumentException e) {
46                 e.printStackTrace();
47             }
48         }
49 }
```

在上述代码中，讲解了如何利用DOM4j来解析XML资源文件，其中第19行新建SAXReader解析器，第22行解析指定文件，第24~42行利用获得的Document不断地循环遍历出节点的属性和内容。上述代码的运行结果如图14.5所示。

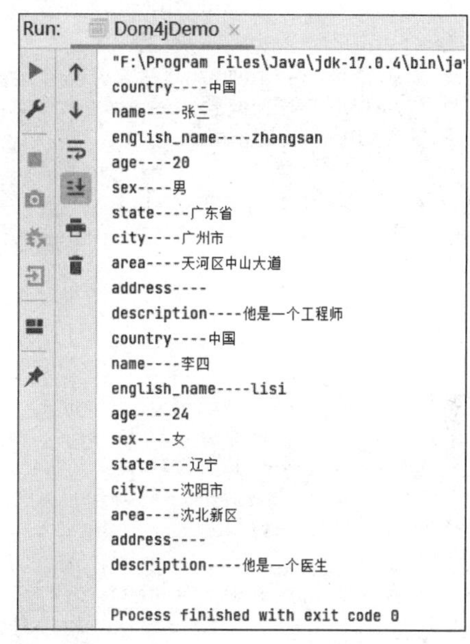

图14.5　DOM4j解析XML文件效果图

通过上述代码可以发现，使用DOM4j解析XML时主要有以下步骤：

01 创建SAXReader实例：

```
SAXReader saxReader = new SAXReader();
```

02 利用SAXReader获取XML的Document：

```
Document document =
saxReader.read(file);
```

从上述步骤可以看出，利用DOM4j解析XML文件十分便捷且极易使用。要想了解更多的DOM4j内容，可查询DOM4j的API和相关例子。

注意：上述例子中，采用的解析器是SAXReader，并通过它来获取XML文件的Document，在DOM4j中还可以利用“DocumentHelper.parseText(text);”来获取XML的Document。

14.5 XML与Java类映射JAXB

上一节讲解了XML的解析方法，分别使用DOM、SAX和DOM4j这3种方法解析XML，从而从XML数据文件中获得想要的数据，这非常有用，但获取数据需要编写大量的代码，工作量巨大。那有没有更加简单的方法获得XML数据和生成XML呢？答案是肯定的，利用JAXB的API就可以解决这个问题。

14.5.1 什么是XML与Java类映射

所谓的映射就是一一对应的关系。例如，有一个XML数据文件：

```
<?xml version="1.0" encoding="UTF-8"?>
<user>
   <name>张三</name>
   <english_name>zhangsan</english_name>
    <age>20</age>
```

```
    <sex>男</sex>
    <address>广东省广州市</address>
    <description>他是一个工程师</description>
</user>
```

另有一个User的Java类：

```
------------------------ User.java--------------------------
public class User {
    String name;                //姓名
    String english_name;        //英文名
    String age;                 //年龄
    String sex;                 //性别
    String address;             //地址
    String description;         //描述

    public User() {
        super();
        // TODO Auto-generated constructor stub
    }
    public User(String name, String englishName, String age, String sex,
            String address, String description) {
        super();
        this.name = name;
        english_name = englishName;
        this.age = age;
        this.sex = sex;
        this.address = address;
        this.description = description;
    }
    //省略get、set方法
}
```

在开发过程中，要将XML中的name元素与User类中的name属性对应、english_name元素与User类中的english_name属性对应、age元素与User类中的age属性对应等，这种XML数据与Java类的对应关系就是一种映射。

14.5.2 JAXB的工作原理

JAXB映射主要由4个部分构成：Schema、JAXB映射类、XML文档、Java对象。其工作原理如图14.6所示。

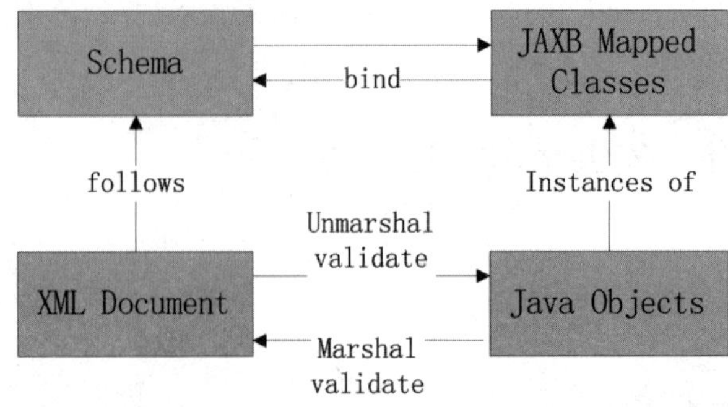

图14.6 JAXB的工作原理

Schema可以理解为表结构，XML Document是数据来源，JAXB提供类的映射方法、Object是Java中对应的类。

14.5.3 将Java对象转换为XML

将Java对象转换为XML的过程为marshal，在JDK6中可以通过注入的方式将Java类映射成XML文件。

【例 14.4】 将 Java 对象转换为 XML。

首先，更改 User 类：

```
import javax.xml.bind.annotation.XmlRootElement;
@XmlRootElement
public class User {
    String name;                    //姓名
    String english_name;            //英文名
    String age;                     //年龄
    String sex;                     //性别
    String address;                 //地址
    String description;             //描述

   //属性的get和set方法省略
}
```

上述代码使用“@XmlRootElement”方式注入 XML 的根元素，那么类中的属性就是 XML 文档中的元素。

其次，编写转换类 JAXBMarshalDemo：

```
----------------------- JAXBMarshalDemo.java--------------------------
import java.io.File;

import javax.xml.bind.JAXBContext;
import javax.xml.bind.JAXBException;
import javax.xml.bind.Marshaller;

import com.eshore.pojo.User;

public class JAXBMarshalDemo {

    public static void main(String[] args) {
        //创建XML对象，将它保存在F盘下
        File file = new File("F:\\user.xml");
        //声明一个JAXBContext对象
        JAXBContext jaxbContext;
        try {
            //指定映射的类，创建JAXBContext对象的上下文
            jaxbContext = JAXBContext.newInstance(User.class);
            //创建转换对象Marshaller
            Marshaller m = jaxbContext.createMarshaller();
            //创建XML文件中的数据
            User user = new User("张三", "zhangsan", "30", "男",
           "广州市天河区", "他是一个工程师");
```

```
24              //将Java类中的User对象转换到XML
25              m.marshal(user, file);
26          } catch (JAXBException e) {
27              e.printStackTrace();
28          }
29      }
30  }
```

从上述代码可以得出转换XML数据的一般步骤：

01 创建JAXBContext上下文对象，参数为映射的类，如代码第18行所示。

02 通过JAXBContext对象的createMarshaller()方法，创建Marshaller转换对象，如代码第20行所示。

03 为转换的类设置内容，如代码第22行所示。

04 通过方法marshal将Java对象输出到指定位置，如代码第25行所示，参数是映射的类和输出文件。

运行上述代码，将在F盘下生成user.xml文件，内容如下：

```
<?xml version="1.0" encoding="UTF-8" standalone="yes" ?>
<user>
  <name>张三</name>
  <english_name>zhangsan</english_name>
  <age>30</age>
  <sex>男</sex>
  <address>广州市天河区</address>
  <description>他是一个工程师</description>
 </user>
```

14.5.4 将XML转换为Java对象

上一小节介绍了如何将Java对象转换为XML数据，本小节将介绍如何使XML数据转换为Java对象，转换的过程与上一小节的过程正好相反。

【例 14.5】 将 XML 数据转换为 Java 对象。

User 类的代码不变，变化的是 JAXBMarshalDemo 中的内容。编写的 JAXBUnmarshalDemo 类的代码如下：

```
-------------------------JAXBUnmarshalDemo.java---------------------------
import java.io.File;

import javax.xml.bind.JAXBContext;
import javax.xml.bind.JAXBException;
import javax.xml.bind.Unmarshaller;

import com.eshore.pojo.User;

public class JAXBUnmarshalDemo {

    public static void main(String[] args) {
        //创建XML对象，将它保存在F盘下
```

```
13          File file = new File("F:\\user.xml");
14          //声明一个JAXBContext对象
15          JAXBContext jaxbContext;
16          try {
17              //指定映射的类，创建JAXBContext对象的上下文
18              jaxbContext = JAXBContext.newInstance(User.class);
19              //创建转换对象Unmarshaller
20              Unmarshaller u = jaxbContext.createUnmarshaller();
21              //转换指定XML文档为Java对象
22              User user = (User)u.unmarshal(file);
23              //输出对象中的内容
24              System.out.println("姓名----"+user.getName());
25              System.out.println("英文名字----"+user.getEnglish_name());
26              System.out.println("年龄----"+user.getAge());
27              System.out.println("性别----"+user.getSex());
28              System.out.println("地址----"+user.getAddress());
29              System.out.println("描述----"+user.getDescription());
30          } catch (JAXBException e) {
31
32              e.printStackTrace();
33          }
34      }
35  }
```

从上述代码中可以发现，代码与上一小节的相比变化不大，将其转换的步骤分解如下：

01 创建JAXBContext上下文对象，参数为映射的类，如第18行代码所示。

02 通过JAXBContext对象的createUnmarshaller ()方法创建Unmarshaller转换对象，如第20行代码所示。

03 利用方法unmarshal将指定的XML文件转换为Java对象，转换时需要强制转换为映射类对象。

运行上述代码，在Java控制台的输出如下：

```
姓名----张三
英文名字----zhangsan
年龄----30
性别----男
地址----广州市天河区
描述----他是一个工程师
```

14.5.5 更为复杂的映射

例14.1中的XML文件就是一个复杂的XML文件，现在要将这个复杂的XML数据转换为Java对象。从数据上可以发现user元素不仅有元素还有属性，子元素address中还包含有子元素city和area，那么原有的类结构已经不适用了，需要进行改造。

【例 14.6】 将复杂 XML 数据转换为 Java 对象。

首先，将 User 类进行改造，使它包含属性 country。代码如下：

```
----------------------User.java--------------------------
import javax.xml.bind.annotation.XmlAccessType;
```

```
02  import javax.xml.bind.annotation.XmlAccessorType;
03  import javax.xml.bind.annotation.XmlAttribute;
04  import javax.xml.bind.annotation.XmlElement;
05
06  @XmlAccessorType(XmlAccessType.FIELD)
07  public class User {
08      @XmlAttribute
09      private String country;                    //国家
10      @XmlElement
11      private String name;                       //姓名
12      @XmlElement
13      private String english_name;               //英文名
14      @XmlElement
15      private String age;                        //年龄
16      @XmlElement
17      private String sex;                        //性别
18      @XmlElement
19      private Address address;                   //地址
20      @XmlElement
21      private String description;                //描述
22      public User() {
23          super();
24
25      }
26      public User(String name, String englishName, String age, String sex,
27              Address address, String description,String country) {
28          super();
29          this.name = name;
30          english_name = englishName;
31          this.age = age;
32          this.sex = sex;
33          this.address = address;
34          this.description = description;
35          this.country = country;
36      }
37      @Override
38      public String toString() {
39          String str=name+"来自"+country+"英文名: "+english_name+"性别:"+sex+
40          " 年龄: "+age+" 现住"+address.toString()+","+description;
41          return str;
42      }
43  }
```

在上述代码中，利用“@XmlAttribute”表示元素的属性，“@XmlElement”表示元素的子元素，第 38~41 行代码是重写类的 toString 方法。

注意：“@XmlAccessorType(XmlAccessType.FIELD)”代表 User 类中每个非静态字段都会绑定到 XML 中，即将类中的属性定义为 private，因为在默认情况下每个公共字段和每个 get、set 方法都会自动绑定到 XML 中。

Address 类的代码如下：

```
------------------------Address.java----------------------------
import javax.xml.bind.annotation.XmlAccessType;
import javax.xml.bind.annotation.XmlAccessorType;
```

```
03  import javax.xml.bind.annotation.XmlAttribute;
04  import javax.xml.bind.annotation.XmlElement;
05
06  @XmlAccessorType(XmlAccessType.FIELD)
07  public class Address {
08      @XmlAttribute
09      private String state;                     //国家
10      @XmlElement
11      private String city;                      //城市
12      @XmlElement
13      private String area;                      //地区
14      public Address() {
15          super();
16      }
17      public Address(String state, String city, String area) {
18          super();
19          this.state = state;
20          this.city = city;
21          this.area = area;
22      }
23      @Override
24      public String toString() {
25          String str=state+" "+city+" "+area;
26          return str;
27      }
28  }
```

在上述代码中，第 08、09 行定义了 Address 元素的 state 属性，第 10~13 行定义了 city 和 area 元素，第 24~27 行重写 Address 类的 toString()方法。

因为 XML 文件中有多个 user 元素，所以需要定义一个容器和根元素的类 Users，代码如下：

```
-----------------------Users.java--------------------------
@XmlRootElement(name="users")
@XmlAccessorType(XmlAccessType.FIELD)
public class Users {

    @XmlElement(name="user")
    private List<User> list = new ArrayList<User>();

    public Users() {
        super();

    }

    public Users(List<User> list) {
        super();
        this.list = list;
    }

    public void setList(List<User> list) {
        this.list = list;
    }

    public List<User> getList() {
```

```
23            return list;
24        }
25
26    }
```

在上述代码中，第 01 行指定元素节点名称为“users”，第 05、06 行定义 list 容器并指定元素名称为“user”，第 13~24 行给出 get 和 set 方法。

其次，建立转换类 JAXBComplexUnmarshalDemo，代码如下：

```
----------------------- JAXBComplexUnmarshalDemo.java-----------------------
01    public class JAXBComplexUnmarshalDemo {
02        public static void main(String[] args) {
03            //创建XML对象，将它保存在F盘下
04            File file = new File("F:\\users.xml");
05            //声明一个JAXBContext对象
06            JAXBContext jaxbContext;
07            try {
08                //指定映射的类，创建JAXBContext对象的上下文
09                jaxbContext = JAXBContext.newInstance(Users.class);
10                //创建转换对象Unmarshaller
11                Unmarshaller u = jaxbContext.createUnmarshaller();
12                //转换指定XML文档为Java对象
13                Users users = (Users)u.unmarshal(file);
14                List<User> list = users.getList();
15                for(User user:list){
16                    //输出对象中的内容
17                    System.out.println("输出----"+user.toString());
18                }
19
20            } catch (JAXBException e) {
21
22                e.printStackTrace();
23            }
24        }
25    }
```

在上述代码中，第15~18行代码输出对象数据，其余代码与类JAXBUnmarshalDemo相似。运行上述代码，在Java控制台中的输出结果如下：

```
输出----张三 来自 中国 英文名：zhangsan 性别:男 年龄：25 现住广东省 广州市 天河区中山大道，他是一个工程师
输出----李四 来自 中国 英文名：lisi 性别:女 年龄：30 现住辽宁省 沈阳市 沈北新区，他是一个医生
```

14.6 小　　结

本章详细介绍了XML的相关内容，包括什么是XML、XML的用途以及技术框架，以及XML的基本语法（包括元素命名规则、元素的定义规范等）。在了解了上述内容之后，又介绍了3种XML解析方法，即DOM、SAX和DOM4j，最后讲解了Java与XML映射的使用方法。

14.7 习　　题

（1）什么是XML?它的作用有哪些？

（2）简述XML中的元素以及属性的命名规则。

（3）简述XML中的语法规则。

（4）判断在下面的 XML 元素定义中哪些是错误的。

```
a. <user_name> b. <user name> c. <1address> d. <&and> e. <and-or>
f. <xmlTel> g. <XMLTel> h. <.phone> i. <phone>
```

（5）有如下XML文件，利用DOM、SAX和DOM4j分别进行解析。

```
<?xml version="1.0" encoding="UTF-8"?>
<user>
 <property name="name" value="张三"/>
 <property name="english_name" value="zhangsan"/>
 <property name="age" value="20"/>
 <property name="sex" value="男"/>
 <property name="address" value="广东省广州市"/>
 <property name="description" value="他是一个工程师"/>
</user>
```

（6）将上述的 XML 文件映射为 Java 的 User 类。

第 15 章 资源国际化

通常情况下，一个Web应用是运行在互联网中的，从理论上讲它可以被全球所有的互联网在线用户访问到。但是不同国家地区的访问者都有自己的语言，Web应用需要根据访问者的语言和习惯来自动调整页面的显示内容，这时就需要用到资源国际化编程。本章将介绍资源国际化编程，从而使读者掌握简单的国际化编程和本地化编程，以开发出适应性更强的Web网站。

本章主要涉及的知识点有：

- 资源国际化简介
- 资源国际化编程
- I18N与L10N的区别
- Servlet的资源国际化

15.1 资源国际化简介

资源国际化就是要解决不同国家和地区之间的文化差异问题，包括语言的差异、生活习惯的差异等。在Java语言中，提供了相关的方法来支持资源国际化，例如资源绑定类ResourceBundle、地区类Locale等。

资源国际化一般有两种编程：国际化编程（I18N）和本地化编程（L10N）。

1. 国际化编程 I18N

I18N是Internationlization的缩写，其含义是指让软件产品随着国家和地区中语言的不同自动显示相适应的内容，而不是用代码将这些不同的信息写在程序中。通常需要进行国际化编程的信息包括数字、货币信息、日期与时间等。

2. 本地化编程 L10N

L10N为资源本地化，全称为Localization。资源本地化编程就是要向使用不同语言和处于不同地区的访问者提供适合的页面。

一个Web应用程序只有当需要实现多种不同语言版本时才有必要进行本地化编程，当然某些应用程序当前只需要一种语言，但考虑到以后可能具有多语言的需求，因此也有必要利用本地化编程的方式来实现，以方便将来进行程序上的拓展。

15.2 资源国际化编程

从Java语言诞生开始，就为资源国际化做了准备，在Java中提供了一些类，用于对程序进行国际化，使得实现国际化变得容易；又因为它是基于Unicode编码设计的编程语言，所以Java程序可以支持目前世界上所有的语言。

15.2.1 资源国际化示例

先看一个原始的例子，即没有利用资源国际化的方式：

```
01  <%@ page contentType="text/html; charset=UTF-8"%>
02  <!DOCTYPE HTML>
03  <html>
04    <head>
05        <title>资源国际化编程示例</title>
06    </head>
07      <body>
08            <p>您好，资源国际化编程示例</p>
09      </body>
10  </html>
```

在上述代码中，第8行代码中的字符串“您好，资源国际化编程示例”是写在程序中的，当某用户访问该页面时，它只会显示中文，若是在没有中文字库的系统中则会显示乱码。现在把字符串用键一值对方式写入properties资源文件中：

```
    message_zh_CN.properties:
    helloInfo= \u60A8\u597D\uFF0C\u8D44\u6E90\u56FD\u9645\u5316\u7F16\u7A0B\u793A
\u4F8B
    message.properties:
    helloInfo = Hello, Internationalization Example
```

一般情况下我们用*_zh_CH.properties表示中文资源文件，*.properties表示默认的资源文件，“*”表示的内容即为资源文件名称。这两个文件经过编译都位于Web项目的classpath中，在IDE中一般是位于src包下面。同时把程序变更为：

```
    <%@ page contentType="text/html; charset=UTF-8"%>
    <%@taglib prefix="fmt" uri="http://java.sun.com/jsp/jstl/fmt" %>
    <!DOCTYPE HTML>
    <html>
```

```
    <head>
      <title>资源国际化编程示例</title>
    </head >
      <body>
          <fmt:bundle basename="message">
                <fmt:message key=" helloInfo"/>
          </fmt:bundle>
      </body>
  </html>
```

在上述代码中引用到了JSTL的fmt标签库中的<fmt:bundle>标签，其中指定绑定的资源名为message。在中文系统中用户访问时会显示“您好，资源国际化编程示例”，效果如图15.1所示；非中文系统则显示“Hello, Internationalization Example”，效果如图15.2所示。这样就实现了资源国际化。

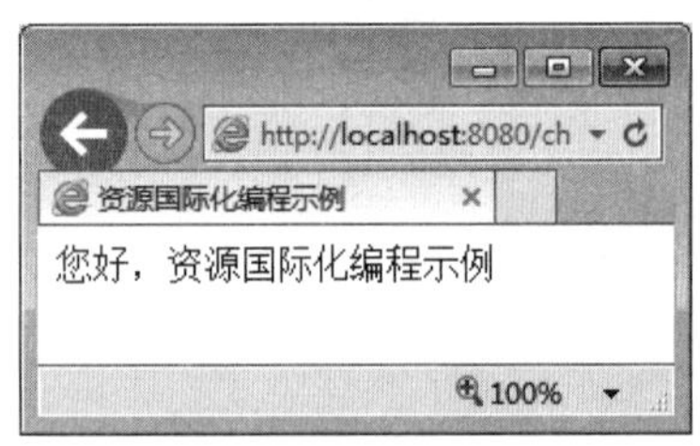

图15.1　显示中文

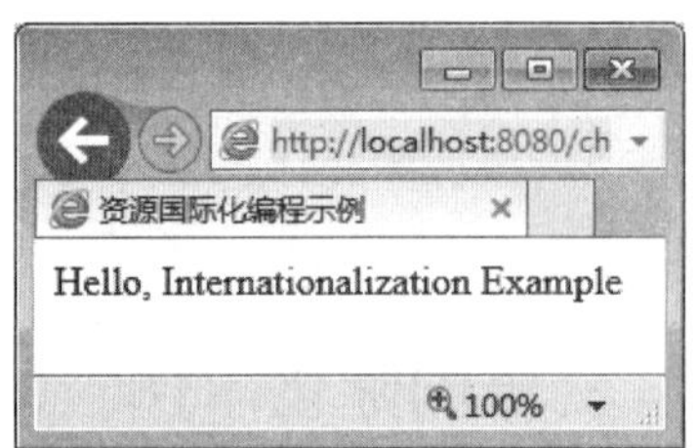

图15.2　显示英文

注意：<fmt:bundle>标签的具体用法请参见第11.3.2节。在调试时，可以通过浏览器中的“显示语言”选项卡进行设定。

15.2.2　资源文件编码

上一小节介绍了资源国际化的简单示例，让读者了解了资源国际化是什么样的，本小节将介绍有关资源文件的编码内容。

一般而言我们采用properties文件来保存资源文件。properties文件是以键－值对的形式来保存文件的。messages_zh_CN.properties中保存的是经过UTF-8编码之后的ASCII字符，Unicode字符中不允许出现中文、日文等其他字符的文字。经过编码之后的文件就可以直接在程序中被引用。

注意：Unicode是为了解决传统的字符编码不统一而产生的，它为每种语言中的每个字符设定了统一并且唯一的二进制编码。在字节数上，Unicode字符占用两个字节，而ASCII字符只占1个字节。

在IntelliJ IDEA中，可以在File | Settings | File Encodings设置中勾选Transparent native-to-ascii conversion复选框，如图15.3所示，然后在编辑器中直接输入中文即可。虽然此时在编辑器中看到的资源文件内容都是中文，但是实际上IntelliJ IDEA已对中文做了转码。

如果想验证文件是否已被成功转码，可以在输入中文后再回到设置页面，将之前勾选的选项取消勾选，此时若查看资源文件则会看到输入的中文已全部被转为ASCII码。所以运用此项设置后，省略了开发人员转码的步骤，使开发更便捷。

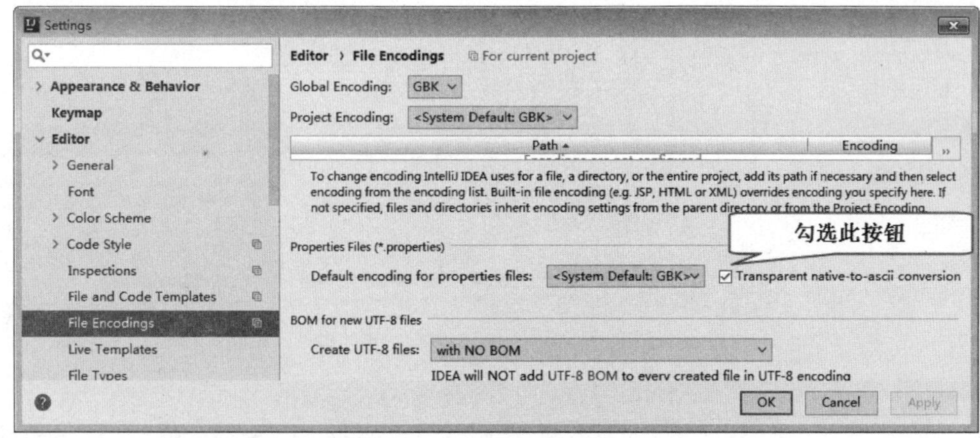

图15.3　IntelliJ IDEA设置自动转码

在JDK 8及以下的版本中，安装目录的bin文件夹下自带native2ascii.exe工具，通过该工具也可实现转码，JDK 9及以后的版本中该工具已不存在，因为本书使用的版本为JDK 17，所以在此不再详述。

15.2.3　显示所有Locale代码

在Java中，与国际化编程关系比较密切的类为Locale。Locale的一个实例代表了一个特定的语言编码。它提供了语言参数构造方法，例如public Locale(String language)。

通过以下构造方法可以产生一个美国英语的 Locale 对象：

```
Locale loc = new Locale("en","US");
```

在运行 Java 程序时，Locale 类由 Java 虚拟机提供；运行 Web 程序时，Locale 由浏览器提供。Locale 里面记录着客户的地区与语言信息。Locale 的代码格式形如 zh_CN，其中小写的 zh 为语种，表示简体中文，大写的 CN 为国家（或地区），表示中国。

下面的程序显示了Java支持的所有Locale。

【例 15.1】　输出所有的 Locale 代码。

showAllLocale.jsp 用于显示所有的 Locale，源代码如下：

```
------------------------ showAllLocale.jsp--------------------------
<%@ page contentType="text/html; charset=UTF-8"%>
<%@page import="java.util.Locale" %>
<%@taglib prefix="c" uri="http://java.sun.com/jsp/jstl/core" %>
<!DOCTYPE HTML>
<html>
  <head>
    <title>资源国际化显示所有的Locale代码</title>
  </head>
  <%
     Locale[] availableLocales = Locale.getAvailableLocales();
     request.setAttribute("availableLocales",availableLocales);
  %>
    <body>
```

```
14          <table border="1" width="100%" cellpadding="2" cellspacing="1">
15             <tr>
16                   <td>名称</td>
17                   <td>国家</td>
18                   <td>国家名称</td>
19                   <td>语言</td>
20                   <td>语言名称</td>
21                   <td>别名</td>
22               </tr>
23               <c:forEach items="${availableLocales}" var="locale">
24                  <tr>
25                    <td>${locale.displayName}</td>
26                    <td>${locale.country}</td>
27                    <td>${locale.displayCountry}</td>
28                    <td>${locale.language}</td>
29                    <td>${locale.displayLanguage}</td>
30                    <td>${locale.variant}</td>
31                  </tr>
32               </c:forEach>
33          </table>
34      </body>
35  </html>
```

在上述代码中，第 09~12 行代码用于获得所有本地的 Locale 类，并将它们传入页面中；第 23~32 行代码利用 forEach 标签遍历输出 Locale 类中所包含的名称、国家、国家名称、语言、语言名称以及别名等信息。运行结果如图 15.4 所示。

图15.4 showAllLocale.jsp运行结果图

将语言代码与国家代码用下画线连接，即为该Locale的代码。例如，中国大陆是zh_CN、美国是en_US、日本是ja_JP。一般情况下，默认的资源文件名为*.properties，例如message.properties；而国家或者地区的资源文件名为*_Locale.properties，例如中国大陆的资源文件名为message_zh_CN.properties，美国的资源文件名为message_en_US.properties。

注意：若一个系统支持多个 Locale 功能，则它一般具有多个资源文件。例如，若某 Locale 对应的 properties 文件存在，则会优先显示该 properties 文件里的对应内容；若 properties 文件不存在或者该文件存在，但是对应的键-值对不存在，则会获取默认资源文件里的内容；若默认资源文件不存在指定的内容，则抛出异常。

15.2.4 带参数的资源

带参数的资源即资源内容是动态的，部分内容是可以变化的，变化的部分利用参数指定，资源的参数通过{0}、{1}等指定。

【例 15.2】 输出带参数的资源。

showParam.jsp 是输出带参数资源的页面，其页面源代码如下：

```
----------------------- showParam.jsp--------------------------
01  <%@ page contentType="text/html; charset=UTF-8"%>
02  <%@page import="java.util.Date" %>
03  <%@taglib prefix="fmt" uri="http://java.sun.com/jsp/jstl/fmt" %>
04  <%@taglib prefix="c" uri="http://java.sun.com/jsp/jstl/core" %>
05  <!DOCTYPE HTML>
06  <html>
07    <head>
08      <title>显示带参数的资源</title>
09    </head>
10
11      <body>
12          带参数的资源示例<br/>
13          <fmt:bundle basename="param">
14          <c:set var="todayT" value="<%=new Date()%>"/>
15            <fmt:message key="message">
16                <!-- 输出地址 -->
17                 <fmt:param value="${pageContext.request.remoteAddr }" />
18                 <!-- 输出Locale -->
19                 <fmt:param value="${pageContext.request.locale }" />
20                 <!-- 输出浏览器显示语言 -->
21                 <fmt:param value="${pageContext.request.locale.
displayLanguage }" />
22                 <!-- 输出日期 -->
23                 <fmt:param value="${todayT}" />
24            </fmt:message>
25          </fmt:bundle>
26      </body>
27  </html>
```

在上述代码中，第 15~24 行输出参数中的信息，如请求的 IP 地址、浏览器语言、日期等。运行结果如图 15.5 所示。

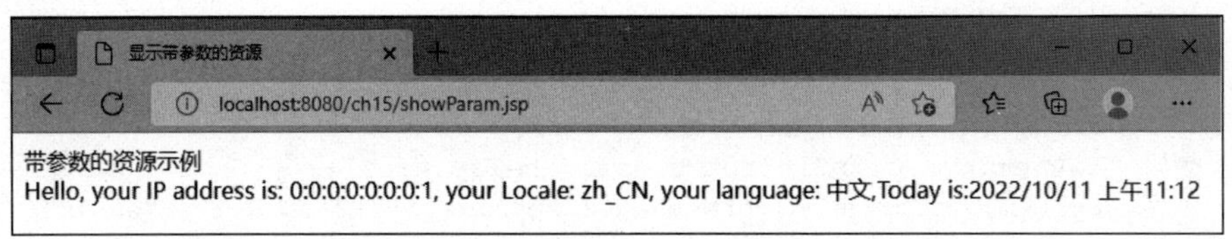

图15.5 showParam.jsp运行结果图

15.2.5 ResourceBundle类

本地化编程需要用到名称为 ResourceBundle 的类，它位于 java.util 之下，这个类实际上只是一个抽象的父类，真正操作的是其子类 ListResourceBundle 和 PropertyResourceBundle。ResourceBundle类中最重要的方法是getBundle，通过指定参数调用getBundle方法获取当前Locale对应的本地化资源。ResourceBundle提供了6个getBundle方法，其中最为常用的是包含baseName和locale参数的方法。例如：

```
publc static final ResourceBundle getBundle(String baseName,Locale locale)
```

参数baseName用于指定资源的路径，参数locale用于指定当前的Locale。

【例15.3】 利用Java获取资源内容。

```
----------------------JavaGetResourceBundle.java--------------------------
package com.eshore;

import java.text.MessageFormat;
import java.util.Date;
import java.util.MissingResourceException;
import java.util.ResourceBundle;

public class JavaGetResourceBundle {
    // 资源名称
    private static final String BUNDLE_NAME = "message";
    // 资源绑定
    private static final ResourceBundle RESOURCE_BUNDLE =
ResourceBundle.getBundle(BUNDLE_NAME);
    // 返回不带参数的资源
    public static String getMessage(String key) {

        try {
            return RESOURCE_BUNDLE.getString(key);
        } catch (MissingResourceException e) {
            return key;
        }
    }
    // 返回带任意个参数的资源
    public static String getMessage(String key, Object... params) {
        try {
            String value = RESOURCE_BUNDLE.getString(key);
            return MessageFormat.format(value, params);
        } catch (MissingResourceException e) {
            return key;
        }
    }

    public static void main(String args[]) {
        System.out.println(getMessage("helloInfo"));
        System.out.println(getMessage("message", "127.0.0.1", "en_US", "英文",
new Date()));
    }
}
```

在上述代码中，ResourceBundle类会绑定message的资源，然后根据用户的不同Locale选择显示不同的properties文件内容。

15.2.6 Servlet的资源国际化

Servlet的资源国际化就是在Servlet类中运用ResourceBundle类绑定资源文件，并输出到页面中，例如将例15.3进行改造，将JavaGetResourceBundle类改造成Servlet类并输出到页面。

【例 15.4】 Servlet 实现资源国际化。

使用 ResourceBundle 类实现资源国际化，当浏览器语言发生变化时，网页内容自动切换为对应语种。

```
-----------------------GetResourceBundleServlet.java--------------------------
package com.eshore;
import java.io.IOException;
import java.io.PrintWriter;
import java.util.Locale;
import java.util.ResourceBundle;
import jakarta.servlet.ServletException;
import jakarta.servlet.http.HttpServlet;
import jakarta.servlet.http.HttpServletRequest;
import jakarta.servlet.http.HttpServletResponse;
public class GetResourceBundleServlet extends HttpServlet {
    private static final long serialVersionUID = 1L;
    public GetResourceBundleServlet() {
        super();
    }
    public void destroy() {
        super.destroy(); // 销毁父类
    }
    public void doGet(HttpServletRequest request, HttpServletResponse
response) throws ServletException, IOException {
         doPost(request,response);
    }
    public void doPost(HttpServletRequest request, HttpServletResponse
response) throws ServletException, IOException {
        //设定页面请求的Locale
        Locale loc = request.getLocale();
        //绑定welcome资源文件
        ResourceBundle rb = ResourceBundle.getBundle("welcome", loc);
        //获取文件上的welcomeinfo内容
        String welcomeinfo = rb.getString("welcomeinfo");
        //获取文件上的message内容
        String message = rb.getString("message");
        response.setContentType("text/html;charset=utf-8");
        PrintWriter out = response.getWriter();
        out.println("<!DOCTYPE HTML>");
```

```
44          out.println("<HTML>");
45          out.println("  <HEAD><TITLE>welcomeinfo</TITLE></HEAD>");
46          out.println("  <BODY>");
47          out.println(" <h2>"+welcomeinfo+"</h2>");
48          out.println("  http://blog.sina.com.cn/u/1268307652</br>");
49          out.println(" <h4>"+message+"</h4>");
50          out.println("  </BODY>");
51          out.println("</HTML>");
52          out.flush();
53          out.close();
54      }
55      public void init() throws ServletException {
56
57      }
58  }
```

在上述代码中，第 34 行代码用于取得页面请求的 Locale，第 36 行代码利用指定的 Locale 获取绑定的资源文件，第 43~53 行代码将文件内容输出到页面中。完成上述代码并编译与部署后，在浏览器中访问该 Servlet，可以发现当浏览器发送不同的 Locale 时它的页面将分别显示中文和英文文字。运行效果如图 15.6 和图 15.7 所示。

图 15.6　Servlet 资源国际化中文显示效果

图 15.7　Servlet 资源国际化英文显示效果

15.2.7　显示所有Locale的数字格式

一般情况下，Web程序中需要格式化的数据包括数字、日期、时间、百分数、货币等。不同地区或国家的显示方法会稍有不同，也有可能截然不同。例如，“2,000”，美国人会认为是2000，但欧洲人会认为是2。如果显示不对，差别就会很大，因此JSP中尽量使用JSTL标签，支持资源国际化，根据用户的Locale自动选择合适的数据格式。

【例 15.5】　显示所有 Locale 的数字格式。

showNumber.jsp 页面用于显示所有地区的 Locale 数字格式，其源代码如下：

```
-----------------------showNumber.jsp--------------------------
<%@ page contentType="text/html; charset=UTF-8"%>
<%@ taglib uri="http://java.sun.com/jsp/jstl/core" prefix="c"%>
<%@ taglib uri="http://java.sun.com/jsp/jstl/fmt" prefix="fmt"%>
<%@page import="java.util.*"%>
<%
    request.setAttribute("availableLocales", Locale.getAvailableLocales());
%>
<!DOCTYPE HTML>
<html>
  <head>
```

```
11        <title>显示所有的日期格式</title>
12      </head>
13        <body>
14              <table border="1" width="100%" cellpadding="2" cellspacing="1">
15                  <tr>
16                      <td>Locale码</td>
17                      <td>语言</td>
18                      <td>日期时间</td>
19                      <td>数字</td>
20                      <td>货币</td>
21                      <td>百分比</td>
22                  </tr>
23                  <c:set var="date" value="<%=new Date()%>"/>
24                  <c:forEach var="locale" items="${availableLocales}">
25                      <fmt:setLocale value="${ locale }" />
26                      <tr>
27                          <td align="left">
28                              ${ locale.displayName }
29                          </td>
30                          <td align="left">
31                              ${ locale.displayLanguage }
32                          </td>
33                          <td>
34                              <fmt:formatDate value="${date}" type="both" />
35                          </td>
36                          <td>
37                              <fmt:formatNumber value="100000.5" />
38                          </td>
39                          <td>
40                              <fmt:formatNumber value="100000.5" type="currency" />
41                          </td>
42                          <td>
43                              <fmt:formatNumber value="100000.5" type="percent" />
44                          </td>
45                      </tr>
46                  </c:forEach>
47              </table>
48          </body>
49      </html>
```

在上述代码中，第 05~07 行是向页面存放本地的 Locale 集合。JSP 运用<c:forEach>标签循环遍历输出各地的日期、货币、百分比格式，运行结果如图 15.8 所示。

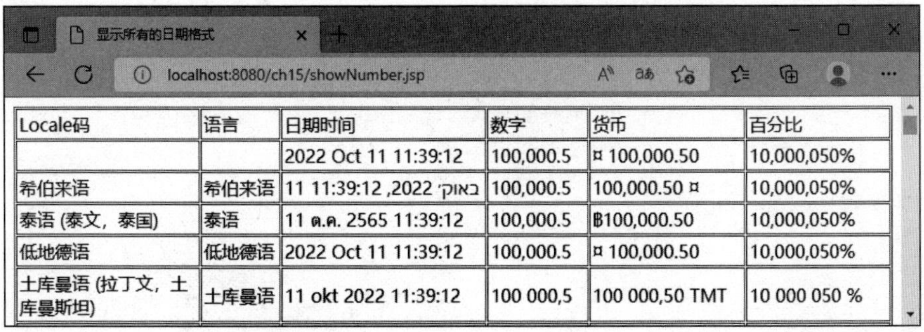

显示所有的日期格式

localhost:8080/ch15/showNumber.jsp

Locale码	语言	日期时间	数字	货币	百分比
		2022 Oct 11 11:39:12	100,000.5	¤ 100,000.50	10,000,050%
希伯来语	希伯来语	11 11:39:12 ,2022 באוק׳	100,000.5	100,000.50 ¤	10,000,050%
泰语 (泰文，泰国)	泰语	11 ต.ค. 2565 11:39:12	100,000.5	฿100,000.50	10,000,050%
低地德语	低地德语	2022 Oct 11 11:39:12	100,000.5	¤ 100,000.50	10,000,050%
土库曼语 (拉丁文，土库曼斯坦)	土库曼语	11 okt 2022 11:39:12	100 000,5	100 000,50 TMT	10 000 050 %

图15.8　showNumber.jsp页面运行结果

15.2.8 显示全球时间

前面已经介绍了如何使用<fmt:timeZone>标签显示全球时间。TimeZone类代表时区，用来表示不同地区间的时间差异。地球上共有24个时区，每两个相邻的时区间相差1个小时。

【例 15.6】 显示全球时间。

ShowTimeZone.jsp 页面用于显示全球时间，源代码如下：

```
-----------------------showTimeZone.jsp--------------------------
<%@ page contentType="text/html; charset=UTF-8"%>
<%@page import="java.util.*"%>
<%@taglib prefix="c" uri="http://java.sun.com/jsp/jstl/core"%>
<%@taglib prefix="fmt" uri="http://java.sun.com/jsp/jstl/fmt"%>
<!DOCTYPE HTML>
<html>
    <head>
        <title>显示全球时间</title>
    </head>
    <%
        Map<String, TimeZone> hashMap = new HashMap<String, TimeZone>();
        for (String id : TimeZone.getAvailableIDs()) { // 所有可用的TimeZone
            hashMap.put(id, TimeZone.getTimeZone(id));
        }
        request.setAttribute("timeZoneIds", TimeZone.getAvailableIDs());
        request.setAttribute("timeZone", hashMap);
        request.setAttribute("date",new Date());//当前时间
    %>
    <body>
        <fmt:setLocale value="zh_CN" />
        现在时刻：<%=TimeZone.getDefault().getDisplayName()%>
        <fmt:formatDate value="${date}" type="both" />
        <br />
        <table border="1">
            <tr>
                <td>时区ID</td>
                <td>时区</td>
                <td>现在时间</td>
                <td>时差间隔</td>
            </tr>
            <c:forEach var="id" items="${ timeZoneIds }" varStatus="status">
                <tr>
                    <td>${id}</td>
                    <td>${timeZone[id].displayName}</td>
                    <td>
                <!-- 用fmt标签格式化日期输出 -->
                <fmt:timeZone value="${id}">
                <fmt:formatDate value="${date}" type="both"
timeZone="${id}" />
                </fmt:timeZone>
                    </td>
                    <td>
                        ${ timeZone[id].rawOffset / 60 / 60 / 1000 }
                    </td>
                </tr>
```

```
45                </c:forEach>
46            </table>
47        </body>
48    </html>
```

在上述代码中，第 10~18 行向页面传递 TimeZone 类中所包含的全球时间内容，第 31~45 行代码遍历输出 TimeZone 类中的内容。运行结果如图 15.9 所示。

时区ID	时区	现在时间	时差间隔
Africa/Abidjan	格林尼治标准时间	2022年10月11日 上午3:40:38	0.0
Africa/Accra	格林尼治标准时间	2022年10月11日 上午3:40:38	0.0
Africa/Addis_Ababa	东部非洲时间	2022年10月11日 上午6:40:38	3.0
Africa/Algiers	中欧标准时间	2022年10月11日 上午4:40:38	1.0
Africa/Asmara	东部非洲时间	2022年10月11日 上午6:40:38	3.0
Africa/Asmera	东部非洲时间	2022年10月11日 上午6:40:38	3.0
Africa/Bamako	格林尼治标准时间	2022年10月11日 上午3:40:38	0.0
Africa/Bangui	西部非洲标准时间	2022年10月11日 上午4:40:38	1.0
Africa/Banjul	格林尼治标准时间	2022年10月11日 上午3:40:38	0.0

图15.9　显示全球时间效果图

上述情况不是直接输出日期时间，而是根据不同的Locale显示不同的日期时间。JSP中可以使用<fmt:timeZone>标签输出日期，它会根据用户所在的时区自动输出当地时间。

15.3　小　　结

本章介绍了资源国际化编程的基础知识，随着Web应用范围的不断扩展，了解并运用资源国际化编程在Web开发中已经越来越重要了。在学习完本章的内容之后，读者可以根据本章示例尝试开发出一个可以显示几个国家语言的Web页面。

15.4　习　　题

（1）在Java中，实现资源国际化编程的关键类名称是什么？

（2）在Web中，一般是对哪些内容进行资源国际化编程？

（3）利用编码获取带参数的资源文件内容。

第 16 章

家校通门户网站

在Web开发中，一个门户网站是十分常见的，门户网站的建设难度随着客户的要求而不断增加，如果只是简单地展示产品信息，那就非常简单，如果是像新浪网、腾讯、网易等那样的门户网站就要相对复杂很多。本章将介绍简单的门户网站的制作。

本章主要涉及的知识点有：

- 开发一个简单的家庭学校网站
- 了解一个网站的制作流程

16.1 网页首页的布局

开发一个门户网站，主要涉及的是样式布局问题，可以利用Dreamweaver工具生成一个页眉布局，随后再进行样式的编排。

首页 index.jsp 从整体分析网站的布局，源代码如下：

```
----------------------index.jsp--------------------------
<%@ page import="java.util.*" pageEncoding="UTF-8"%>
<!DOCTYPE HTML>
<html >
    <head>
        <title>门户网站一首页</title>
      <jsp:include page="head.jsp"/>
    </head>

    <body>
        <!-- 头部开始 -->
        <jsp:include page="menu_top.jsp"/>
        <!-- 头部结束 -->

        <div class="hr01"></div>
```

```

          <!--主体内容开始 -->
          <div class="content">
              <div class="content_left">
                  <!-- 登录信息开始 -->
                  <jsp:include page="login.jsp"/>
                  <!-- 登录信息结束 -->

                  <!-- 简单帮助开始 -->
                  <jsp:include page="simpleHelp.jsp"/>
                  <!-- 简单帮助结束 -->
              </div>
              <div class="content_right">
                  <!-- 中间内容开始 -->
                  <jsp:include page="home.jsp"/>
                  <!-- 中间内容结束 -->

                  <!-- 公告开始 -->
                  <jsp:include page="notice.jsp"/>
                  <!-- 公告结束 -->

                  <!-- 友情链接开始 -->
                  <jsp:include page="friendlyLink.jsp"/>
                  <!-- 友情链接结束 -->
              </div>
          </div>

          <!--主体内容结束 -->

          <!--页脚开始 -->
          <jsp:include page="bottom.jsp"/>
      </body>
  </html>
  <!--页脚结束 -->
```

从上述代码中可以分析出，页面由 3 个部分组成：页头、正文内容、页脚。其中正文内容又分为登录内容、中间内容、公共内容和友情链接 4 个部分。

16.2 导入样式页面

head.jsp 是导入样式页面，源代码如下：

```
----------------------- head.jsp--------------------------
<link rel="stylesheet" href="css/website.css" type="text/css"/>
 <script type="text/javascript" src="js/jquery-3.6.0.js"></script>
 <script type="text/javascript">
  function showNotice(url,w,h){
     var iWidth,iHeight;
     iWidth=w||400;  iHeight=h||400;
     var iTop=(window.screen.availHeight-30-iHeight)/2;
     var iLeft=(window.screen.availWidth-10-iWidth)/2;
     window.open (url,'newwindow',
```

```
10              'height='+iHeight+',width='+iWidth+',top='+iTop+',left='+iLeft+
11              ',directories=no,toolbar=no,menubar=no,scrollbars=no,'+
12              'resizable=no,location=no, status=no')
13     }
14   </script>
```

在上述代码中，第 01 行导入页面样式，第 02 行导入 jQuery.js，第 03~14 行调整弹出窗口的位置。

16.3 显示页面头内容

menu_top.jsp 用于显示页面头内容，源代码如下：

```
----------------------- menu_top.jsp--------------------------
01   <script type="text/javascript">
02     /**
03      *根据menuId高亮显示顶部菜单
04      * @param menuId
05      */
06     function hightLightTopMenu(menuId) {
07        $("#topMenu>ul>li").removeClass("over");//所有菜单高亮关闭
08        $("#" + menuId).addClass("over");// 高亮
09     }
10   </script>
11   <div class="top"><img src="image/portal/logo_title.gif" width="243"
12   height="52" border="0"/></div>
13   <div id="topMenu" class="menu">
14     <ul>
15       <li id="topMenu_home"><a href="website?action=index">首页</a></li>
16       <li id="topMenu_teacher"><a href="website?action=index">教师</a></li>
17       <li id="topMenu_parent"><a href="website?action=index">家长</a></li>
18       <li id="topMenu_student" ><a href="website?action=index">学生</a></li>
19       <li id="topMenu_help"><a href="website?action=index" target="_blank">帮
助</a></li>
20     </ul>
21     <span class="question"><a href="website?action=index"  >
22       <img src="image/portal/ico01.gif" width="16" height="16" border="0"/>问
题反馈</a>
23     </span>
24  </div>
```

在上述代码中，第 01~09 行是用于菜单高亮显示的 JavaScript 代码，第 13~23 行代码利用 ul、li 方式列出菜单项。

16.4 用户登录页面

login.jsp 为登录页面，主要用于显示登录者的信息，源代码如下：

```
------------------------ login.jsp---------------------------
<%@ page import="java.util.*" pageEncoding="UTF-8"%>
<!-- 登录信息开始 -->
<div class="login_bg">
  <div class="login_title">
<img src="images/portal/login_title.gif" width="95" height="26"/></div>
  <div class="login_content">
      <div class="infor04">尊敬的百恼同学，您好！</div>
      <br/>
          <div class="infor05">积分：1000</div>
          <div class="infor06">
          [<a href="website?action=showInfo">我的资料</a>]
          [ <a href="website?action=modifyPass">密码管理</a>]
          [ <a href="website?action=portalLogout">退出</a>]
          </div>
  </div>
</div>
<!-- 登录信息结束 -->
```

在上述代码中，为了简化操作，将登录者的信息写成固定的，当然也可以创建一个输入框，让用户在输入用户名与密码后登录，随后显示用户名，这个时候就要用到数据库中的知识点。

16.5 帮助页面

simpleHelp.jsp 是帮助页面的信息显示页面，源代码如下：

```
------------------------ simpleHelp.jsp---------------------------
<%@ page import="java.util.*" pageEncoding="UTF-8"%>
<!-- 简单帮助开始 -->
<div class="left01_bg">
  <div class="left01_title_help"></div>
  <div class="left01_content_ul">
    <ul>
      <li><a href="#">家长手机号码换了怎么办？ </a></li>
      <li><a href="#">教师如何向家长和学生留言？ </a></li>
      <li><a href="#">教师如何发表专栏文章？</a></li>
      <li><a href="#">我的手机号码换了怎么办？</a></li>
      <li><a href="#">家长如何向教师留言？</a></li>
      <li><a href="#">家长如何发表专栏文章？</a></li>
      <li><a href="#">学生如何向教师留言？ </a></li>
      <li><a href="#">学生如何发表专栏文章？ </a></li>
    </ul>
  </div>
</div>
<!-- 简单帮助结束 -->
```

在上述代码中，帮助列表中的内容是固定的，其实这些内容可以从数据库中查询出来，随着库中内容的变化而变化。

16.6 网页主体内容

home.jsp 是中间内容的信息显示页面，中间内容同样可以通过数据库查询的方式显示出来，也可以通过固定的内容进行显示，源代码如下：

```
------------------------ home.jsp---------------------------
<script type="text/javascript">
  $(document).ready(function() {
    hightLightTopMenu("topMenu_home");// 高亮显示顶部菜单
  })
</script>
<%@ page import="java.util.*" pageEncoding="UTF-8"%>
<div class="center01">
      <div class="center01_top"></div>
      <div class="center01_content">
        <div class="detail">
          <img src='images/portal/show1.jpg' width="64" height="64"/>
          <div class="ritdel">
            <div class="ritdel_title">家校短信</div>
            <div class="ritdel_contet"><a href="websiteIndex?show1">
            通过“家校短信”能够将学校通知、家庭作业、
            学生表现等各种信息以短信的方式及时通知给家长，
            极大地方便了教师和家长之间的沟通交流.....</a>.</div>
          </div>
        </div>
        <div class="hr02"></div>
        <div class="detail">
          <img src='images/portal/ico_6402.gif' width="64" height="64"/>
          <div class="ritdel">
            <div class="ritdel_title">考勤短信（考勤报安）</div>
            <div class="ritdel_contet"><a href="#">
               每位学生配备一张智能校园卡
               以加密方式存储数据。此卡广泛用于......</a></div>
          </div>
        </div>
        <div class="hr02"></div>
        <div class="detail">
          <img src='images/portal/ico_6403.gif' width="64" height="64"/>
          <div class="ritdel">
            <div class="ritdel_title">亲情电话</div>
            <div class="ritdel_contet"><a href="#">
                    学生可以通过配置的智能校园卡在校园的智能
                    计算机上拨打亲情电话......</a></div>
          </div>
        </div>
        <div class="hr02"></div>
        <div class="detail">
          <img src='images/portal/show2.jpg' width="64" height="64"/>
          <div class="ritdel">
            <div class="ritdel_title">留言互动</div>
            <div class="ritdel_contet"><a href=" ">家长、学生通过
```

```
                "留言互动"可以向任课教师进行电子留言以及回复留言，教师也可以通过
                 "留言互动"向学生、家长进行电子留言以及回复留言......</a> </div>
            </div>
          </div>
          <div class="hr02"></div>
          <div class="detail">
            <img src='images/portal/show3.jpg' width="64" height="64"/>
            <div class="ritdel">
              <div class="ritdel_title">发表文章</div>
              <div class="ritdel_contet"><a href="">教师、家长、学生分别
              可以在教师专栏、家长专栏、学生专栏发表文章，将自己教学、家庭教育、
              学习中的心得与他人进行分享......</a></div>
            </div>
          </div>
          <div class="hr02"></div>
          <div class="detail">
            <img src='images/portal/show4.jpg' width="64" height="64"/>
            <div class="ritdel">
              <div class="ritdel_title">成绩短信</div>
              <div class="ritdel_contet"><a href=" ">教师可以在"家校通"系统中
              录入学生的成绩，通过"成绩短信"能够将学生的成绩发送到家长手机上，
              家长可以及时了解到学生的学习情况......</a></div>
            </div>
          </div>
          <div class="hr02"></div>
        </div>
        <div class="center01_bottom"></div>
      </div>
```

上述代码通过 div 的方式列出内容，再通过样式控制其在页面上的显示位置。

16.7　网页公告内容

notice.jsp 是公告显示页面，用于显示一些公告内容，也可以通过数据库查询的方式进行显示，源代码如下：

```
----------------------- notice.jsp--------------------------
<%@ page import="java.util.*" pageEncoding="UTF-8"%>
<div class="right01">
      <div class="right01_top"></div>
      <div class="right01_content">
        <div class="right01_title"><img src='images/portal/title_ad.gif'
                    width="83" height="27"/></div>
        <div class="right01_content_ul">
          <ul>
              <li>
                <a href="website?showNotice" target="_blank" >
                    平潭一中喜中状元...
                </a>
                <br/>
                <span>发布时间：2022-11-01 20:38</span>
              </li>
```

```
              <li>
                <a href="website?showNotice" target="_blank" >
                    高考数学难度加大...
                </a>
                <br/>
                <span>发布时间：2022-10-30 21:38</span>
              </li>
              <li>
                <a href="website?showNotice" target="_blank" >
                    语文试题答案...
                </a>
                <br/>
                <span>发布时间：2022-10-30 00:38</span>
              </li>
              <li>
                <a href="website?showNotice" target="_blank" >
                    理综试题答案...
                </a>
                <br/>
                <span>发布时间：2022-10-26 20:38</span>
              </li>
          </ul>
          <ul><li><a target="_blank" href="#" >更多>></a></li></ul>
        </div>
      </div>
      <div class="right01_bottom"></div>
    </div>
```

在上述代码中，同样是用 div 来显示内容，再通过 ul 来固定顺序。

16.8 友情链接页面

friendlyLink.jsp 为友情链接页面，源代码如下：

```
----------------------- friendlyLink.jsp--------------------------
<%@ page import="java.util.*" pageEncoding="UTF-8"%>
<div class="right02">
  <div class="right02_top"></div>
  <div class="right02_content">
    <img src='images/portal/title_link.gif' width="245" height="27"/>
    <div class="link" style="height:212px;">
      <a href="http://gd.ct10000.com/zq/" target="_blank">中国电信</a> <br/>
      <a href="http://edu.qq.com/" target="_blank">腾讯教育</a>   <br/>
      <a href="http://www.chinaedu.edu.cn/" target="_blank">中国教育信息网</a> <br/>
      <a href="#" target="_blank">平潭教育信息网</a> <br/>
      <a href="#" target="_blank">平潭第一中学</a>   <br/>
      <a href="#" target="_blank">永泰第一中学</a>   <br/>
      <a href="#" target="_blank">福州第一中学</a>   <br/>
     <a href="#">
     <img src='images/portal/link_img01.gif' width="217" height="52"/></a>
    </div>
```

```
        <div class="right02_bottom"></div>
      </div>
    </div>
```

上述代码中显示了一些学校教育机构相关的超链接内容。

16.9　网页底部的版权信息内容

bottom.jsp 为底部页面，主要用于显示版权信息内容，源代码如下：

```
------------------------ bottom.jsp---------------------------
<%@ page import="java.util.*" pageEncoding="UTF-8"%>
<div class="bottom"><img src='images/portal/logo_bottom.gif' width="75"
height="31"/>
  <span>JSP 版权所有 Copyright ©2022 建议最佳分辨率：1024px*768px</span>
</div>
```

上述代码直接输出版权信息，当然也可以通过自定义标签等方式显示，读者可以自行修改。

16.10　家校通门户网站预览效果

部署项目后在浏览器中输入项目名称，运行效果如图16.1所示。

图16.1　家校通门户网站

16.11 小　结

本章主要介绍了如何开发一个企业门户网站，并给出具体的代码示例。一般而言，开发一个企业门户网站的注意事项如下：

- 了解企业的真实需求：是想开发一个简单的企业介绍静态页面还是相对复杂的门户网站，是否包括用户登录、后台数据显示与管理等。
- 安全性与稳定性：企业的门户网站对于稳定性的要求相对较高，不能有宕机的情况出现。同时，又要防止因恶意的网站攻击而导致的系统崩溃。
- 美工的优化：在门户的建设中，美工的优化也是相当重要的，清晰、操作方便的界面风格会提高企业门户的访问量。

企业门户网站建设的基本步骤如下：

01 企业的域名选择。域名的选择要简洁，方便用户记忆。

02 网站的需求分析。在网站建设之初，要跟企业进行明确而详细的沟通，进而提供解决方案。

03 网页设计。网页的首页设计至关重要，要突出企业的业务主线，尽量突显企业的重点和特点。

04 网页推广。网站建设完成后，要加强企业的网站宣传力度，让更多的用户了解、熟知，从而发挥网络营销的优势。

第 17 章 在线购物系统

目前，出现了形形色色的购物网站，人们也开始习惯于网上购物带来的便利，因此网站的建设要求也越来越复杂。本章将与读者一起完成一个简易的网上购物系统。本系统采用JSP+Servlet+Java Bean技术，JSP页面负责展示数据，业务逻辑则在Servlet中实现，Java Bean负责数据的处理。这是JSP的小型项目常用的分层思想，也是现在诸多框架（Struts、Spring、Spring MVC、Hibernate、MyBatis）常用的技术，希望读者能够熟练掌握这种分层技术，这将对以后学习大型项目的开发框架起到事半功倍的作用。

本章主要涉及的知识点有：

- 以“在线购物系统”的开发与实现为主线，从系统需求、系统总体架构、数据库设计、系统详细设计这4个方面进行深入分析，详细讲解该系统的实现过程。
- 复习前面各章节所介绍的知识点。

17.1 系统需求分析

某电商为加快人们对自己商品的了解，想开发一个在线购物系统，通过该系统人们可在网上查询商品的相关信息，并对自己满意的商品进行下单购买。

系统分成4大模块，即用户登录模块、用户管理模块、购物车模块、支付模块，如图17.1所示。

1. 用户登录模块

用户登录模块的主要作用是判断用户是否登录与进行用户退出系统的操作。用户只有登录系统才可以进行购买和支付操作。

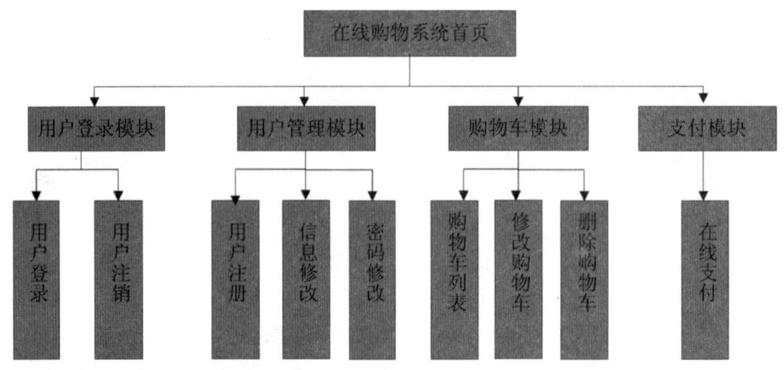

图17.1　系统模块结构图

2. 用户管理模块

用户管理模块负责对注册的用户进行管理，可以进行用户注册、信息修改、密码修改等操作。

3. 购物车模块

购物车模块负责对已登录用户购买的商品进行管理，可以进行查看购物车列表、修改购物车列表、删除购物车列表等操作。

4. 支付模块

支付模块负责对购买商品进行支付操作。

注意：本系统只对购物车商品进行状态处理，没有实现银行的支付接口。

17.2　系统总体架构

根据用户的需求，目前只是做一个简单的网上购物系统，包括4个模块的功能，业务逻辑比较简单，容易理解，采用目前比较流行的三层架构即可。

系统的层次结构如图17.2所示。

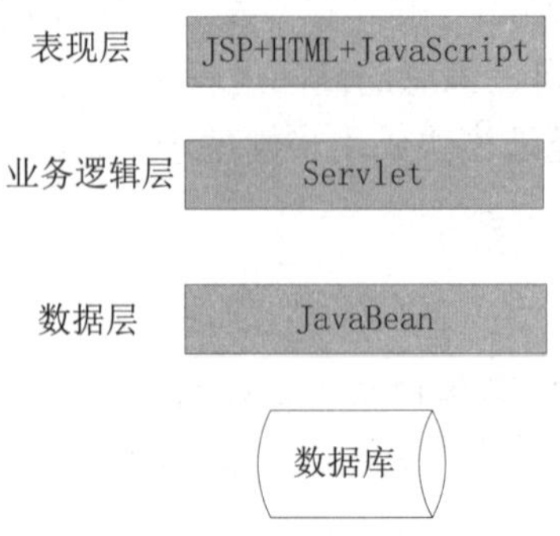

图17.2　系统分层结构图

系统的流程如图17.3所示。

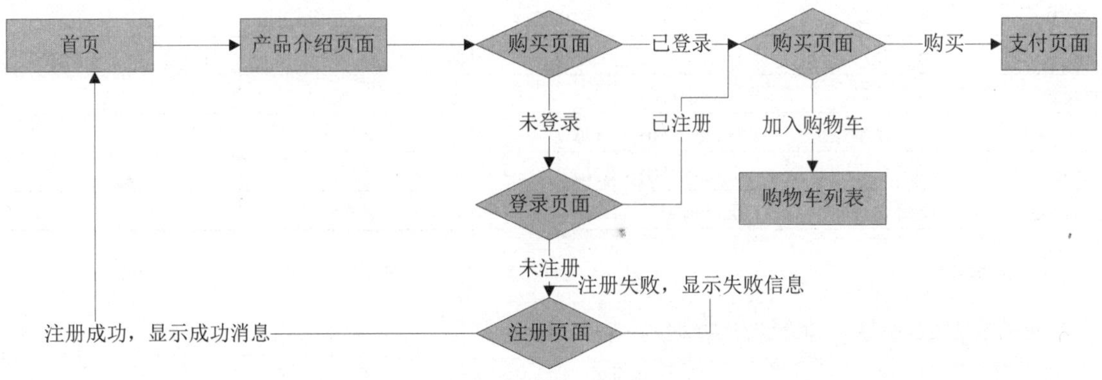

图17.3　系统流程图

17.3　数据库设计

在数据库设计时，一般先构建E-R图（Entily Relationship Diagram，实体一关系图），再根据E-R图创建数据库表、视图等。当然，也可以借助一些OOA工具（例如PowerDesigner、Rose等）进行数据库的设计。

17.3.1　E-R图

本系统仅为简易在线购物系统，所以涉及的表结构很简单，只有3张表：用户、购物车和商品。其E-R图如图17.4所示。

图17.4　系统E-R图

一个用户只能有一个购物车，从购物车支付商品之后就清空这个用户的购物车数据，购物车里可以拥有多个商品ID，同样一个商品ID可以存在多个购物车中，所以购物车与商品之间为多对多的关系。

17.3.2　数据物理模型

根据图17.4所示的E-R图，可以设计出系统的数据物理模型。

下面对关系图中表的设计进行简要分析。

1. 用户表

用户表（user表）如表17.1所示。

表 17.1 users 表

字段名称	含　义	数据类型	是否主键	是否为空	是否外键	其他约束
uid	用户 ID	int	Yes	No	No	AUTO_INCREMENT
uname	用户姓名	varchar(20)	No	No	No	
passwd	用户登录密码	varchar(50)	No	No	No	
email	用户 E-mail	varchar(50)	No	No	No	
lastlogin	最后登录时间	datetime	No	No	No	

2. 商品表

商品表（goods表）如表17.2所示。

表 17.2 goods 表

字段名称	含　义	数据类型	是否主键	是否为空	是否外键	其他约束
gid	物品 ID	int	Yes	No	No	AUTO_INCREMENT
kinds	商品类别	varchar(50)	No	No	No	
gname	商品名称	varchar(100)	No	No	No	
gphoto	商品照片	varchar(100)	No	No	No	
types	商品型号	varchar(100)	No	No	No	
producer	生产商	varchar(50)	No	No	No	
price	商品价格	float(10,2)	No	No	No	
carriage	商品运费	float(10,2)	No	No	No	
pdate	生产日期	datetime	No	No	No	
paddress	生产地址	varchar(100)	No	No	No	
described	商品描述	varchar(200)	No	No	No	

3. 购物车表

购物车表（shoppingcart表）如表17.3所示。

表 17.3 shoppingcart 表

字段名称	含　义	数据类型	是否主键	是否为空	是否外键	其他约束
id	购物车 ID	int	Yes	No	No	AUTO_INCREMENT
uid	用户 ID	int	No	No	Yes	
gid	物品 ID	int	No	No	Yes	
status	购物车状态	int	No	No	No	0：未支付商品 1：已支付商品
number	物品数量	int	No	No	No	

17.4 系统详细设计

下面详细列出本系统的所有源代码和所有文件。

17.4.1　系统包的介绍

在线购物系统采取三层架构的模式设计，所以读者可以照搬第9章中的目录结构来建立该系统。系统开发的Java Bean和Servlet类包如图17.5所示。

其中，action包中存放的是系统所有操作的相关类；dao包中存放的是系统数据操作的相关类；db包中存放的是系统数据库链接的相关类；factory包中存放的是数据实现的相关类；filters包中存放的是系统的Servlet过滤器相关类；pojo包中存放的是系统的Java Bean相关类；service包中存放的是数据真实操作的相关类；tag包中存放的是系统的自定义标签相关类；utils包中存放的是系统的工具类。

有关页面中的相关代码存放在web目录下，其中common包中存放的是共有的页面；css包中存放的是页面样式；js包中存放的是页面用到的JavaScript；其他目录按照各功能模块存放相应的页面。

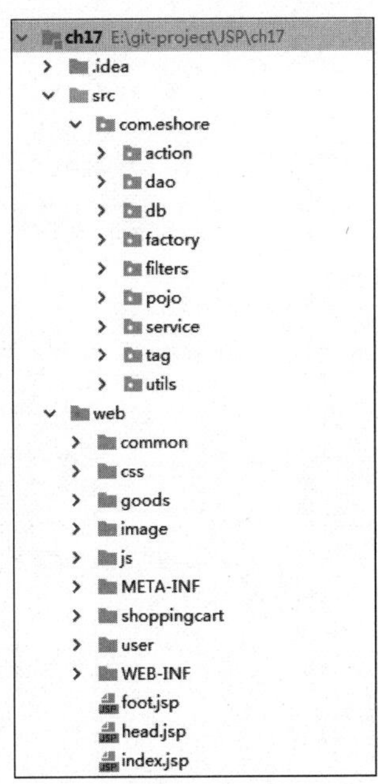

图17.5　系统的包结构

17.4.2　系统的关键技术

1. 数据库连接

数据库连接直接引用第13章中的数据库连接代码。代码的说明详见第13章。

2. 系统分页技术

在本系统中运用自定义标签的方式进行分页。首先，定义标签基本类，为的是以后可以扩展它。源代码如下：

```
----------------------- BaseTagSupport.java--------------------------
01  import jakarta.servlet.ServletRequest;
02  import jakarta.servlet.jsp.tagext.TagSupport;
03  //基本的标签类
04  public class BaseTagSupport extends TagSupport{
05      private static final long serialVersionUID = 1L;
06      protected ServletRequest getRequest(){
07          return pageContext.getRequest();
08      }
09  }
```

在上述代码中，第 04 行继承自 TagSupport 类，第 07 行获取 ServletRequest 请求对象，以后的标签类只要继承 BaseTagSupport 就可以获取到请求对象。

其次，编写分页标签类 PageTag.java，部分源代码如下（详细源代码可参看本书配套资源中的相关文件）：

```
------------------------PageTag.java----------------------------
01  public class PageTag extends BaseTagSupport {
02      private static final Logger log = Logger.getLogger(PageTag.class);
03      private PageObject object;                  //分页对象
04      private String link;                        //分页链接
05      private String script;                      //页面JavaScript方法名
06      //参数的get和set方法
07      ...
08      public int doStartTag() throws JspException {
09          int[] iparams={0,0,0};
10          String[] sparams={"",""};
11          if(object!=null && object.getData()!=null){
12              iparams[0]=object.getDataCount();
13              iparams[1]=object.getPageCount();
14              iparams[2]=object.getCurPage();
15              if(link!=null && link!=""){
16                  sparams[0]=link;
17              }
18              if(script!=null && script!=""){
19                  sparams[1]=script;
20              }
21          }
22          getRequest().setAttribute("iPageObjectTag", iparams);
23          getRequest().setAttribute("sPageObjectTag", sparams);
24          return EVAL_BODY_INCLUDE;
25      }
26      public int doEndTag() throws JspException {
27          getRequest().removeAttribute("iPageObjectTag");
28          getRequest().removeAttribute("sPageObjectTag");
29          return EVAL_PAGE;
30      }
31  }
```

在上述代码中，第03~05行代码定义标签中用到的属性，包括object、link、script；第09~21行代码利用int类型数组分别存放分页中的数据、分页的页数、当前页，String类型数组分别存放跳转链接和页面的方法。分页对象PageObject的部分源代码如下：

```
------------------------PageObject.java----------------------------
public class PageObject {

    private final int DEFAULT_PAGE_SIZE = 10;           //默认显示记录数
    private final int DEFAULT_CUR_SIZE = 1;             //默认当前页
    private List data;                                  //数据列表
    private int dataCount;                              //数据总数
    private int pageSize;                               //显示记录数
    private int pageCount;                              //总页数
    private int curPage;                                //当前页

    public int getCurPage() {                           //获得当前页
        if (curPage < DEFAULT_CUR_SIZE) {
            curPage = DEFAULT_CUR_SIZE;
        }
        return curPage;
    }

```

```
    //参数的get和set方法省略
    ...
    public int getPageCount() {                              //获得页数
        if (dataCount > 0) {
            pageCount = dataCount % pageSize == 0 ?
            (dataCount / pageSize) : (dataCount / pageSize + 1);
        }
        return pageCount;
    }
    public int getPageSize() {                               //获得每页的显示数量
        if (pageSize < 1) {
            pageSize = DEFAULT_PAGE_SIZE;
        }
        return pageSize;
    }

    public void reqProperty(HttpServletRequest request) {
        String curPage = null, pageSize = null, dataCount = null;

        curPage = request.getParameter("curPage");      //设定当前页数
        if (curPage != null && curPage != "") {
            try {
                this.curPage = Integer.valueOf(curPage).intValue();
            } catch (NumberFormatException ex) {
            }
        }

        pageSize = request.getParameter("pageSize");  //设定每页的显示数量
        if (pageSize != null && pageSize != "") {
            try {
                this.pageSize = Integer.valueOf(pageSize).intValue();
            } catch (NumberFormatException ex) {
                ex.printStackTrace();
            }
        }

        dataCount = request.getParameter("dataCount");     //设定总数量
        if (dataCount != null && dataCount != "") {
            try {
                this.dataCount = Integer.valueOf(dataCount).intValue();
            } catch (NumberFormatException ex) {
                ex.printStackTrace();
            }
        }
    }

    public int getBeginPoint() {                                 //获取开始的数据点
        return (getCurPage() - 1) * getPageSize();
    }
    //获得PageObject对象
    public static PageObject getInstance(HttpServletRequest request) {
        PageObject pageObject = new PageObject();
        pageObject.reqProperty(request);
        return pageObject;
    }
}
```

在上述代码中设定了分页中需要用到的属性，例如数据列表、数据总数、总页数、当前页等，如第 05~09 行代码所示；第 11~16 行代码获得当前页；第 20~26 行代码获得页数；第 45~61 行代码设定分页中的数据总数、总页数、显示数量等值；第 67~72 行代码获取分页对象。

最后，定义一个标签文件 lms.tld，存放在 WEB-INF 目录下。文件的内容如下：

```
----------------------lms.tld-------------------------
<?xml version="1.0" encoding="UTF-8"?>
<taglib xmlns="https://jakarta.ee/xml/ns/jakartaee"
        xmlns:xsi="http://www.w3.org/2001/XMLSchema-instance"
        xsi:schemaLocation="https://jakarta.ee/xml/ns/jakartaee
        https://jakarta.ee/xml/ns/jakartaee/web-jsptaglib_3_1.xsd"
        version="2.1">
    <tlib-version>1.0</tlib-version>
    <short-name>lms</short-name>
    <uri>/lms-tags</uri>
    <tag>
        <description>分页</description>
        <name>page</name>
        <tag-class>com.eshore.tag.PageTag</tag-class>
        <body-content>JSP</body-content>
        <attribute>
            <description>PageObject对象</description>
            <name>object</name>
            <required>true</required>
            <rtexprvalue>true</rtexprvalue>
            <type>com.eshore.tag.PageObject</type>
        </attribute>
        <attribute>
            <description>转向URL如:fileQuery.jsp?guid=123456</description>
            <name>link</name>
            <rtexprvalue>true</rtexprvalue>
            <type>java.lang.String</type>
        </attribute>
        <attribute>
            <description>onclick的js function如:doSubmit,必须带一个参数
                pageNo(要转向的页码)</description>
            <name>script</name>
            <rtexprvalue>true</rtexprvalue>
            <type>java.lang.String</type>
        </attribute>
    </tag>
</taglib>
```

如上述代码所示，标签定义了3个属性，即object、link、script，并设置它们的属性类型。具体说明可以参考第12章的内容。分页页面page.jsp的源代码如下：

```
----------------------page.jsp-------------------------
<%@ page pageEncoding="UTF-8"%>
<%@taglib prefix="c" uri="http://java.sun.com/jsp/jstl/core" %>
<style type="text/css">
.page_bg td{ background:#d9ecf2; height:24px; border-bottom:1px solid #abc4de;}
</style>
```

```
<table width="100%" align="center" cellpadding="0" cellspacing="0">
        <tr class="page_bg">
            <td style="width:10%; " height="35" nowrap="nowrap"
>  共有 <strong><span id="count">
<c:out value="${iPageObjectTag[0]}"/></span></strong>条记录 </td>
            <td style="width:12%;" nowrap="nowrap"> 当前第 <strong>
<c:out value="${iPageObjectTag[2]}"/></strong> 页/共 <strong>
<c:out value="${iPageObjectTag[1]}"/></strong> 页</td>
            <td style="width:63%; line-height:35px; vertical-align:middle;
padding-top:5px; padding-right:5px;" valign="middle" align="right"
nowrap="nowrap">
             <img title="第一页"
 src="${pageContext.request.contextPath}/common/images/dg_btn_lt_end.gif"
border="0" onclick="toPage(1)" style="cursor:hand"/> 
             <img title="上一页"
 src="${pageContext.request.contextPath}/common/images/dg_btn_lt.gif"
border="0" onclick="toPage(${iPageObjectTag[2]-1})"
style="cursor:hand"/> 
             <img title="下一页"
src="${pageContext.request.contextPath}/common/images/dg_btn_rt.gif"
border="0" onclick="toPage(${iPageObjectTag[2]+1})"
style="cursor:hand"/> 
             <img title="最后一页"
 src="${pageContext.request.contextPath}/common/images/dg_btn_rt_end.gif"
border="0" onclick="toPage(${iPageObjectTag[1]})"
style="cursor:hand"/></td>
             <td style="width:9%;" nowrap="nowrap">到第
<input id="tagCurPage" size="2" maxlength="3"
onkeypress="return myKeyPress(event);" style="cursor:hand;
position:relative;
top:2px; height:16px; padding:0px; margin:0px;"/> 页</td>
             <td style="width:5%;" colspan="6" nowrap="nowrap">
 <img
src="${pageContext.request.contextPath}/common/images/turnpage_go.gif"
 width="24" height="20" border="0" title="GO" onclick="toPage(-1)"
style="cursor:hand; position:relative; top:2px;"/></td>
            </tr>
</table>
```

上述代码就是分页页面的通用代码，在运用的时候可以直接引入到要分页的页面中。

3. 自定义版权标签

自定义版权标签和标签文件copyright.tld可以参见第12章中自定义版权标签的介绍，这里就不再给出它的源代码。

4. 业务操作类

在开发系统中，尽量采用解耦方式开发程序，这样可以提高系统的可读性和可维护性。同样，在本系统开发中，应尽可能分清数据的操作和数据库的操作，因此先定义一个业务操作类DAOFactory，它的作用是获取各个业务的数据操作方法，部分源代码如下：

```
----------------------- DAOFactory.java---------------------------
public class DAOFactory {
    //取得商品业务操作类
    public static GoodDao getGoodDAOInstance()throws Exception {
```

```
04          return new GoodService();
05      }
06      //取得购物车操作类
07      public static ShoppingCartDao getShoppingCartDAOInstance()throws
Exception {
08          return new ShoppingCartService();
09      }
10      //取得用户操作类
11      public static UserDao getUserDAOInstance()throws Exception {
12          return new UsersService();
13      }
14  }
```

在上述代码中，第 03~12 行代码用于获得商品操作类 GoodService、购物车操作类 ShoppingCartService、用户操作类 UsersService。

17.4.3 过滤器

1. 字符过滤器

在第7章中，介绍了Servlet的过滤器的使用，在本系统中也将运用字符过滤器与登录过滤器。其中，字符过滤器代码与第7章中的过滤器代码一样，具体说明请参见第7章的介绍。

2. 登录过滤器

登录过滤器的作用是防止用户没有登录系统就对商品进行支付或者查看购物车中的商品。在过滤的过程中判断 session 中是否存在该用户，如果没有，就说明用户没有登录，系统返回到登录页面。LoginFilter 的部分源代码如下：

```
-----------------------LoginFilter.java---------------------------
@WebFilter(
     description = "登录过滤",
     filterName = "loginFilter",
     urlPatterns = { "/user/*","/shoppingcart/*" }
)
public class LoginFilter implements Filter {
    private static Logger log = Logger.getLogger("LoginFilter");
    private String filterName="";//过滤器名称
    public void destroy() {
        log.debug("请求销毁");
    }
    public void doFilter(ServletRequest req, ServletResponse res,
        FilterChain chain) throws IOException, ServletException {
        HttpServletRequest request = (HttpServletRequest) req;
        HttpServletResponse response = (HttpServletResponse) res;
        log.debug("请求被"+filterName+"过滤");
        String uname =(String) request.getSession().getAttribute("uname");
        //请求过滤，如果用户为空，则返回登录页面
        if (uname == null) {
            request.setAttribute("status", "请先登录");
            request.getRequestDispatcher("/login.jsp")
.forward(request, response);
```

```
        } else {
            chain.doFilter(req, res);
        }
    }
    public void init(FilterConfig filterConfig) throws ServletException {
        ...
    }
}
```

17.5　系统首页与公共页面

系统首页 index.jsp 是展示商品的页面，源代码如下：

```
-----------------------index.jsp--------------------------
<%@ page import="java.util.*" pageEncoding="UTF-8"%>
<!DOCTYPE HTML>
<html>
    <head>
        <title>淘淘网—开心淘！</title>
        <jsp:include page="common/common.jsp"/>
        <script type="text/javascript" src="js/common/index.js"></script>
    </head>

    <body>
        <div align="center">
            <div id="top">
                <jsp:include page="head.jsp"></jsp:include>
            </div>
            <p>
            <div id="logoselect">
                <jsp:include page="logo_select.jsp"></jsp:include>
            </div>
            <input id="status" type="hidden" name="status"
value="${status}">
            <div id="main">
                <div>
                    <br>
                    <table border="1" id="list">
                        <tr class="goodlist">
                            <td>
                                <br/>
                                数
                                <br/>
                                <br/>
                                码
                                <br/>
                            <td>
                            <td>
                                <a href="goods?keyWord=cellphone&
keyClass=2&action=index-select">品牌手机</a>
                                <br>
                                <a href="goods?keyWord=huawei&
```

```
38  keyClass=1&action=index-select">华为</a>
39                                  <a href="goods?keyWord=iPhone&
40  keyClass=1&action=index-select">iphone</a>|
41                              </td>
42                           ...<!—其他代码省略，跟<td>中类似—>
43                          </tr>
44                      </table>
45                  </div>
46              </div>
47              <div id="foot">
48                  <jsp:include page="foot.jsp"></jsp:include>
49              </div>
50          </div>
51      </body>
52  </html>
```

如上代码由4个部分构成：第12~14行是首页的顶部内容；第16~18行是首页的搜索框内容；第20~45行是首页的正文内容，利用超链接来显示各个商品；第46~48行是首页的底部内容，用于显示版权信息。第06行代码引入公共样式的页面，该页面的代码如下：

```
01  <link rel="stylesheet" type="text/css" href="css/styles.css">
02  <script type="text/javascript" src="js/jquery.js"></script>
03  <script type="text/javascript" src="js/jquery.validate.js"></script>
04  <script type="text/javascript" src="js/messages_cn.js"></script>
```

上述代码的第 01 行导入系统用到的样式文件，第 02 行引入 jquery.js。head.jsp 页面中包含用户是否已经登录的判断，以及显示超链接等信息。

底部 foot.jsp 页面的源代码如下：

```
-----------------------foot.jsp------------------------
01  <%@ taglib prefix="linl" uri="/copyright-tags" %>
02  <!DOCTYPE HTML>
03  <html>
04      <head>
05          <title>淘淘网—开心淘！</title>
06      </head>
07      <body>
08          <div align="center">
09          <hr>
10              <font size="2" color="black">
11              <linl:copyright startY="2014" user="JSP"/>
12                          <a
13                  href="swarding99@163.com">联系我们</a> </font>
14          </div>
15      </body>
16  </html>
```

在上述代码中，第 01 行代码使用自定义版权标签，第 11~13 行代码利用版权标签显示内容。首页的效果如图 17.6 所示。

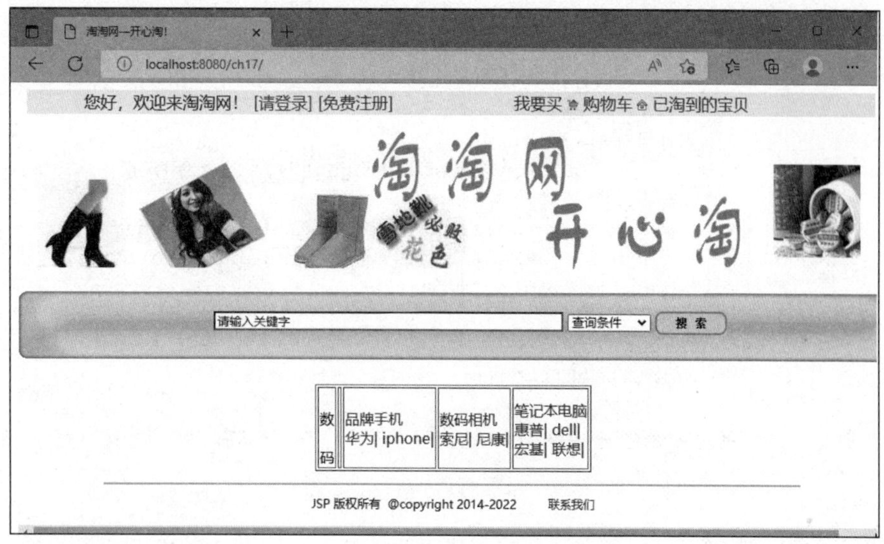

图17.6　在线购物系统首页

17.6　用户登录模块

用户登录页面login.jsp提供用户名与密码输入框，输入用户名与密码后进行提交，验证成功后跳转到系统首页，否则提示相应的错误信息。前面章节已有类似的代码，这里由于篇幅原因不再讲解。

form 表单中用于跳转的 Servlet 类为 LoginServlet.java，源代码如下：

```
----------------------LoginServlet.java-------------------------
package com.eshore.action;

@WebServlet(
    urlPatterns = { "/login" },
    name = "loginServlet"
)
public class LoginServlet extends HttpServlet {
    public void doPost(HttpServletRequest request, HttpServletResponse
response) throws ServletException, IOException {
        String uname = request.getParameter("uname");      //获取用户名
        String passwd = request.getParameter("passwd");    //获取用户密码
        String action = request.getParameter("action");    //获取action类型
        String path = null;
        try{
            if (action.equals("login")) {                      //如果是登录
                Users user = DAOFactory.getUserDAOInstance().
                    queryByName(uname);                         //根据用户名查询用户
                if (passwd.equals(user.getPasswd())) {  //输入的密码与数据库
中的一致
                    request.getSession().setAttribute("uname", uname);
                    request.getSession().setAttribute("uid",
user.getUid());
```

```
21                  path = "index.jsp";
22              } else {
23                  request.setAttribute("status", "用户名或密码错误！");
24                  path = "login.jsp";
25              }
26          } else if (action.equals("logout")) {  //用户退出，注销session中的用户
27              request.getSession().removeAttribute("uname");
28              request.getSession().removeAttribute("uid");
29              path = "login.jsp";
30          }
31      }catch(Exception e){
32          e.printStackTrace();
33      }
34      request.getRequestDispatcher(path).forward(request, response);
35    }
36 }
```

在上述代码中，第03~06行利用注入的方式声明了Servlet；第15~30行判断页面中的action参数是登录还是退出，如果是登录操作，就验证密码是否输入正确并保存到session中，如果是退出操作，则注销session中的用户。登录页面如图17.7所示。

图17.7　登录页面效果图

17.7　用户管理模块

用户管理模块包括用户注册、用户信息修改、用户信息查看、用户密码修改等内容，下面将逐一介绍。

17.7.1 用户注册

用户注册页面 register.jsp，需要用户提供用户名、密码、邮箱等信息，其中用户名与邮箱必须是在本系统中未被注册过的。利用 form 表单提供用户名、密码、邮箱输入框且都是必输项，action 路径是 register，提交方法为 POST，页面的底部公共页即为版权显示页面。由于篇幅原因，页面源代码可以参考前面章节中的部分代码。注册提交的 RegisterServlet 类源代码如下：

```
----------------------- RegisterServlet.java---------------------------
01 @WebServlet(
02     urlPatterns = { "/register" },
03     name = "registerServlet"
04 )
05 public class RegisterServlet extends HttpServlet {
06 
07     public void doPost(HttpServletRequest request, HttpServletResponse
08 response) throws ServletException, IOException {
09         //获取页面参数，包括用户名、密码、邮箱
10         String uname = request.getParameter("uname");
11         String passwd = request.getParameter("passwd");
12         String email = request.getParameter("email");
13         String path = null;
14         //为用户设定属性值
15         Users user = new Users(uname,passwd,email);
16         if (DAOFactory.getUserDAOInstance().
17                 queryByName(uname).getUid() == 0) {           //用户名可用
18             if (DAOFactory.getUserDAOInstance().
19                     queryByEmail(email).getUid() == 0) {      //邮箱可用
20                 if (DAOFactory.getUserDAOInstance().addUser(user) == 1) {
21                     request.getSession().setAttribute("uname", uname);
22                     request.getSession().setAttribute("uid",
23             DAOFactory.getUserDAOInstance().queryByName(uname).getUid());
24                     path = "index.jsp";
25                     request.setAttribute("status", "恭喜您，注册成功！");
26                 } else {
27                     path = "register.jsp";
28                     request.setAttribute("status", "注册失败，请重试……");
29                 }
30             } else {
31                 path = "register.jsp";
32                 request.setAttribute("status", "电子邮箱已被注册");
33             }
34         }else{
35             path = "register.jsp";
36             request.setAttribute("status", "用户名已被注册");
37         }
38         request.getRequestDispatcher(path).forward(request, response);
39     }
40 }
```

在上述代码中，第01~04行代码用于声明Servlet，并配置url为register；第10~13行代码用于接收页面传递的用户名、密码、邮箱等参数；第16~36行代码用于判断用户名和邮箱是否被注册过，如果被注册过，就提示错误消息，如果都未被注册过，就提示“注册成功”并跳转到首页。DAOFactory获得用户操作类UsersService，源代码如下：

```
----------------------UsersService.java--------------------------
01  public class UsersService implements UserDao {
02
03      private DBConnection dbconn = null;             //定义数据库连接类
04      private UserDao dao = null;                     //声明DAO对象
05      // 在构造方法中实例化数据库连接，同时实例化DAO对象
06      public UsersService() throws Exception {
07          this.dbconn = new DBConnection();
08          this.dao = new UserDaoImpl(this.dbconn.getConnection());
09        // 实例化GoodDao的实现类
10      }
11      public int addUser(Users user) throws Exception {
12          ...
13      }
14  ...
15  }
```

在上述代码中，第 03、04 行代码用于声明一个数据库连接类和数据操作 UserDao 对象；第 06、07 行代码在构造方法中初始化数据库连接和实例化 UserDao 对象。UsersService 类的主要作用是启动和关闭数据库连接，有关数据的具体操作可在 UserDaoImpl 中进行。

UserDao 接口的源代码如下：

```
---------------------- UserDao.java--------------------------
public interface UserDao {
    //添加用户
    public int addUser(Users user) throws Exception;
    //修改用户信息
    public int editInf(int uid,String uname,String email) throws Exception;
    //修改用户密码
    public int editPasswd(int uid,String passwd) throws Exception;
    //根据用户id删除用户
    public int deleteUser(int uid) throws Exception;
    //根据用户名查询用户
    public Users queryByName(String uname) throws Exception;
    //根据用户E-mail查询用户
    public Users queryByEmail(String email) throws Exception;
}
```

在上述接口的代码中，定义了用户注册中需要用到的方法。具体对数据的操作位于实现类UserDaoImpl中，部分源代码如下：

```
---------------------- UserDaoImpl.java--------------------------
public class UserDaoImpl implements UserDao {

    private Connection conn = null;                 //数据库连接对象
    private PreparedStatement pstmt = null;         // PreparedStatement对象
    ResultSet rs = null;

    // 通过构造方法取得数据库连接
    public UserDaoImpl(Connection conn) {
        this.conn = conn;
    }
    public int addUser(Users user) throws Exception{
        String sql = "insert into users(uname,passwd,email,lastlogin)
                     values(?,?,?,sysdate())";
```

```
14          int result = 0;
15          pstmt = this.conn.prepareStatement(sql);//获取PreparedStatement对象
16          pstmt.setString(1, user.getUname());        //设定用户名
17          pstmt.setString(2, user.getPasswd());       //设定用户密码
18          pstmt.setString(3, user.getEmail());        //设定用户邮箱
19          result = pstmt.executeUpdate();             //执行数据库操作
20          pstmt.close();
21          return result;
22      }
23  ...
24  }
```

在上述代码中，第03、04行用于声明数据库的连接对象和PreparedStatement对象；第11~22行实现新增用户的方法，其中第12、13行写出SQL语句、第15行获得PreparedStatement对象、第16~18行向SQL语句中填入参数值、第19行执行数据库操作方法并返回结果值。用户实体类User.java的源代码如下：

```
-----------------------User.java--------------------------
01  public class Users {
02      private int uid;                                //用户ID
03      private String uname;                           //用户名
04      private String passwd;                          //用户密码
05      private String email;                           //用户邮箱
06      private Date lastLogin;                         //最后的登录时间
07      //省略get和set方法
08  }
```

在上述代码中，第 03~07 行代码列出 User 对象所用到的 Java Bean 属性，并给出属性的 get 和 set 方法。注册页面效果如图 17.8 所示。

图17.8　注册页面效果图

17.7.2 用户信息修改

用户信息修改页面editinfo.jsp的实现相对简单，form表单提供用户名、邮箱等输入框，action则跳转到UserServlet类中。页面的内容可以参考用户注册页面register.jsp，这里省略源代码。

用户管理模块中有很多操作方法，例如修改用户信息、修改用户名和密码、查看用户信息等，为了降低耦合度并提高代码的复用性和可读性，本系统建立一个UserServlet，以页面传递的action为标识来判断具体执行的是哪个操作。UserServlet.java的源代码如下：

```
----------------------- UserServlet.java--------------------------
01  @WebServlet(
02      urlPatterns = { "/user" },
03      name = "userServlet"
04  )
05  public class UserServlet extends HttpServlet {
06
07      public void doPost(HttpServletRequest request, HttpServletResponse
08  response) throws ServletException, IOException {
09
10          String action = request.getParameter("action");
11          Action targetAction =null;
12          String path = null;
13          if (action.equals("show")) {                         //查看用户列表
14              targetAction = new ShowUserAction();
15              path=targetAction.execute(request, response);
16          } else if (action.equals("editinf")) {               //修改用户信息
17              targetAction = new EditinfUserAction();
18              path=targetAction.execute(request, response);
19          }...
20          request.getRequestDispatcher(path).forward(request, response);
21      }
22  }
```

在上述代码中，代码的逻辑是十分清晰的，第01~04行代码声明一个Servlet；第10行代码获得页面传送的action参数值；第13~19行代码根据action值判断执行的是哪一步操作，显然代码执行的是第16行的修改用户信息操作，所以程序跳转到EditinfUserAction方法中并返回页面的跳转信息。EditinfUserAction类是具体执行修改用户信息的方法，源代码如下：

```
----------------------- EditinfUserAction.java--------------------------
public class EditinfUserAction implements Action{
    public String execute(HttpServletRequest request,
        HttpServletResponse response) throws ServletException, IOException {
        //获取用户的ID值
        int uid = Integer.parseInt(String.valueOf(
        request.getSession().getAttribute("uid")));
        //获取用户的用户名
        String uname = request.getParameter("uname");
        //获取用户的邮箱
        String email = request.getParameter("email");
        //根据用户名查询用户
        Users user=DAOFactory.getUserDAOInstance().queryByName(
                String.valueOf(request.getSession().getAttribute ("uname")));
        if(user.getUname().equals(uname)||
```

```
15                  DAOFactory.getUserDAOInstance().
16                  queryByName(uname).getUid()==0){          //用户名未注册
17              if(user.getEmail().equals(email)||
18                      DAOFactory.getUserDAOInstance().
19                      queryByEmail(email).getUid()==0){     //邮箱未被注册
20                  if(DAOFactory.getUserDAOInstance().
21                          editInf(uid, uname, email)==1){   //用户信息修改成功
22                      request.getSession().setAttribute("uname", uname);
23                      request.setAttribute("status", "信息修改成功！");
24                  }else{                                   //用户信息修改失败
25                      request.setAttribute("status", "修改操作失败，请重试！");
26                  }
27              }else{                                       //邮箱已经被注册
28                  request.setAttribute("status","电子邮箱账号已被注册,请换一个！");
29              }
30          }else{                                           //判断用户名已经存在
31              request.setAttribute("status", "用户名已存在，请换一个！");
32          }
33          return "shoppingcart?action=lookbus";
34      }
35  }
```

在上述代码中，第 05、06 行代码先从 session 中取得登录用户的 ID 值；第 07~10 行代码取得页面输入的用户名与邮箱。然后利用 UserService 中的方法对用户进行判断，如果用户名已经存在，就提示用户名已存在，如第 31 行代码所示；如果邮箱已经被注册，就提示邮箱账号已经被注册，如第 28 行代码所示；如果用户修改信息成功，就提示信息修改成功，如第 23 行代码所示。

接口 Action 中的方法很简单，即提供一个 String 的返回值，传入 HttpRequest、HttpResponse，源代码如下：

```
------------------------ Action.java---------------------------
/**
 * 业务操作的接口类
 */
public interface Action {
    public String execute(HttpServletRequest request,
            HttpServletResponse response)
            throws ServletException, IOException;
}
```

跳转至修改页面的action方法EditAction.java的源代码如下：

```
------------------------ EditUserAction.java---------------------------
public class EditUserAction implements Action {

    public String execute(HttpServletRequest request,
HttpServletResponse response) throws ServletException, IOException {
        //根据用户名查询用户
        Users user = DAOFactory.getUserDAOInstance().queryByName(
        String.valueOf(request.getSession().getAttribute("uname")));
        System.out.println(user.getEmail());
        request.setAttribute("email", user.getEmail());
        return "user/editinf.jsp";
    }
}
```

在上述代码中，第 06~08 行代码从 session 中查询已经登录的用户，然后根据用户名通过 UserService 查询出该用户，获得该用户的邮箱并输出到页面中。

修改用户信息的页面效果如图17.9所示。

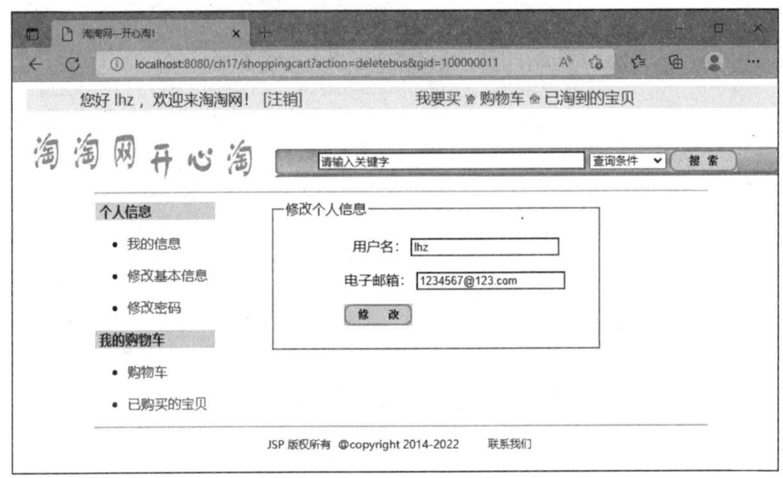

图17.9 修改用户信息页面效果图

17.7.3 用户信息查看

查看用户信息的结果页面为 myinf.jsp，其主要功能是显示用户信息。这里为了使操作简单，仅显示用户名与邮箱信息，源代码如下：

```
-----------------------myinf.jsp-----------------------------
<%@ page import="java.util.*" pageEncoding="UTF-8"%>
<!DOCTYPE HTML>
<html>
    <head>
        <title>查看用户</title>
    </head>

    <body>
        <div align="center" style="width: 60%; padding-left: 10%">
            <fieldset>
                <legend>
                    个人信息
                </legend>
                <div align="left" style="padding-left: 20%">
                    <p>
                        <label>
                              用户名:
                        </label>${uname }<br/>
                        <label>
                            电子邮箱:
                        </label>${email }
                    <p>
                </div>
            </fieldset>
        </div>
```

```
26         </body>
27     </html>
```

在上述代码中，第18~21行代码利用EL标签来显示用户名和邮箱。业务中的方法类ShowUserAction.java的源代码如下：

```
----------------------ShowUserAction.java-------------------------
public class ShowUserAction implements Action{

    public String execute(HttpServletRequest request,
HttpServletResponse response) throws ServletException, IOException {
        //根据用户名查询用户
        Users user = DAOFactory.getUserDAOInstance().queryByName(
        String.valueOf(request.getSession().getAttribute ("uname")));
        request.setAttribute("email", user.getEmail());
        return "user/myinf.jsp";
    }
}
```

在上述代码中，从 session 中查询已经登录的用户，然后根据用户名通过 UserService 方法查询出用户，并将邮箱输出到页面中。

17.7.4　用户密码修改

修改用户密码的页面为editpasswd.jsp，该页面提供form表单，表单中有密码输入框，根据比对密码的合法性来判断是否修改成。页面跳转到EditPasswdAction类方法中，源代码可参见register.jsp页面。用户密码修改页面效果如图17.10所示。

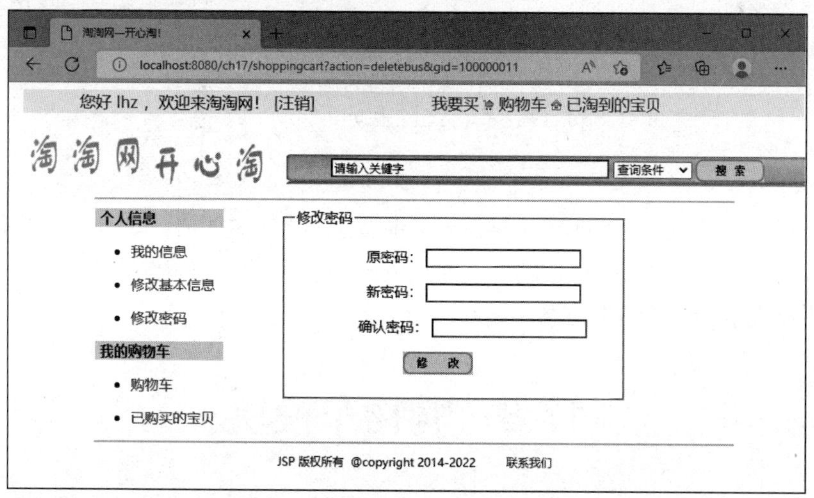

图17.10　editpasswd.jsp页面效果图

EditPasswdAction 类的源代码如下：

```
----------------------EditPasswdAction.java-------------------------
public class EditPasswdAction implements Action{
    public String execute(HttpServletRequest request,
            HttpServletResponse response) throws ServletException,
IOException {        //获取用户的ID值
```

```
05          int uid = Integer.parseInt(String.valueOf(
06                  request.getSession().getAttribute("uid")));
07          //获取旧密码
08          String oldPasswd = request.getParameter("oldPasswd");
09          //获取新密码
10          String passwd = request.getParameter("passwd1");
11          String confirdPasswd = request.getParameter("passwd2");
12          //根据用户名查询用户
13          Users user =DAOFactory.getUserDAOInstance().
14              queryByName(String.valueOf(
15                      request.getSession().getAttribute("uname")));
16          //判断输入的旧密码与原来的旧密码是否一致
17          //如果一致就进行修改
18          if(user.getPasswd().equals(oldPasswd)){
19              if(isValidPassword(passwd,confirdPasswd)){     //验证密码
20                  request.setAttribute("status", "密码为空或者密码不一致！");
21              }
22              if(DAOFactory.getUserDAOInstance().
23                      editPasswd(uid, passwd)==1){       //密码修改成功
24                  request.setAttribute("status", "密码修改成功！");
25              }else{                                     //密码修改失败
26                  request.setAttribute("status", "密码修改操作失败，请重试！");
27              }
28          }else{                                         //输入密码错误
29              request.setAttribute("status", "原密码错误，你不能修改密码！");
30          }
31          return "shoppingcart?action=lookbus";
32      }
33      //验证密码，如果密码为空或长度小于6或与确认密码不一致
34      //返回true
35      public boolean isValidPassword(String passwd,String confirdPasswd){
36          return passwd==null||confirdPasswd==null
37          ||passwd.length()<6||confirdPasswd.length()<6
38          ||!passwd.equals(confirdPasswd);
39      }
40  }
```

在上述代码中，第 05 行代码获得 session 中已登录的用户 ID 值；第 08~11 行代码获取页面中获得的输入密码值；第 19~30 行代码通过 UserService 方法验证密码的合法性，若验证通过，则修改成功，否则提示相应的错误信息。

17.8 购物车模块

购物车模块包括添加购物车、删除购物车、查看购物车、修改购物车等功能。

17.8.1 添加购物车

为了代码的可维护性和可读性，只建立一个 ShoppingCartServlet，利用 action 来标识跳转到哪个方法中。ShoppingCartServlet.java 的源代码如下：

```
----------------------- ShoppingCartServlet.java---------------------------
01  @WebServlet(
02      urlPatterns = { "/shoppingcart" },
03      name = "shoppingCartServlet"
04  )
05  public class ShoppingCartServlet extends HttpServlet {
06
07      public void doPost(HttpServletRequest request, HttpServletResponse
08  response) throws ServletException, IOException {
09          String uids = String.valueOf(request.getSession().
getAttribute("uid"));
10          String action = String.valueOf(request.getParameter("action"));
11          Action targetAction =null;
12          String path = null;
13          try{
14              if (uids == null || uids.equals("null")) {
15                  path = "login.jsp";
16              } else {
17                  if (action.equals("deletebus")) {       //删除购物车
18                      targetAction = new DeletShoppingCartAction();
19                      path=targetAction.execute(request, response);
20                  } else if (action.equals("intobus")) {//单击加入购物车时处理
21                      targetAction = new InsertShoppingCartAction();
22                      path=targetAction.execute(request, response);
23                  }...
24              }
25          }catch(Exception e){
26              e.printStackTrace();
27          }
28          request.getRequestDispatcher(path).forward(request, response);
29      }
30  }
```

在上述代码中，第10行代码用于获得页面传入的action标识，根据action值跳转到相应的方法中。InsertShoppingCartAction.java代表添加到购物车中，源代码如下：

```
----------------------- InsertShoppingCartAction.java-------------------------
public class InsertShoppingCartAction implements Action {

    public String execute(HttpServletRequest request,
HttpServletResponse response) throws ServletException, IOException {
        //获取商品的ID
        int gid = Integer.parseInt(String.valueOf(request
                .getParameter("gid")));
        //获取商品的数量
        int number = Integer.parseInt(String.valueOf(request
                .getParameter("number")));
        //获取登录用户的ID
        String uids = String.valueOf(request.getSession().
getAttribute("uid"));
        int uid = Integer.parseInt(uids);
        ShoppingCart bus = DAOFactory.getShoppingCartDAOInstance().
                getGoodsId(uid, gid, 0);
        if (bus.getId() == 0) {        // 如果购物车中不存在，则加入购物车
            DAOFactory.getShoppingCartDAOInstance().addBus(gid, uid,
number);
```

```
18
19          } else {                          // 否则修改未付款的商品数量
20              DAOFactory.getShoppingCartDAOInstance().
updatebus(bus.getId(),
21                      bus.getNumber() + number, 0);
22          }
23          request.setAttribute("status", "已将该宝贝添加到您的购物车");
24          return "goods?sid=" + gid
25                  + "&action=goodslist-select";
26      }
27  }
```

在上述代码中，第06~12行代码分别获取商品的ID、商品的数量、登录用户的ID；第16~22行代码判断该商品是否存在于购物车中，如果不存在则直接添加，否则修改购物车中该商品的数量；最后返回商品的列表页面。DAOFactory获得ShoppingCartService类，其作用就是连接数据库和关闭数据库，源代码如下：

```
----------------------- ShoppingCartService.java---------------------------
01  public class ShoppingCartService implements ShoppingCartDao {
02
03      private DBConnection dbconn = null;         // 定义数据库连接类
04      private ShoppingCartDao dao = null;         // 声明DAO对象
05      // 在构造方法中实例化数据库连接，同时实例化DAO对象
06      public ShoppingCartService() throws Exception {
07          this.dbconn = new DBConnection();
08          this.dao = new ShoppingCartDaoImpl(this.dbconn.getConnection());
09          // 实例化GoodDao的实现类
10      }
11      ...
12      public PageObject getPageObject(String curPage, PageObject
13  pageObject, List<Object> listObject) {
14          pageObject = this.dao.getPageObject(curPage, pageObject,
listObject);
15          return pageObject;
16      }
17  }
```

在上述代码中，第07、08行代码用于新建数据库连接，而后再实现具体的数据操作。具体的数据执行动作位于ShoppingCartDaoImpl类中，源代码如下：

```
-----------------------ShoppingCartDaoImpl.java---------------------------
public class ShoppingCartDaoImpl implements ShoppingCartDao {

    private Connection conn = null;                     //数据库连接对象
    private PreparedStatement pstmt = null;             //数据库操作对象

    ResultSet rs = null;
    Vector<ShoppingCart> busVector = new Vector<ShoppingCart>();

    // 通过构造方法取得数据库连接
    public ShoppingCartDaoImpl(Connection conn) {
        this.conn = conn;
    }
    //删除指定的购物车信息
    public int deleteGoods(int gid, int uid,int status) throws Exception{
```

```
        String sql = "delete from shoppingcart where uid=? and gid=? and status=?";
        int result = 0;
        this.pstmt = this.conn.prepareStatement(sql); //获取PreparedStatement对象
        this.pstmt.setInt(1, uid);
        this.pstmt.setInt(2, gid);
        this.pstmt.setInt(3, status);
        result = pstmt.executeUpdate();          //执行数据库操作
        this.pstmt.close();                      //关闭PreparedStatement操作

        return result;
    }
    ...
}
```

上述代码就是对购物车中的数据进行具体的操作。接口ShoppingCartDao的源代码如下：

```
----------------------- ShoppingCartDao.java-------------------------
public interface ShoppingCartDao {

    ///根据购物车状态、用户ID查询购物车
    public Vector<ShoppingCart> getAppointedGoods(int uid, int status)throws Exception;
    //根据用户ID获取所有的商品
    public Vector<ShoppingCart> getAllGoods(int uid)throws Exception;
    //根据购物车状态、商品ID、用户ID查询购物车
    public ShoppingCart getGoodsId(int uid, int gid, int status)throws Exception;
    //根据购物车状态、商品ID、用户ID删除购物车
    public int deleteGoods(int gid, int uid, int status)throws Exception;
    //根据用户ID、购物车状态删除购物车
    public int deleteAll(int uid, int status)throws Exception;
    //添加购物车
    public int addBus(int gid, int uid, int number)throws Exception;
    //修改购物车信息
    public int updatebus(int id, int number, int status)throws Exception;
    //更新购物车信息
    public int updateShopcarts(String ids,int status) throws Exception;
    //购物车的分页对象
    public PageObject getPageObject(String curPage,PageObject
        pageObject,List<Object> listObject);
}
```

上述代码用于定义接口所用到的方法。

17.8.2　删除购物车

DeletShoppingCartAction 类是删除购物车的方法，首先获取商品的 ID 值和登录用户的 ID，从库中查询出指定的商品，然后再删除，源代码如下：

```
----------------------- DeletShoppingCartAction.java--------------------------
public class DeletShoppingCartAction implements Action {

    public String execute(HttpServletRequest request,
HttpServletResponse response) throws ServletException, IOException {
        //获取商品的ID
```

```
06            int gid = Integer.parseInt(String.valueOf(request
07                    .getParameter("gid")));
08            //获取登录用户的ID
09            String uids = String.valueOf(request.getSession().getAttribute
("uid"));
10            int uid = Integer.parseInt(uids);
11            if (DAOFactory.getShoppingCartDAOInstance().
12                    deleteGoods(gid, uid, 0) == 1) { //删除购物车中指定的商品
13                request.setAttribute("status", "已从购物车中删除商品");
14            } else {                              //删除失败
15                request.setAttribute("status", "删除商品操作失败，请重试");
16            }
17            return "shoppingcart?action=lookbus";
18        }
19    }
```

在上述代码中，第06行代码获得商品的ID值；第09行代码获得登录用户的ID值；第11~14代码行根据ShoppingCartService删除指定商品。

17.8.3 查看购物车

ShowShoppingcartAction类用于查看购物车列表类，根据登录用户的ID获取购物车列表，然后进行遍历，获取相关的信息，源代码如下：

```
-----------------------ShowShoppingcartAction.java--------------------------
public class ShowShoppingcartAction implements Action {

    public String execute(HttpServletRequest request,
HttpServletResponse response) throws ServletException, IOException {
        //新建TempGoods对象
        Vector<TempGoods> tempVector = new Vector<TempGoods>();
        //获取登录用户的ID
        String uids = String.valueOf(request.getSession().
getAttribute("uid"));
        int uid = Integer.parseInt(uids);
        float countPrice = 0.0f;
        //获取用户所有未支付的购物车列表
        Vector<ShoppingCart> busVector =
        DAOFactory.getShoppingCartDAOInstance().
                getAppointedGoods(uid, 0);
        for (int i = 0; i < busVector.size(); i++) {
            ShoppingCart cart = new ShoppingCart();
            cart = (ShoppingCart) busVector.get(i);    //获取购物车
            Goods good=new Goods();
            TempGoods tempGoods = new TempGoods();
            Vector<Goods> gVector=DAOFactory.getGoodDAOInstance().
                queryGoodBySid(cart.getGid());          //获取指定商品
            if(gVector.size()>0&&gVector!=null)
                good =(Goods)gVector.get(0);
            //组合TempGoods对象
            tempGoods.setGood(good);
            tempGoods.setNumber(cart.getNumber());
            tempVector.add(tempGoods);
            countPrice+=cart.getNumber()*good.getPrice(); //计算价格
        }
        request.setAttribute("goods", tempVector);
```

```
31            request.setAttribute("countPrice",countPrice);
32            return "shoppingcart/bus.jsp";
33        }
34    }
```

在上述代码中，第17行代码获得购物车列表集合；第19~28行代码遍历购物车中的列表集合，并获得指定商品；第30~32行代码组合TempGoods对象。在该方法中，还用到了两个类TempGoods和ShoppingCart的Java Bean。ShoppingCart的源代码如下：

```
----------------------- ShoppingCart.java-------------------------
01    public class ShoppingCart {
02    
03        private int id;                                        //购物车ID
04        private int gid;                                       //商品ID
05        private int uid;                                       //用户ID
06        private int number;                                    //物品的数量
07        private int status;                                    //1：已付款；0：未付款
08        //省略get和set方法
09    }
```

第 01~09 行代码列出了 ShoppingCart 对象的属性，并给出它们的 get 和 set 方法。

临时购物车对象 TempGoods 的源代码如下：

```
01    public class TempGoods {
02        private Goods good;                                    //商品对象
03        private int number;                                    //购买的数量
04        //省略get和set方法
05    }
```

第 01~05 行代码列出 TempGoods 对象的属性并给出它们的 get 和 set 方法。

物品对象 Goods 的源代码如下：

```
01    public class Goods {
02    
03        private int gid;
04        private String kinds;          // 类型
05        private String gname;          // 名字
06        private String gphoto;         // 实物图片
07        private String types;          // 型号
08        private String producer;       // 生产商
09        private String paddress;       // 出产地
10        private String described;      // 描述
11        private Date pdate;            // 生产日期
12        private float price;           // 单价
13        private float carriage;        // 运费
14        private int keyclass;          //小类别
15        private int big_keyclass; //大类别
16        private String keyword;        //类别名称
17        //省略get和set方法
18    }
```

第 01~18 行代码列出 Goods 对象的属性，并给出它们的 get 和 set 方法。

17.8.4 修改购物车

EditShoppingCartAction 类是修改购物车类，先获取商品的 ID 值和商品的数量，再根据登录用户的 ID 获取购物车列表，取得购物车列表信息，源代码如下：

```
public class EditShoppingCartAction implements Action {
    public String execute(HttpServletRequest request,
HttpServletResponse response) throws ServletException, IOException {
        //获取商品的ID
        int gid = Integer.parseInt(String.valueOf(request
                .getParameter("gid")));
        //获取商品的数量
        int number = Integer.parseInt(String.valueOf(request
                .getParameter("number")));
        //获取登录用户的ID
        String uids = String.valueOf(request.getSession()
.getAttribute("uid"));
        int uid = Integer.parseInt(uids);
        //获得指定的购物车列表
        ShoppingCart bus = DAOFactory.getShoppingCartDAOInstance().
                    getGoodsId(uid, gid, 0);
        DAOFactory.getShoppingCartDAOInstance(). //更新购物车中的商品数量
                    updatebus(bus.getId(), number, 0);
        return "shoppingcart?action=lookbus";
    }
}
```

在上述代码中，首先分别获取页面中商品的 ID 和商品的数量，然后对指定的购物车进行修改，修改成功后跳转到查看购物车方法中。

1. 删除购物车中的所有商品

DeleteallAction 类用于删除购物车中的所有商品，先根据登录用户的 ID 值获取其购物车对象，再进行删除操作，源代码如下：

```
01  public class DeleteallAction implements Action {
02      public String execute(HttpServletRequest request,
03  HttpServletResponse response) throws ServletException, IOException {
04          //获取登录用户的ID
05          String uids = String.valueOf(request.getSession()
.getAttribute("uid"));
06          int uid = Integer.parseInt(uids);
07          if (DAOFactory.getShoppingCartDAOInstance().
08                  deleteAll(uid, 0) > 0) {              //删除购物车商品
09              request.setAttribute("status", "您的购物车中没有商品。");
10          } else {                                      //删除失败
11              request.setAttribute("status", "删除商品操作失败，请重试。");
12          }
13          return "shoppingcart?action=lookbus";
14      }
15  }
```

在上述代码中，第 07~09 行代码用于删除购物车中的商品。

2. 显示购物车中的商品信息

查看购物车页面 bus.jsp 主要用于显示购物车中的商品信息，包括商品的价格、购买的数量等，源代码如下：

```
-----------------------bus.jsp--------------------------
<!DOCTYPE HTML>
<html>
    <head>
        <title>淘淘网—开心淘！</title>
        <jsp:include page="../common/common.jsp"/>
        <script type="text/javascript" src="js/shopcart/bus.js"></script>
    </head>
    <body>
        <div id="top">
            <jsp:include page="../head.jsp"/>
        </div>
        <p>
        <div>
            <jsp:include page="../logo_select1.jsp"/>
        </div>
        <input id="status" type="hidden" name="status" value="${status }">
        <div align="center">
            <div style="width: 80%; height: 78%;">
                <div id="left" align="left">
                    <div style="padding-top: 2px;">
                        <div id="title">
                        我的购物车
                    </div>
                    <ul>
                        <li>
                        <a href="shoppingcart?action=lookbus">购物车
                                        </a>
                            <p>
                        <li>
                        <a href="shoppingcart?action=paid">已购买的宝贝
                                        </a>
                            <p>
                    </ul>
                </div>
            </div>
            <div id="right" align="left">
                <div
                    style="padding-right: 3%; padding-left: 5%;
width: 92%; height:100%;">
                    <div align="center">
                        <div id="title" align="left">
                            <table width="100%">
                                <tr style="text-align: center">
                                    <td width="100px" >图片</td>
                                    <td width="180px">宝贝详细</td>
                                    <td width="90px">单价（元）</td>
                                <td width="150px">数量</td>
                                <td width="100px">总计（元）</td>
                                <td colspan="2" width="150px">操作
                                    </td>
                            </tr>
                        </table>
                    </div>
```

```
                    <form action="shoppingcart" method="post" id="bus">
                        <table width="100%" border="0" >
                            <input type="hidden" name="action"
                                    value="editbus">
                                ...
            <div id="foot">
                <jsp:include page="../foot.jsp"/>
            </div>
        </div>
    </body>
</html>
```

页面效果如图 17.11 所示。

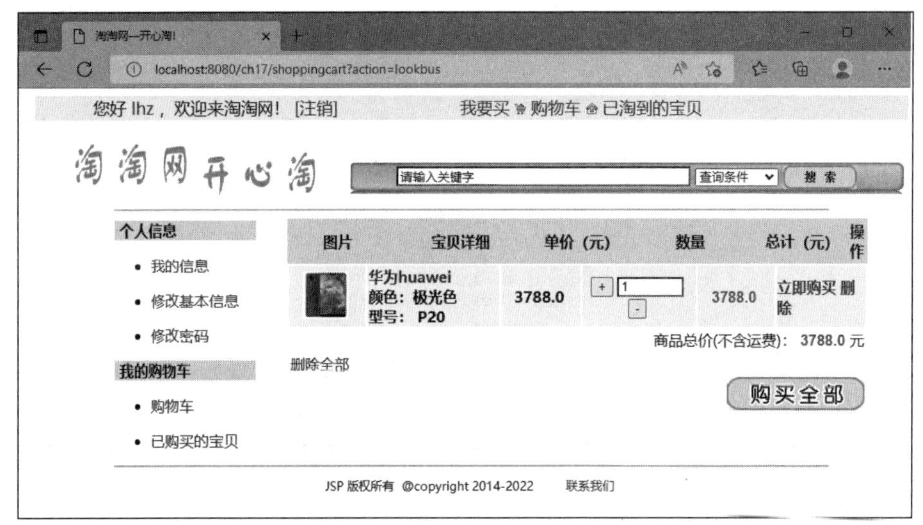

图17.11　bus.jsp页面效果图

17.9　商 品 模 块

商品模块包括查看商品列表、查看单个商品功能等。

17.9.1　查看商品列表

GoodServlet 类是商品的 Servlet 类，同样可通过 action 的标识来决定是查询商品列表还是查询单个商品，源代码如下：

```
------------------------ GoodServlet.java---------------------------
@WebServlet(
    urlPatterns = { "/goods" },
    name = "goodsServlet"
)
public class GoodServlet extends HttpServlet {
    public void doPost(HttpServletRequest request, HttpServletResponse
response) throws ServletException, IOException {
```

```
08          //判断action类型
09          String action=request.getParameter("action");
10          String path=null;
11          Vector<Goods> gVector=new Vector<Goods>();
12
13          try{
14             if(action.equals("index-select")){                          //查询商品列表
15                String keyWord=request.getParameter("keyWord"); //获取查询的输入值
16                String keyClass=request.getParameter("keyClass");//获取查询类别
17                gVector=DAOFactory.getGoodDAOInstance().         //获得所有商品
18                      queryAll(keyWord, keyClass);
19                request.setAttribute("goods", gVector);
20                path="goods/goodslist.jsp";
21             }else if(action.equals("goodslist-select")){//指定商品列表
22                Goods good=new Goods();
23                String sid=request.getParameter("sid");  //获得商品的ID
24                gVector=DAOFactory.getGoodDAOInstance(). //获得指定的商品对象
25                queryGoodBySid(Integer.valueOf(sid));
26                if(gVector.size()>0&&gVector!=null)
27                   good =(Goods)gVector.get(0);
28                request.setAttribute("good", good);
29                path="goods/good.jsp";
30             }
31          }catch(Exception e){
32             e.printStackTrace();
33          }
34          request.getRequestDispatcher(path).forward(request, response);
35      }
36  }
```

在上述代码中，第09行代码获取action标识，根据action值跳转到相应的方法中。index-select表示查询商品列表，通过GoodService查询所有商品的方法，跳转到goodslist.jsp页面中。GoodService类的部分源代码如下：

```
-----------------------GoodService.java---------------------------
public class GoodService implements GoodDao {
    private DBConnection dbconn = null;                    //定义数据库连接类
    private GoodDao dao = null;                            //声明DAO对象
    // 在构造方法中实例化数据库连接，同时实例化DAO对象
    public GoodService() throws Exception {
        this.dbconn = new DBConnection();
        // 实例化GoodDao的实现类
        this.dao = new GoodDaoImpl(this.dbconn.getConnection());
    }
    public PageObject getPageObject(String curPage, PageObject pageObject,
List<Object> listObject) {
        pageObject = this.dao.getPageObject(curPage, pageObject, listObject);
        return pageObject;
    }
    public Vector<Goods> queryGoodBySid(int sid) throws Exception {
        ...
    }
    ...
}
```

在上述代码中，GoodService类主要是实现GoodDao接口，并在构造方法中实例化GoodDao对象，如第07行代码所示，在方法实现中调用GoodDao的相应方法。GoodDaoImpl是具体的商品数据操作类，源代码如下：

```
----------------------- GoodDaoImpl.java---------------------------
public class GoodDaoImpl implements GoodDao {
    private Connection conn = null;            //数据库连接对象
    private PreparedStatement pstmt = null;    //数据库操作对象
    ResultSet rs = null;
    // 通过构造方法取得数据库连接
    public GoodDaoImpl(Connection conn) {
        this.conn = conn;
    }
    //分页显示商品列表
    public PageObject getPageObject(String curPage,PageObject pageObject,
        List<Object> listObject){
        SetPageObject setPageObject = SetPageObject.getInstance();
        //获取分页对象PageObject
        pageObject = setPageObject.setPageObjectData(curPage, pageObject,
        listObject);
        return pageObject;
    }
    //根据商品ID查询指定商品
    public Vector<Goods> queryGoodBySid(int sid) throws Exception {
        ...
    }
}
```

在上述代码中，主体的思想都是获得PreparedStatement对象，然后执行数据库操作方法，得到结果返回值，并将返回值设置到容器中。接口GoodDao中的方法定义了Goods中用到的各个方法，源代码如下：

```
----------------------- GoodDao.java-------------------------
public interface GoodDao {
    //添加商品
    public int addGood(Goods good) throws Exception;
    //删除指定商品
    public int deleteGood(int gid) throws Exception;
    //更新指定商品
    public int updateGood(Goods good) throws Exception;
    //根据商品的ID查找商品
    public Vector<Goods> queryGoodBySid(int sid) throws Exception;
    //根据类型、输入关键字查询商品列表
    public Vector<Goods> queryAll(String keyWord, String keyClass) throws
Exception;
    //分页显示商品列表
    public PageObject getPageObject(String curPage,PageObject
                pageObject,List<Object> listObject);
}
```

通过上述代码可以清楚地了解接口中的方法。

goodslist.jsp 是显示商品列表的页面，利用 JSTL 中的 forEach 遍历结果集，源代码如下：

```
---------------------- goodslist.jsp--------------------------
<%@ page import="java.util.*" pageEncoding="UTF-8"%>
<%@taglib prefix="c" uri="http://java.sun.com/jsp/jstl/core" %>
<!DOCTYPE HTML>
<html>
    <head>
        <title>淘淘网—开心淘！</title>
        <link rel="stylesheet" type="text/css" href="css/styles.css">
    </head>
    <body>
        <div align="center">
            <div id="top">
                <jsp:include page="../head.jsp"/>
            </div>
            <p>
            <div id="logoselect">
                <jsp:include page="../logo_select1.jsp"/>
            </div>
            <div>
                <div style="background-color: #E1F0F0; width: 1000px;
                        height: 35px; font-size: 25px; color: red">
                    <table width="1000px">
                        <tr>
                            <td width="13%">图片</td>
                            <td width="21%">产品</td>
                            <td width="18%">单价</td>
                            <td width="19%">运费</td>
                            <td width="18%">型号</td>
                            <td >出产地</td>
                        </tr>
                    </table>
                </div>
                <div id="main">
                    <table width="1000px" border="0" id="list">
                       <c:choose>
                        <c:when test="${empty goods}">
                            <div align="left">
                                    <span>抱歉，没有找到符合您条件的商品，
                                                请看看别的</span>
                                    <br>
                                <jsp:include page="../recommend.jsp"/>
                                </div>
                          </c:when>
                          <c:otherwise>
                            <c:forEach items="${goods}" var="good">
                    <tr height="100px">
                        <td width="13%">
        <a href="goods?sid=${good.gid}&action=goodslist-select">
        <img src="${good.gphoto}"
                        width="100px" height="100px" border="0">
        </a>
                        </td>
                        <td width="21%">
        <a href="goods?sid=${good.gid}&action=goodslist-select">
        ${good.gname}</a>
                                         <br>
```

```
                                    ${good.described}
                                    <br>
                                    出厂日期：${good.pdate}
                        </td>
                        <td width="18%">${good.price}¥
                        </td>
                        <td width="19%">${good.carriage}¥
                        </td>
                        <td width="18%">${good.types}</td>
                        <td>${good.paddress}</td>
                    </tr>
                            </c:forEach>
                        </c:otherwise>
                      </c:choose>
                    </table>
                </div>
                <div id="foot">
                    <jsp:include page="../foot.jsp"/>
                </div>
            </div>
        </div>
    </body>
</html>
```

在上述代码中，页面只负责显示商品的列表，而业务中的操作交给后台处理，页面中不出现JSP代码。

17.9.2 查看单个商品

查询单个商品的Servlet与查询商品列表是一样的，将goodslist.jsp页面中循环遍历的代码删除就是good.jsp页面的代码。

17.10 支付模块

支付模块包括支付商品、查看已支付商品、支付中的页面等。

17.10.1 支付商品

PayAction类是一个支付单个商品的方法，源代码如下：

```
-----------------------PayAction.java--------------------------
public class PayAction implements Action {
    public String execute(HttpServletRequest request,
HttpServletResponse response) throws ServletException, IOException {
        //获得参数flag标识
        String flag = request.getParameter("flag");
        //获取登录用户的ID
        String uids = String.valueOf(request.getSession(). getAttribute("uid"));
        int uid = Integer.parseInt(uids);
        if(flag.equals("payall")){
```

```
10              //支付全部
11              String shopcartId = request.getParameter("shopcartId");
12              //获得所有购物车的ID
13              try{
14                  if (DAOFactory.getShoppingCartDAOInstance().
15                          updateShopcarts(shopcartId, 1) != 0) {
16                      //更新购物车中的状态
17                      request.setAttribute("status","交易成功!您可以继续选购宝贝。");
18                  }
19              }catch(Exception e){
20                  e.printStackTrace();
21              }
22          }else{                                          //支付单个商品
23              int gid = Integer.parseInt(String.valueOf(request
24                      .getParameter("gid")));             //获得商品的ID
25              int number = Integer.parseInt(String.valueOf(request
26                      .getParameter("number")));          //获得购买商品的数量
27              try{
28                  ShoppingCart bus = DAOFactory.getShoppingCartDAOInstance().
29                          getGoodsId(uid, gid, 0);        //获得指定的购物车
30                  if (DAOFactory.getShoppingCartDAOInstance().
31                      updatebus(bus.getId(), number, 1)!= 0){//更新购物车中的状态
32                      request.setAttribute("status","交易成功!您可以继续选购宝贝");
33                  }
34              }catch(Exception e){
35                  e.printStackTrace();
36              }
37          }
38          return "index.jsp";
39      }
40  }
```

在上述代码中，第 05 行代码利用 flag 标识来判断是支付单个商品还是支付所有商品。如果是支付所有商品，就获取所有商品的 ID 值，调用 ShoppingCartService 类中支付所有商品的方法；如果是支付单个商品，就获取该商品的 ID 值，调用 ShoppingCartService 类中支付单个商品的方法。

17.10.2 查看已支付商品

ShowPaidAction 类是查看已支付商品的方法，源代码如下：

```
----------------------- ShowPaidAction.java---------------------------
public class ShowPaidAction implements Action {

    public String execute(HttpServletRequest request,
HttpServletResponse response) throws ServletException, IOException {

        Vector tempVector = new Vector();
        //获取登录用户的ID
        String uids = String.valueOf(request.getSession()
.getAttribute("uid"));
        int uid = Integer.parseInt(uids);
        //获取分页对象PageObject
        PageObject pageObject = PageObject.getInstance(request);
```

```
            //获取该用户已经支付的购物车列表
            Vector<ShoppingCart> busVector = DAOFactory.
                    getShoppingCartDAOInstance().
                getAppointedGoods(uid, 1);
            //遍历购物车列表
            for (int i = 0; i < busVector.size(); i++) {
                ShoppingCart cart = new ShoppingCart();
                cart = (ShoppingCart) busVector.get(i);
                Goods good=new Goods();
                TempGoods tempGoods = new TempGoods();
                Vector<Goods> gVector=DAOFactory.getGoodDAOInstance().
                    queryGoodBySid(cart.getGid());        //获取指定商品
                if(gVector.size()>0&&gVector!=null)
                    good =(Goods)gVector.get(0);
                //设置TempGoods对象值
                tempGoods.setGood(good);
                tempGoods.setNumber(cart.getNumber());
                tempVector.add(tempGoods);
            }
            String curPage = request.getParameter("curPage"); //获取当前页
            pageObject = DAOFactory.getGoodDAOInstance().
getPageObject(curPage, pageObject, tempVector);//向页面传送分页内容
        request.setAttribute("pageObject", pageObject);
        return  "shoppingcart/paidbus.jsp";
    }
}
```

上述代码用于遍历已经支付的商品，将商品对象存放到分页对象中，页面用分页的方式显示。

查看已支付商品列表页面paidbus.jsp，其源代码与bus.jsp类似，只是在paidbus.jsp中增加分页标签。

17.10.3 支付中的页面

支付中的页面 pay.jsp 用于显示快递的地址、支付的金额等信息，部分源代码如下：

```
-----------------------pay.jsp--------------------------
<html>
    <head>
        <title>淘淘网—开心淘！</title>
    </head>
    <body>
        <div id="top">
            <jsp:include page="../head.jsp"/>
        </div>
        <p>
        <div>
            <jsp:include page="../logo_select1.jsp"/>
        </div>
        <div align="center">
            <div style="width: 80%; height: 100%;">
    <table width="100%" align="center" border="0">
        <tr>
            <td width="30%">
                <div align="center" style="border: 1px solid #c1eae8;">
```

```
                    <div id="title" align="left">
                                    宝贝信息
                    </div>
    <a id="img-link" href="goods?sid=${good.gid}&action=goodslist-select">
    <img src="${good.gphoto}" width="115" height="115" border="0">
    </a>
                    <div style="padding-left: 5%" align="left">
                                    <p>
                                        宝贝名称：
    <a href="goods?sid=${good.gid}&action=goodslist-select">
    ${good.gname}<br>${good.described}</a>
                                    <p>
                                        宝贝单价：
                <font id="price" color="blue">${good.price}</font>元
                                    <p>
                                        宝贝运费：
                <font id="carriage" color="blue">${good.carriage}</font>元
                                    <p>
                                        出产地：
                <font color="blue">${good.producer}</font>
                                    <p>
                                        出厂日期：
                <font color="blue">${good.pdate}</font>
                    </div>
                </div>
            </td>
            <td width="70%">
        <form action="shoppingcart" method="post" id="pay" name="pay">
                <input type="hidden" name="gid" value="${good.gid}">
                <input type="hidden" name="action" value="pay">
                <div style="padding-left: 3%; padding-top: 3px; ">
                    <div>
                        <div id="title">
                                    确认收货地址
                        </div>
                        <br>
                        ...
        <input type="submit" value="确认无误,购买" style="background-image:
        url(image/button1.jpg); width: 150px; height: 35px;
        border-style: none; font-weight: bold;">
            </td>
        </tr>
    </table>
    </body>
</html>
```

在上述代码的页面中，可以显示购买的地址、购买的总金额等信息以及“确认无误，购买”按钮，页面效果如图 17.12 所示。

到目前为止，系统中的实现代码基本介绍完毕，剩余的只是页面的样式和JavaScript代码，用于增加页面的美观性和易操作性。如果读者要弄懂这些样式和JavaScript代码，可以找相关图书继续深入学习。

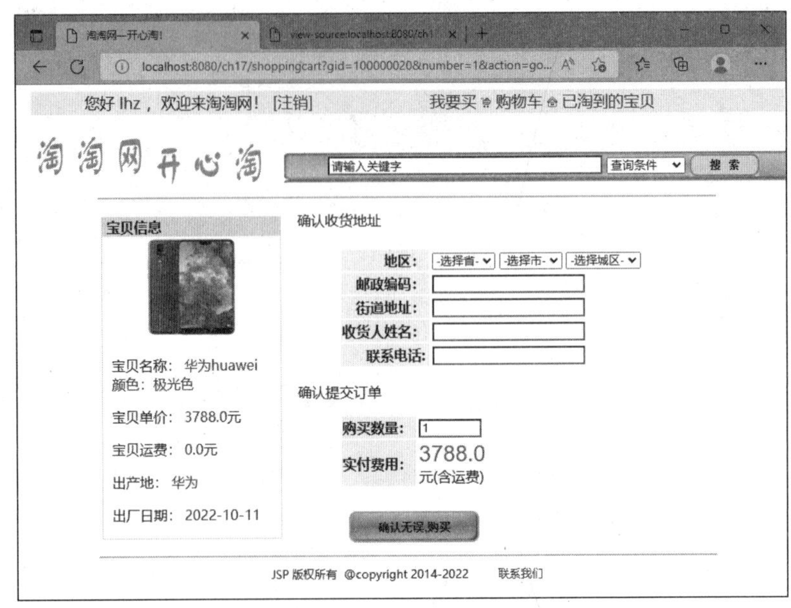

图17.12　pay.jsp页面效果图

17.11　小　　结

本章以“在线购物系统”的开发和实现为主线，从系统需求、系统总体框架、数据库设计、系统详细设计这4个方面逐步进行深入分析，详细讲解了该系统的实现过程。

读者在学习本章时最需要注意的地方就是如何将前面学习的内容串联起来，为将来的实战开发打下较好的基础。如果想进一步掌握Spring框架的应用开发，推荐学习《Spring Boot从零开始学（视频教学版）》《Spring Boot整合开发案例实战》《Spring Boot企业级开发实战（视频教学版）》等书。